THE NUCLEUS

A Multi-Volume Treatise

THE NUCLEOSOME

Volume 1 • 1995

THE NUCLEUS

A Multi-Volume Treatise

THE NUCLEOSOME

Editor: ALAN P. WOLFFE
  Laboratory of Molecular Embryology
  National Institute of Child Health and
    Human Development
  Bethesda, Maryland

VOLUME 1 • 1995

JAI PRESS INC.

Greenwich, Connecticut   London, England

Transferred to digital printing in 2008.

CONTENTS

LIST OF CONTRIBUTORS

Christopher C. Adams

Department of Biochemistry and
Molecular Biology
Pennsylvania State University
University Park, Pennsylvania

Anthony T. Annunziato

Department of Biology
Boston College
Chestnut Hill, Massachusetts

Trevor K. Archer

London Regional Cancer Centre
London, Ontario, Canada

David J. Clark

Laboratory of Molecular Biology
National Institutes of Health
Bethesda, Maryland

Jacques Cote

Department of Biochemistry and
Molecular Biology
Pennsylvania State University
University Park, Pennsylvania

David S. Gross

Department of Biochemistry and
Molecular Biology
Pennsylvania State University
University Park, Pennsylvania

Jeffrey J. Hayes

Laboratory of Molecular Biology
National Institutes of Health
Bethesda, Maryland

Li-Jung Juan

Intercollege Graduate Program
Pennsylvania State University
University Park, Pennsylvania

Randall H. Morse   David Axelrod Institute
Wadsworth Center for Laboratories and
Research
New York State Department of Health
Albany, New York

Joseph S. Mymryk   London Regional Cancer Centre
London, Ontario, Canada

Craig L. Peterson   Program in Molecular Medicine
University of Massachusetts Medical
Center
Worcester, Massachusetts

Dmitry Pruss   Laboratory of Molecular Biology
National Institutes of Health
Bethesda, Maryland

Sharon Y. Roth   Department of Biochemistry and
Molecular Biology
University of Texas M.D.
Anderson Cancer Center
Houston, Texas

Rhea Utley   Department of Biochemistry and
Molecular Biology
Pennsylvania State University
University Park, Pennsylvania

Michelle Vettese-Dadey   Department of Biochemistry and
Molecular Biology
Pennsylvania State University
University Park, Pennsylvania

Phillip P. Walter   Department of Biochemistry and
Molecular Biology
Pennsylvania State University
University Park, Pennsylvania

Alan P. Wolffe   Laboratory of Molecular Embryology
National Institute of Child and Human
Development
Bethesda, Maryland

Jerry L. Workman   Department of Biochemistry and
Molecular Biology
Pennsylvania State University
University Park, Pennsylvania

PREFACE

The Nucleosome is the first in a series of volumes concerning the properties of the eukaryotic nucleus. Contributions from several of the most active laboratories are brought together to present a focussed overview of a selected aspect of nuclear structure and function. Research on the role of nucleosomal structures in transcriptional regulation is making rapid progress. Experimental work on the accessibility of genes to the transcriptional machinery has clearly established that chromatin structure has a role in the repression of genes. However this repression is highly selective, dependent upon the formation of specific nucleosomes that contain key regulatory elements. More importantly, genetic analysis and biochemical reconstitution have suggested that nucleosomal structures have an essential role in the activation of transcription, and that in certain instances can actually facilitate the initiation process. Mutation or deficiencies in individual histones and their structural domains exert very specific effects on the transcription of individual genes. These functional studies have driven a renaissance of interest in the nucleosome, the histones and their interaction with DNA.

The Nucleosome contains nine chapters describing *in vitro* and *in vivo* experiments that examine the functional consequences of histone-DNA interactions. Chapter 1 details the basic elements of histone and DNA structure within the nucleosome, the sequence variation and modification of the histone proteins and some of their implications for higher order chromatin structure. The assembly of the nucleosome *in vivo* is the major theme of Chapter 2. The role of histone acetylation in the assembly process, the potential structural and functional consequences together with their epigenetic implications is described. Chapter 3 and 4 concern the interplay between histones and transcription factors in various model systems *in vivo* and *in vitro*. Important issues discussed include the impact of nucleosome positioning and modification on the accessibility of different trans-acting factors to cis-acting elements within chromatin. These model experiments lead a functional appreciation of the specific chromatin structures within which particular genes are found *in vivo*. Chapter 5 describes how a yeast repressor protein, α2/MCM1 acts to silence the transcription of a particular class of genes by mediating the assembly of specific nucleosomal structures. This result has major significance for the interrelationship of trans-acting factors with histones, since specific trans-acting factors direct the assembly of nucleosomes with defined functional roles. Chapter 6 and 7 strengthen the evidence for functional nucleosomes in transcriptional regulation. The architecture of promoters is seen to be central to regulatory events. Finally, Chapters 8 and 9 describe genetic and biochemical evidence for molecular machines influencing nucleosomal structure. The activities of the SWI/SNF general activator complex has major consequences for nucleosomal integrity, as does the processivity of RNA polymerases during the transcription process.

The Nucleosome provides a state of the art analysis of a rapidly expanding field. There is currently a flux of ideas concerning the mechanism of nucleosome construction and disruption. Many more contributions will be necessary before we completely understand how histones influence regulatory events. It is hoped that the systems and results described within this volume will encourage other scientists to enter this exciting research area.

Alan Wolffe
April 15th, 1994

Part I

Nucleosome Structure and Assembly

Chapter 1

Histone and DNA Contributions to Nucleosome Structure

DMITRY PRUSS, JEFFREY J. HAYES, and
ALAN P. WOLFFE

The Nucleus
Volume 1, pages 3–29.
Copyright © 1995 by JAI Press Inc.
ISBN: 1-55938-940-0

ABSTRACT

Recent structural studies allow a new functional appreciation of nucleosome structure. The principles of histone-histone and histone-DNA interaction point to a conformational flexibility that provides insight into how *trans*-acting factors and molecular machines might utilize chromatin templates.

I. INTRODUCTION

The nucleosome is the fundamental repeating unit of chromatin. Each nucleosome contains a varying length of DNA dependent upon organism, tissue and cell examined. This nucleosomal repeat length is generally determined as the average distance between micrococcal nuclease cleavage sites within chromatin after very slight digestion. Most of the DNA (~ 160 bp) within the nucleosomal repeat (~ 180–190 bp) is tightly wrapped in two superhelical turns around the core histones. A single molecule of linker histone (e.g. H1) also associates with each nucleosomal repeat. The exact binding site of the linker histone and the path of the linker DNA, that is not wrapped around the core histones is not yet known. However considerable progress has been made in characterizing both DNA and histone structure, these results lead to a new view of nucleosome structure that is the focus of this chapter.

II. THE ORGANIZATION OF THE HISTONE OCTAMER

The four core histones H2A, H2B, H3, and H4 have very selective interactions with each other, together they assemble the octamer $(H2A, H2B, H3, H4)_2$ around which DNA is wrapped. Histone H2A forms a heterodimer with H2B, and H3 forms a heterodimer with H4. The interface between the histones in each heterodimer is very similar (Figure 1) and is described as a 'handshake' motif (Arents et al., 1991). Each core histone has a N-terminal tail domain that reaches outside of the two superhelical turns of DNA within the nucleosome (see later) and a C-terminal domain that is involved in histone–histone interactions inside nucleosomal DNA. The C-terminal domains of each core histone are predominantly α-helical with a long central helix bordered on each side by a loop segment and a shorter helix. This overall structure has been termed the 'histone fold' (Arents et al., 1991). Previous work had found that two molecules of each of the four core histones were present within each nucleosome and that this histone octamer had a tripartite organization, in which a central histone tetramer $(H3, H4)_2$ interacts with two histone dimers (Kornberg, 1974; Kornberg and Thomas, 1974). The tetramer $(H3, H4)_2$ and dimer (H2A, H2B) are stable at physiological ionic strength, whereas the octamer structure is only stable at high ionic strength (~ 2M NaCl) or in the presence of polyanions such as DNA (Eickbusch and Moudrianakis, 1978). Although the area of the interface between the two (H3, H4) heterodimers is less extensive than that between the (H3, H4) and (H2A, H2B) heterodimers (Arents et al., 1991) (Figure

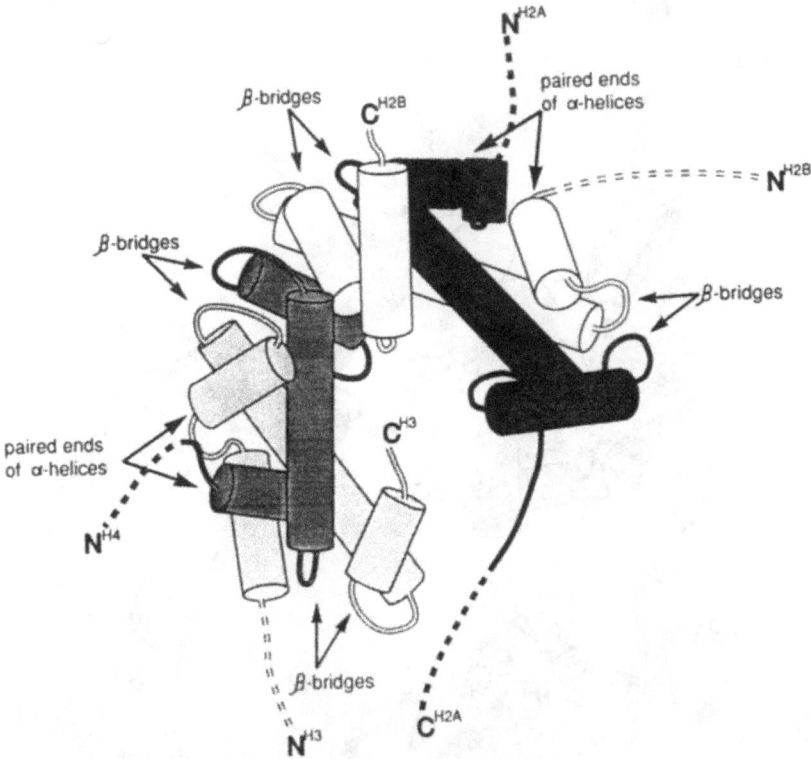

Figure 1. Histone heterodimers and DNA binding motifs. The approximate structures of the H4 and H3 heterodimer, and H2B and H2A are shown. N- and C-terminal polypeptide chains are indicated. The paired ends of α-helices and β-bridge motifs predicted to be involved in DNA binding are indicated. The relative juxtaposition of the two heterodimers in the nucleosome core is shown.

2), the latter contacts break apart first in the absence of counterions. Apparently, the interface between the dimer and the tetramer is more accessible to solvent and is consequently less stable. Disruption of these contacts within the nucleosome is likely to have important consequences for transcription through a nucleosome and for the access of *trans*-acting factors to DNA in the nucleosome (see later).

The shape of the octamer is that of a wedge, in which a central V-shaped tetramer $(H3, H4)_2$ is bordered by two flattened spheres of $(H2A, H2B)$ dimers. The resolved portion of the octamer structure has several grooves and ridges on its surface. These make a left-handed helical ramp onto which DNA may be wrapped (Arents and Moudrianakis, 1993). Within this ramp are eight histone fold motifs containing sixteen loop segments. Due to the dimerization of the histones, loop segments from each half of the dimer are paired to form eight parallel β-bridge segments, two of

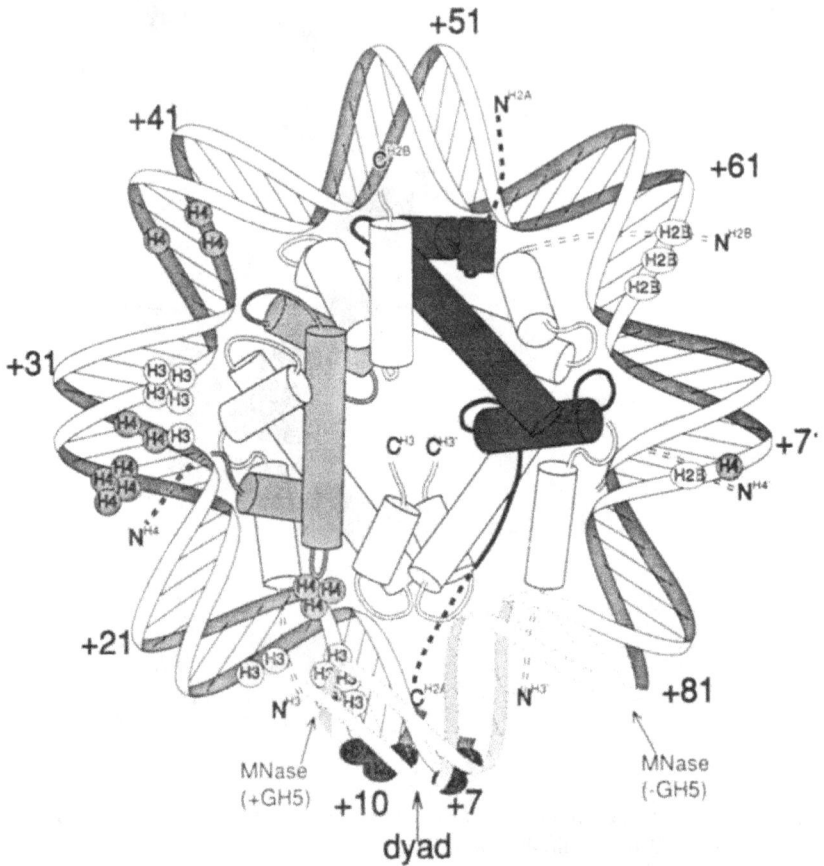

Figure 2. Complexes of the (H3, H4)$_2$ tetramer and (H2A, H2B) dimer with DNA. The approximate structure of the histone (H3, H4)$_2$ tetramer and H2A, H2B dimer bound to 5S DNA is shown. Histones are shaded as in Figure 1. Sites of histone contact revealed by protein-DNA crosslinking are indicated by circles on the double helix. Number refer to the 5S DNA sequence relative to the start site of transcription (+1). The dyad axis of the nucleosome is indicated. The micrococcal nuclease digestion boundaries (MNase) in the presence or absence of the globular domain of histone H5 (GH5) are indicated (see text for details).

which are found within each of the histone dimers (H3, H4) and (H2A, H2B) (Figure 1). Each β-bridge segment is associated with at least two positively charged amino acids which are available to make contact with DNA on the surface of the histone octamer (Arents and Moudrianakis, 1993). The second repeating motif within the nucleosome is assembled from the pairing of the N-terminal ends of the first helical domain of each of the histones in the heterodimers (Figure 1). These four 'paired ends of helices' motifs also appear to contact DNA (Arents and Moudrianakis, 1993). Thus each of the four heterodimers within the core can make at least three pseudo-symmetrical contiguous contacts with three consecutive inward facing minor grooves of DNA (Figure 2). The eight parallel β bridges and the four paired ends of helices motifs provide 12 potential DNA contact sites that are regularly arranged along the ramp on which the double helix is wound (see later).

The N-terminal domains of the core histones are not resolved in the octamer crystal structure, however the positions of the most amino terminal helix within the structured C-terminal domain provides some indication of where the attached N-terminal tail domains might interact with nucleosomal DNA. The formation of such contacts with DNA in the nucleosome has been assayed by histone-DNA cross-linking methodologies (see later).

III. THE INTERACTION OF DNA WITH THE HISTONE OCTAMER

The analysis of chromatin structure with nucleases provided the first footprints of nucleoprotein structures. Of particular importance was the definition of a kinetic intermediate in the digestion of chromatin with micrococcal nuclease known as the nucleosome core, which contains 146 bp of DNA and the histone octamer (van Holde, 1989). The structure of DNA revealed in these pioneering studies has been considerably extended in detail through the use of the hydroxyl radical as a cleavage agent.

The hydroxyl radical cleaves DNA by breaking sugar rings in the phosphodiester backbone. This cleavage reaction occurs without significant sequence specificity, but most importantly, the small size of the hydroxyl radical almost eliminates steric impediments by distantly located components of the structure, such as the neighboring superhelical turns of DNA. Steric hindrance has been a major limitation for interpreting the cleavage patterns obtained using enzymatic probes of DNA structure within chromatin (Klug and Lutter, 1981). Hydroxyl radical cleavage of nucleosome cores reveals that the structure of DNA is different when it is wrapped around the histone octamer compared to when it is free in solution (Hayes et al., 1990; 1991a). Within the nucleosome core, three turns of DNA at the center of nucleosomal DNA (dyad axis) have a helical period of 10.7 bp/turn whereas outside of this region the helical period is 10.0 bp/turn. This DNA structure is ideally suited to match the regularly spaced sites of potential contact on the histone octamer surface, including the three turns of 10.7 bp/turn at the dyad axis (Arents and Moudrianakis, 1993). Thus DNA in the nucleosome core has a tripartite organization comparable to that of the histone octamer. These differences in helical periodicity of DNA in the nucleosome may contribute to determining exactly where along a DNA molecule the histone octamer will prefer to bind.

A

Covalent binding of a protein to a DNA apurine site

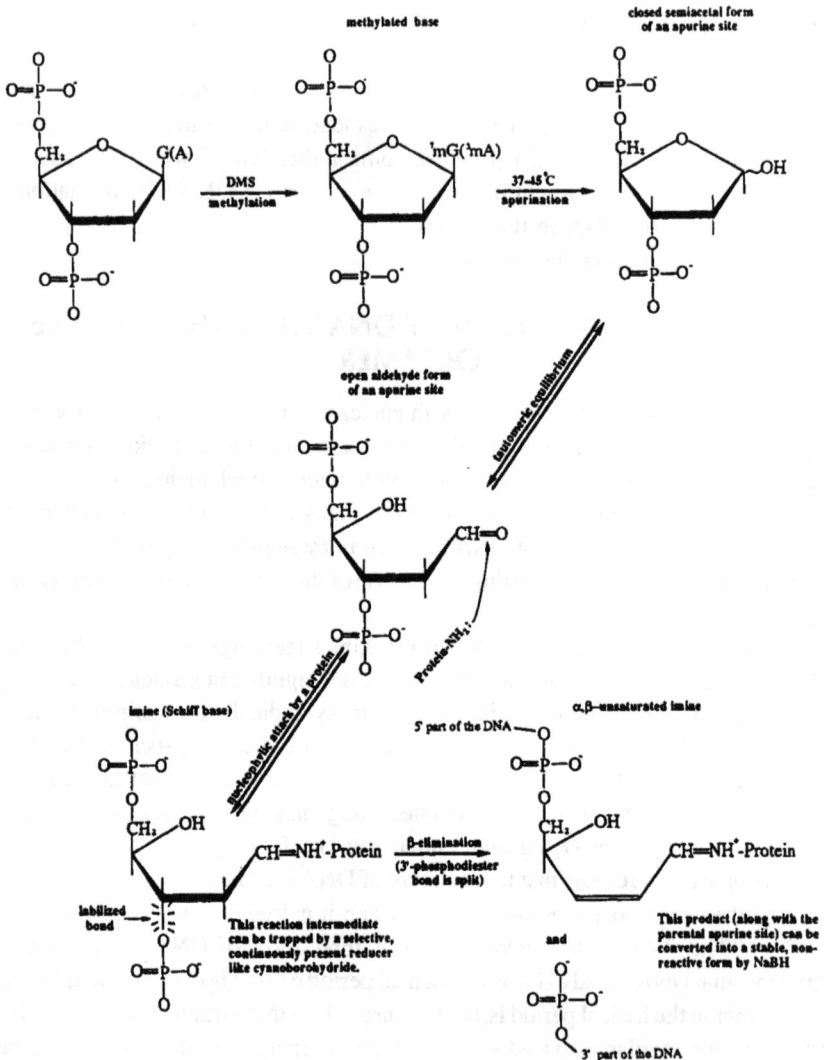

Figure 3. A. The chemistry of protein-DNA crosslinking and **B.** two dimensional mapping of protein-DNA crosslinks. The reaction illustrated in **A** leads to a covalent linkage of protein and DNA coincident with a break in one strand of the double helix. After denaturation, the protein bound to single-stranded DNA is resolved on the first dimension from left to right on the scheme. This resolves nucleoprotein complexes predominantly on the basis of size and charge. The first-dimension gel is then treated

8

(continued)

B

2-Dimensional mapping of the crosslinks

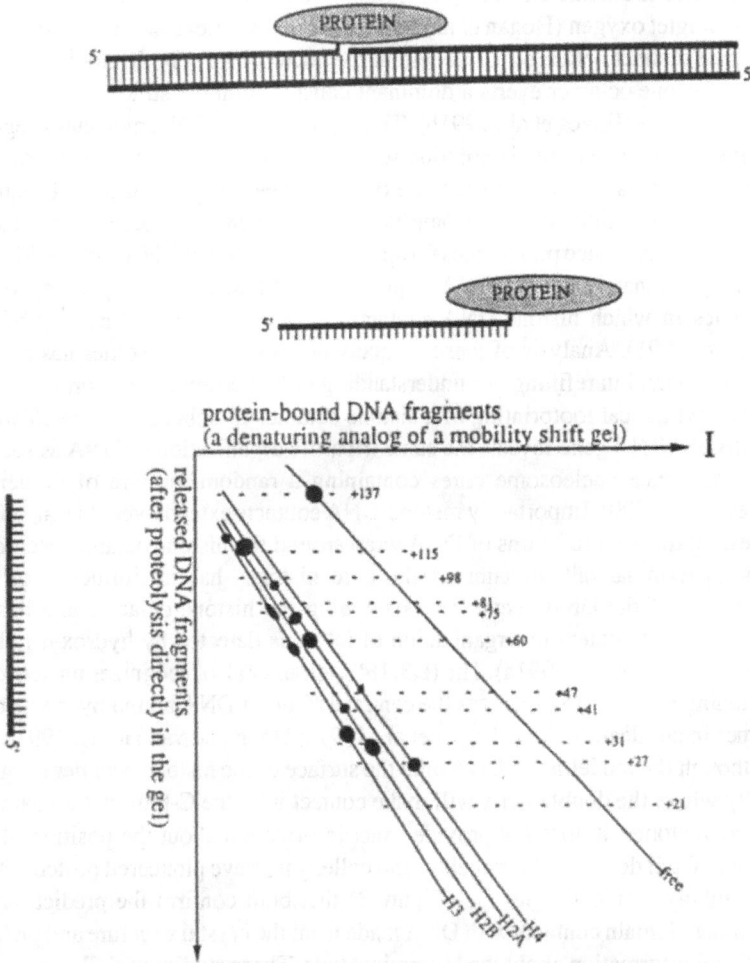

Figure 3. (continued) with protease and the residual DNA resolved on the second dimension from top to bottom on the scheme. This resolves nucleic acid on the basis of size and charge. For the nucleosome core, the dots on the gel represent differently sized DNA molecules, determined on the vertical axis where the DNA is broken, and whose mobility was retarded by differently sized histones. Hence, the dots fall on four diagonals as indicated.

The path of DNA as it wraps around the core histones is not uniform. The low resolution crystal structure of the nucleosome core (Richmond et al., 1984) reveals that the 146 bp of DNA within it are bent in 1.75 turns around the histones. At the dyad axis, the minor groove of DNA is orientated towards solution. Within each 80 bp circle, DNA is more severely distorted about ± 1.5 and ± 3.5 turns from the dyad axis. These sites are preferentially reactive when probed by enzymes sensitive to DNA distortion (Pruss et al., 1994). The sites ± 1.5 turns from the dyad axis are also sensitive to chemical cleavage reagents which recognize unstacked base pairs, such as singlet oxygen (Hogan et al., 1987). These distortions delimit the different regions of distinct helical structure within the nucleosome core. It should be noted that the histone octamer exerts a dominant constraint on the structure of DNA in the nucleosome (Hayes et al., 1991b). This means that all DNA molecules regardless of structure when free in solution will be constrained into the same structure following association with the histone octamer. The complex structural requirements for DNA within the nucleosome have so far prevented the accurate prediction of any unique sequence preferences for specific association of DNA with the histone octamer. Fortunately natural DNA sequences exist that will form positioned nucleosomes in which histone-DNA contacts begin and end at defined sequences (Simpson, 1991). Analysis of these uniquely positioned nucleosomes has proven extremely useful in refining our understanding of nucleosomal structure.

Hydroxyl radical footprinting of a histone octamer associated with the *Xenopus borealis* 5S rRNA gene reveals the same tripartite organization of DNA as seen in mixed sequence nucleosome cores containing a random mixture of sequences (Hayes et al., 1990). Importantly histone-DNA contacts extend over at least 160 bp suggesting that two full turns of DNA wrap around the histone octamer. Removal of the N-terminal tail domains of the core histones has no influence on the recognition of the DNA sequences positioning the histone octamer and has no influence on the extent or organization of DNA as detected by hydroxyl radical cleavage (Hayes et al., 1991a). The (H3, H4)$_2$ tetramer also recognizes nucleosome positioning signals and organizes the central 120 bp of DNA bound by the histone octamer in an identical way (Hayes et al., 1991a; Dong and van Holde, 1991).

Although the modeling of DNA onto the surface of the histone octamer suggests exactly where the double helix will make contact with the C-terminal domains of the core histones, it does not provide exact information about the position of the N-terminal tail domains. Mirzabekov and colleagues have pioneered protein-DNA cross-linking methodologies (see Figure 3) that both confirm the predictions of C-terminal domain contacts with DNA made from the crystal structure and provide additional information about the N-terminal tails. These studies initially employed nucleosome cores containing a mixture of DNA sequences (Mirzabekov et al., 1989), however more recently the same methodologies have been extended to nucleosomes containing defined DNA sequences (Pruss and Wolffe, 1993). Important features of nucleosome structure determined from those studies include the presence of histone H4 cross-linking to DNA over 30 bp to either side of the

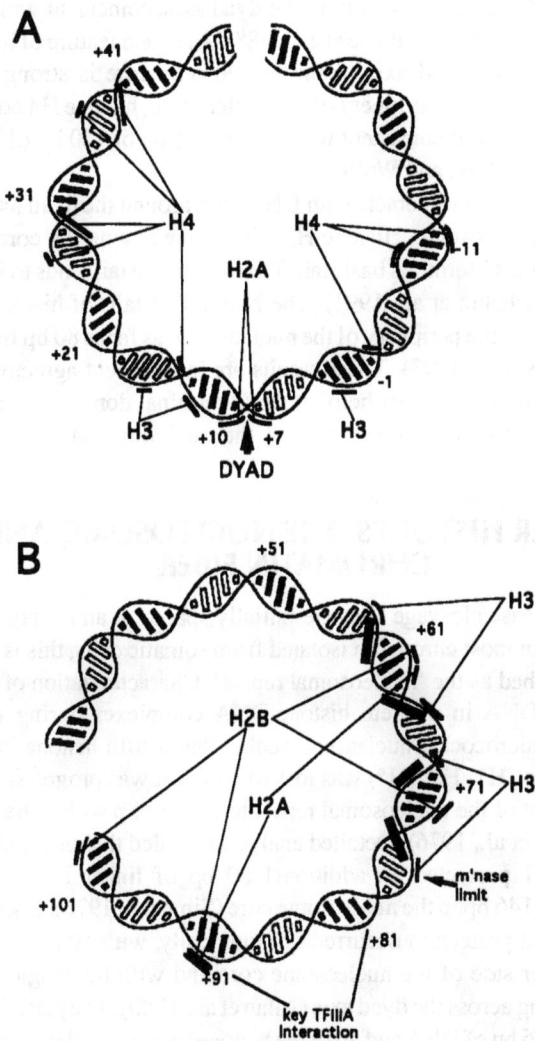

Figure 4. Histone-DNA contacts in the 5S nucleosome core. **A.** Histone-DNA contacts around the dyad axis. DNA is shown in an 80 bp circle, base pairs are indicated by open and closed rectangular blocks. The numbers indicate base pair positions within the *X. borealis* somatic 5S RNA gene relative to the start site of transcription (+1) the dyad axis is indicated. Histone-DNA crosslinks are indicated by closed bars. **B.** Histone-DNA contacts at the 3' boundary of the 5S nucleosome core. Numbers indicate base pair positions within the *X. borealis* somatic 5S RNA gene. Histone H2A and H2B contacts to the coding strand are indicated by small bars and to the non-coding strand by large boxes. Histone H3 contacts are indicated. The major kinetic barrier to micrococcal nuclease caused by the nucleosome core is indicated as is the key TFIIIA recognition element.

11

nucleosomal dyad axis (Figure 4). The N-terminal tail of histone H4 can be cross-linked to DNA at ± 1.5 turns from the dyad axis, coincident with the sites of strong DNA deformation (Ebralidse et al., 1988). A second feature of histone-DNA interaction around the dyad axis of the nucleosome core is strong histone H3 cross-linking. Towards the periphery of the nucleosome, histone H4 contacts DNA over 60 bp from the dyad consistent with the protection of 120 bp of DNA by the $(H3, H4)_2$ tetramer (Pruss and Wolffe, 1993).

The (H2A, H2B) dimer interacts with DNA both around the dyad axis and at the periphery of the nucleosome. Histone H2A is unique among the core histones in having both a N- and C-terminal basic tail. The C-terminal tail binds to DNA around the dyad axis (Gushchin et al., 1991). The N-terminal tails of histones H2A and H2B contact DNA at the periphery of the nucleosome, as far as 80 bp from the dyad axis (Pruss and Wolffe, 1993). These results are in excellent agreement with the proximity of the amino terminal helix of the C-terminal domain of the individual core histones to DNA determined through the modeling studies (Figure 2, see Arents et al., 1991).

IV. LINKER HISTONES, THE NUCLEOSOME, AND THE CHROMATIN FIBER

Micrococcal nuclease cleavage sites are initially spaced at an average distance of 180–190 bp within most chromatin isolated from somatic cells, this is the distance commonly described as the 'nucleosomal repeat'. Characterization of the proteins associated with DNA in discrete histone-DNA complexes during digestion of chromatin with micrococcal nuclease, revealed that a fifth histone known as the linker histone (e.g., H1, H5, H1°) was lost when DNA was progressively reduced in length from that of the nucleosomal repeat to the 146 bp within the nucleosome core (Varshavsky et al., 1976). Detailed analysis revealed that one molecule of the linker histone H1 protected an additional 20 bp of linker DNA immediately contiguous to the 146 bp in the nucleosome core (Simpson, 1978). It was suggested that this additional protection occurred symmetrically, with 10 bp of linker being protected to either side of the nucleosome core and with the single molecule of linker histone lying across the dyad axis (Allan et al., 1980). The particle containing approximately 166 bp of DNA and all of the histones was termed the chromatosome (Simpson, 1978). Linker histones have three domains: N- and C-terminal tails flank a central globular domain. The globular domain alone was found to protect linker DNA. Therefore a model for linker histone interaction was proposed that would require the globular domain to make three contacts with DNA, one with each DNA strand entering and exiting the nucleosome and one with the minor groove of the DNA which is orientated directly away from the nucleosomal surface at the dyad axis (Figure 5A, Allan et al., 1980). Consistent with this hypothesis, DNase I footprinting of dinucleosomes containing linker histones revealed protection from cleavage at the nucleosomal dyad (Staynov and Crane-Robinson, 1988). Moreover

A.

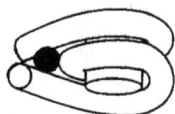

B.

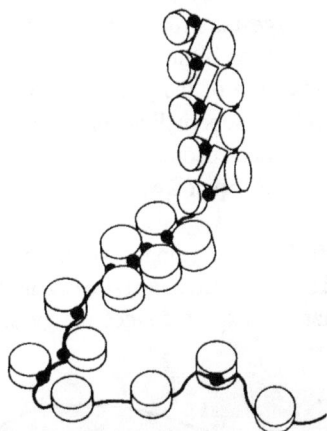

Figure 5. Symmetrical three contact model for linker histone association within the nucleosome and cooperative model for linker histones mediating folding of the nucleosomal array into the chromatin fiber. **A.** The globular domain of the linker histone (black sphere) binds to DNA where it enters and exits wrapping around the histone octamer, and to the central turn of DNA at the dyad axis. **B.** Linker histone molecules positioned as in **A** undergo cooperative interactions within the chromatin fiber, thereby directing the folding up of the nucleosomal array.

the globular domains of the linker histones are known to bind to *naked* DNA molecules cooperatively by constraining two double helices next to each other and stacking linker histones between them (Thomas et al., 1992). The possibility of further interactions of intact linker histones with other linker histones within adjacent nucleosomes, or with the DNA in adjacent nucleosomes, led to the hypothesis that nucleosomal arrays might fold into highly ordered chromatin fibers with linker histones inside the fiber (Figure 5B, Thoma et al., 1979).

The structure of the globular domain of linker histone H5 was recently solved (Ramakrishnan et al., 1993). This structure shows remarkable similarity to that of the DNA binding domain of transcription factor HNF-3 (Clark et al., 1993). Both consist of a bundle of three α-helices attached to a three stranded antiparallel β sheet (see Figure 7 later). HNF-3 binds across the major groove of DNA as a monomer suggesting that the globular domain of H5 will contact nucleosomal DNA in the same way. Since the globular domain of H5 interacts with at least one other DNA molecule (Thomas et al., 1992), a second DNA binding site can be proposed to lie on the opposite side of the protein, separated by 2.5–3nm from the α-helix lying in the major groove. These interactions would be consistent with the known contacts of histone H5 with DNA in the nucleosome (Mirzabekov et al., 1989).

This new structural information on the structure of the globular domain of a linker histone imposes several constraints on existing models of the chromatosome. None of the new structural data suggest a strong interaction with the minor groove of DNA, such as might be envisioned if the linker histone were bound directly over the dyad. Nevertheless interaction of the globular domain with DNA could still occur away from the central DNA superhelical turn. A symmetrical interaction with DNA entering and exiting the nucleosome away from the central turn of DNA at the dyad axis could still protect 20 bp of DNA from micrococcal nuclease digestion (Figure 6A).

An alternative hypothesis is that the single molecule of linker histone interacts asymmetrically with DNA in the nucleosome (Hayes and Wolffe, 1993). Asymmetric protection should perhaps be expected since the globular domain of the linker

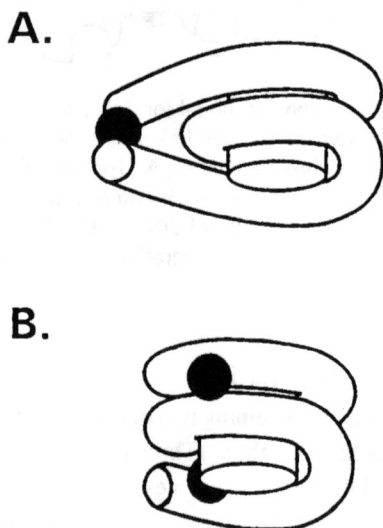

Figure 6. Alternative models for placement of the globular domain of the linker histone within the nucleosome. **A.** The globular domain binds to DNA that enters and exits wrapping around the histone octamer, but away from the central turn of DNA around the octamer. **B.** The globular domain is shown in two possible extremes of an asymmetric position within a single nucleosome, either inside or outside the superhelical turns of DNA.

histone is an asymmetric object and is also unlikely to bind to the minor groove of DNA at the dyad axis (see above). Nucleosomes incorporating specific DNA sequences have once again proven useful in determining possible chromatosome structures. It is possible to show that linker histones bind preferentially to *Xenopus borealis* somatic 5S rRNA gene associated with an octamer of core histones rather than to naked DNA (Hayes and Wolffe, 1993). This preferential binding requires free linker DNA and results in the protection of an additional 20 bp of the linker DNA from micrococcal nuclease digestion. Importantly this additional linker DNA is asymmetrically distributed to either side of the nucleosome core. The 15 bp protected to one side of the 5S nucleosome by H5 are shown in Figure 7 (see later). Formally, asymmetric protection of linker DNA could still occur through interaction of the histone H5 near the nucleosomal dyad. However incorporation of linker histones cause no change to the cleavage of DNA in the 5S nucleosome as detected by hydroxyl radical or DNase I (Hayes and Wolffe, 1993). These results are difficult to reconcile with a model of the nucleosome in which the linker DNA is bound to the outward facing minor groove of 5S DNA on the dyad axis, making equivalent interactions at the entry and exit points of nucleosomal 5S DNA (see Figures 5A and 6A). Moreover neutron scattering studies of chromatosomes place the linker histone near the core histone surface (Lambert et al., 1991) and protein-protein crosslinking data imply contacts between the C-terminal histone fold domain of histone H2A and the globular domain of histone H1 (Boulikas et al., 1980). Thus a distant interaction of the globular domain with DNA away from the octamer surface seems unlikely. The asymmetric protection of linker DNA and lack of protection at the dyad are consistent with the asymmetric linker histone molecule interacting predominantly with one end of nucleosomal 5S DNA close to the surface of DNA and the histone octamer.

More recent observations have shown that the linker histone requires the (H2A, H2B) dimer to be present before stable association with nucleosomal DNA can occur (Hayes et al., 1994). This is consistent with the existence of close linker histone interactions with histone H2A, and indicates a lack of strong independent interactions of the linker histone either with DNA or the $(H3/H4)_2$ tetramer at the dyad axis. The globular domain of histone H5 alone has also been shown to confer asymmetric protection of linker DNA in the 5S nucleosome (Hayes et al., 1994). Moreover, histone-DNA cross-linking reveals a single major contact around 65 bp from the nucleosomal dyad to be made by the globular domain with 5S DNA (Figures 6B and 7). Modeling indicates that this site is largely devoid of spatial constraints imposed by the core histones (Figure 7), hence the globular domain of the linker histone could adopt a variety of positions between 64 to 70 bp from the dyad axis. The position shown in Figure 7A has the globular domain located on the outside of the DNA superhelix at a surprising 3.5nm from the dyad axis. The position shown in Figure 7B locates the globular domain much closer to the dyad axis, but from the inner side of the DNA superhelix. The 'hole' in the octamer filled by the globular domain is framed by the H2A and H3 polypeptide chains. An

A

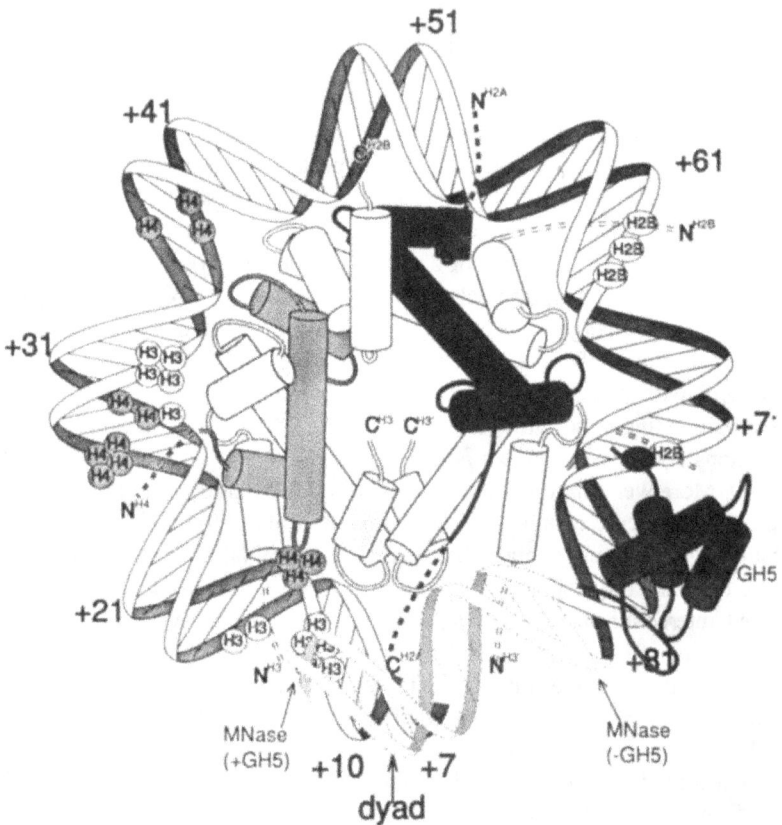

Figure 7. Two possible positions for the linker histone globular domain within the 5S nucleosome. The (H3, H4)$_2$ tetramer and a single (H2A, H2B) dimer are shaded as in Figure 1. Their interactions with DNA are as described in Figure 2. Micrococcal nuclease boundaries with or without the globular domain of histone H5 are indicated

(*continued*)

intermediate location would have the globular domain right at the top edge of the nucleosome (not shown). In all of these instances the globular domain is close to potential sites of interaction with the (H2A, H2B) dimer. If the globular domain is bound at any of these sites it would not be possible for it to interact with DNA at the entry and exit of wrapping around the histone octamer (Figure 6B). However it would be possible for some contact with DNA to be made with the central segment of the nucleosomal DNA, most probably on the central turn of DNA ~ 15 bp from

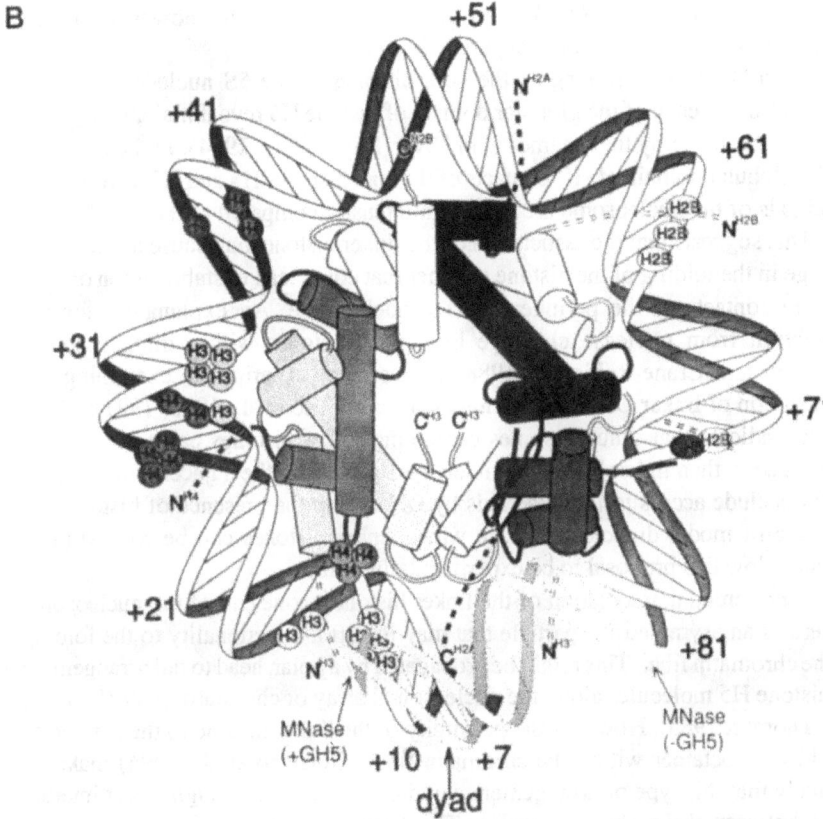

Figure 7. (continued) as in Figure 2. **A.** The globular domain of histone H5 is shown binding to the outside of the superhelical turn of DNA. Contact with the 5S DNA is indicated by the circle on the double helix. **B.** As in **A**, except the globular domain is shown bound inside the superhelical turn of DNA.

the nucleosomal dyad. This model would be consistent with the protein-protein interactions of linker histones with H2A, the lack of protection by linker histones across the dyad axis of the 5S nucleosome and the known dimensions and putative interaction of the globular domain of histone H5 with DNA. Asymmetric binding would presumably be directed by local sequence preferences of the linker histone globular domain (Muyldermans and Travers, 1994). However this model does not offer a simple explanation for the protection of an additional 20 bp of linker DNA including regions at both ends of the nucleosome or for the protection at the dyad axis from DNase I cleavage observed in the footprinting of dinucleosomes (Staynov and Crane-Robinson, 1988). Recent experimental results do however suggest potential explanations for these apparent discrepancies.

Protein-DNA cross-linking of the core histones in the 5S nucleosome in the presence or absence of the globular domain of histone H5 reveal major changes in the contacts made by the octamer with DNA (Hayes et al., 1994). In the presence of the globular domain of H5, contacts of the core histones H3 and H2A around the dyad axis of the nucleosome are altered in intensity compared to those of histone H4. This suggests that the association of the linker histone may cause an allosteric change in the folding of the histone octamer that could lead to stabilization of core histone contacts at the periphery of the nucleosome. An explanation for the protection from nuclease cleavage by linker histones within dinucleosomes (Staynov and Crane-Robinson, 1988) comes from experiments examining the compaction of linker DNA within dinucleosomes (Yao et al., 1990; 1991). These results indicate that dinucleosomes containing linker histones can fold together more readily than linker histone deficient particles. Thus the adjacent nucleosome could occlude access to the dyad axis by DNase I in the presence of histone H1. The use of model dinucleosomes in which linker histones can be reconstituted should allow this proposal to be experimentally tested.

The preferential association of the linker histone to one end of the nucleosome generates an asymmetrical particle that may impart a directionality to the folding of the chromatin fiber. This might be propagated by a polar, head to tail arrangement of histone H5 molecules along the nucleosomal array or chromatin fiber (Lennard and Thomas, 1985). However the proximity of the linker histone to the surface of the histone octamer within the chromatin fiber (Graziano et al., 1994) makes it unlikely that this type of arrangement could be propagated through direct interactions between the globular domains. The asymmetric positioning of the linker histone is also likely to favor interaction of the extended basic C-terminal tail of the linker histone with the linker DNA (Hill and Thomas, 1990). The charge neutralization of the phosphodiester backbone of linker DNA by the basic C-terminal tail of the linker histone has a major role in chromatin compaction (Clark and Kimura, 1990). Moreover the removal of the linker histone away from the dyad axis of the nucleosome would eliminate the need for the DNA at the entry and exit of the nucleosome to come into close proximity. This also eliminates a central requirement of certain models of the chromatin fiber that have histone H1 uniquely localized in the center of the fiber.

The evidence supporting effective compaction of adjacent dinucleosomes, the extended interaction of 160 bp of DNA with the histone octamer and the asymmetric placement of the linker histone all favor the linker DNA being supercoiled between adjacent nucleosomes. This linker DNA organization would support a solenoid model for chromatin higher order folding in which the core DNA together with the linker DNA form a continuous superhelix (Bavykin et al., 1990).

V. NUCLEOSOME CONSTRUCTION AND DISRUPTION

Although the nucleosome is held together by extensive protein-protein and protein-DNA interactions it is clear that the integrity of certain interactions are more important than others. The nucleosome has a central kernel of the $(H3, H4)_2$ tetramer. These are the first histones to be sequestered onto DNA during nucleosome assembly *in vivo* and it is not until 120 bp of DNA is wrapped around this tetramer that two dimers of (H2A, H2B) can associate. Only when 160 bp of DNA is wrapped around the histone octamer can the linker histone also be stably incorporated into the nucleosome (Hayes et al., 1991; Hayes and Wolffe, 1993). Nucleosome assembly depends both on specific histone-histone interactions within and between heterodimers, and between the (H2A, H2B) dimer and the linker histone, and on the progressive deformation of DNA by the $(H3, H4)_2$ tetramer to create a configuration suitable for the association of (H2A, H2B) dimer and finally of the linker histone. Importantly most of the compaction of DNA and the recognition of nucleosome positioning signals are mediated by the $(H3, H4)_2$ tetramer. Thus the selective dissociation of linker histones and (H2A, H2B) dimer can render regulatory elements accessible while retaining DNA in a folded state. This could facilitate transcription through nucleosomes (van Holde et al., 1992; Hansen and Wolffe, 1994) and also the regulated access of transcription factors to DNA in the nucleosome (Hayes and Wolffe, 1992; Chen et al., 1994). It is important to recognize that the whole of chromatin structure is built upon the interaction of the $(H3, H4)_2$ tetramer with DNA. If this interaction is disrupted, contact of the (H2A, H2B) dimers and linker histones within the nucleosome and chromatin fiber are also likely to be disturbed. Therefore modulation of the interaction of $(H3, H4)_2$ tetramer with DNA has the potential to exert a major regulatory role on chromatin integrity and DNA accessibility.

VI. HISTONE MODIFICATIONS, VARIANTS AND MUTANTS: POTENTIAL IMPACT ON NUCLEOSOME STRUCTURE AND FUNCTION

A. Modifications

The core and linker histones are extensively post-translationally modified particularly in the N- and C-terminal tail domains. These reversible covalent modifications include the acetylation and ubiquitination of the core histones, and the phosphorylation of both core and linker histones. The sites of modifications are highly conserved through evolution and are likely to have important structural and functional consequences.

The histone modification that has received the most attention is acetylation of the ε-lysines in the N-terminal tail domains. Acetylation of the N-terminal tails of histones H3 and H4 reduces their net positive charge and weakens their interaction with DNA (Cary et al., 1982). The N-terminal tails of the $(H3, H4)_2$ tetramer make strong contacts with DNA around the dyad axis of the nucleosome, thus modulation of these interactions probably accounts for the global structural changes in the interactions of the histone octamer with DNA associated with this modification (Bauer et al., 1994). Release of the core histone tail domains from interactions with DNA in the nucleosome either directly, or through allosteric structural effects involving the entire histone octamer, facilitates transcription factor access to nucleosomal DNA (Lee et al., 1993). The N-terminal tail of histone H2B is also modified through acetylation. Variants of histone H2B exist in which the N-terminus of H2B is considerably extended in length and interacts stably with linker DNA over 80 bp from the nucleosomal dyad (Hill and Thomas, 1990). This type of interaction would be difficult to achieve unless the linker DNA were folded or supercoiled between adjacent nucleosomes. Thus a role for the basic N-terminal tail of normal H2B in shielding some of the negative charge of the phosphodiester backbone within linker DNA seems probable. Acetylation would reduce the positive charge of the N-terminus of H2B and thus the neutralization of the phosphodiester backbone, and in consequence the folding of linker DNA.

The core histone tails can also have their interactions with nucleosomal DNA altered more severely through removal from chromatin by limited proteolysis using trypsin. This treatment leaves the domains of the core histones within the histone fold intact, and does not disturb the wrapping of DNA around these domains (Hayes et al., 1991a). Nevertheless, DNA associated with 'tailless' core histones in nucleosomal structures becomes more accessible to transcription factors (Lee et al., 1993). Moreover, tailless oligonucleosomes do not fold efficiently into compacted chromatin structures (Garcia-Ramirez et al., 1992), implying that the core histone tail domains have a role in the formation of higher order chromatin structures.

Phosphorylation of the core histone tails will also weaken their interaction with DNA. The histone H4 that is assembled into nascent chromatin is both diacetylated and phosphorylated on its N-terminal tail (van Holde, 1989). The N-terminal tail of histone H2A is also phosphorylated in newly assembled nucleosomes (Kleinschmidt and Steinbeisser, 1991). These modifications of the histones are all associated with the assembly of nascent chromatin following replication. There are two potential roles for these modifications: they may regulate histone interactions with the molecular chaperones required for nucleosome assembly, and once the histones are assembled into nucleosomes, they may retain nucleosomal DNA in a more accessible state for transcription factors to gain access to regulatory elements.

Phosphorylation of the N-terminal tail of histone H3 occurs in response to mitogenic stimulation (Mahadevan et al., 1991), modification of the interaction of this tail domain with DNA at the dyad axis of the nucleosome may again facilitate conformational changes in octamer-DNA interactions rather like acetylation. However, this possibility is yet to be tested. Histone H1 is also phosphorylated in its tail domains in response to mitogenic stimuli, and is heavily phosphorylated at mitosis. Although it is clear that phosphorylation weakens the interaction of histone H1 with naked DNA (Hill et al., 1991) the physiological significance of this modification to the linker histone is unclear. It might be predicted that a reduction in the strength of interaction of H1 with DNA through weakening of electrostatic interactions will facilitate the association of other transacting factors with DNA both within the nucleosome and the chromatin fiber.

In contrast to the widespread modification of the core histones by acetylation, and of the core histones and linker histones by phosphorylation, modification by ubiquitination is directed predominantly to a single histone, H2A. Ubiquitin is a small protein that is linked by its C-terminus to the ε-amino group of lysine 119 of H2A. This is in the immediate vicinity of where the H2A molecule interacts with DNA at the dyad axis of the nucleosome (Figure 2, see Ref. 34). Transcriptionally active *Drosophila* hsp 70 genes are greatly enriched in ubiquitinated H2A suggesting a role for this modification in preventing the formation of repressive chromatin structures (Levinger and Varshavsky, 1982).

Taken together these modifications indicate that the tail domains of the core histones represent the target for the signal transduction pathways that impact on chromatin structure and function. The post-translational modification of histones undoubtedly has a major role in the establishment and maintenance of transcriptional activity. However, the enzymes involved in directing modifications of the histones have not been extensively characterized at the molecular level. There remains a remarkable lack of knowledge concerning the proteins involved and how their activities are regulated and targeted to individual chromatin domains or promoters.

B. Histone Variants

The essential role for the $(H3, H4)_2$ tetramer in nucleosome structure and function is reflected in the remarkable conservation of sequence through evolution of both the N-terminal tail domains, and the C-terminal histone fold domains of the H3 and H4 proteins (Figure 8). The constraints in maintaining interactions with the $(H3, H4)_2$ tetramer and in wrapping DNA on their surface also presumably contribute to the necessity of maintaining sequences within the C-terminal histone fold domains of the (H2A, H2B) dimer (Figure 9). Within the C-terminal domains of H2A and H2B, the β-bridge structures and the paired end of helices motifs show more variation than the histone-histone interface. This suggests that the quality of histone-DNA interactions may change through evolution, but not the interactions between the histones. In contrast the N-terminal domains of histones H2A and H2B and the C-terminal domain of H2A exist with widely variant sequences. Interestingly, some of these H2A variants are evolutionarily conserved.

An evolutionarily conserved variant of H2A that is essential for *Drosophila* development is H2A.vD. This variant is known as H2A.Z in mammals and hv1 in *Tetrahymena* (reviewed by van Daal and Elgin, 1992). H2A.Z has an N-terminal tail that is similar to that of H4, and is post-translationally modified through acetylation to a greater extent than normal H2A. A second conserved variant is

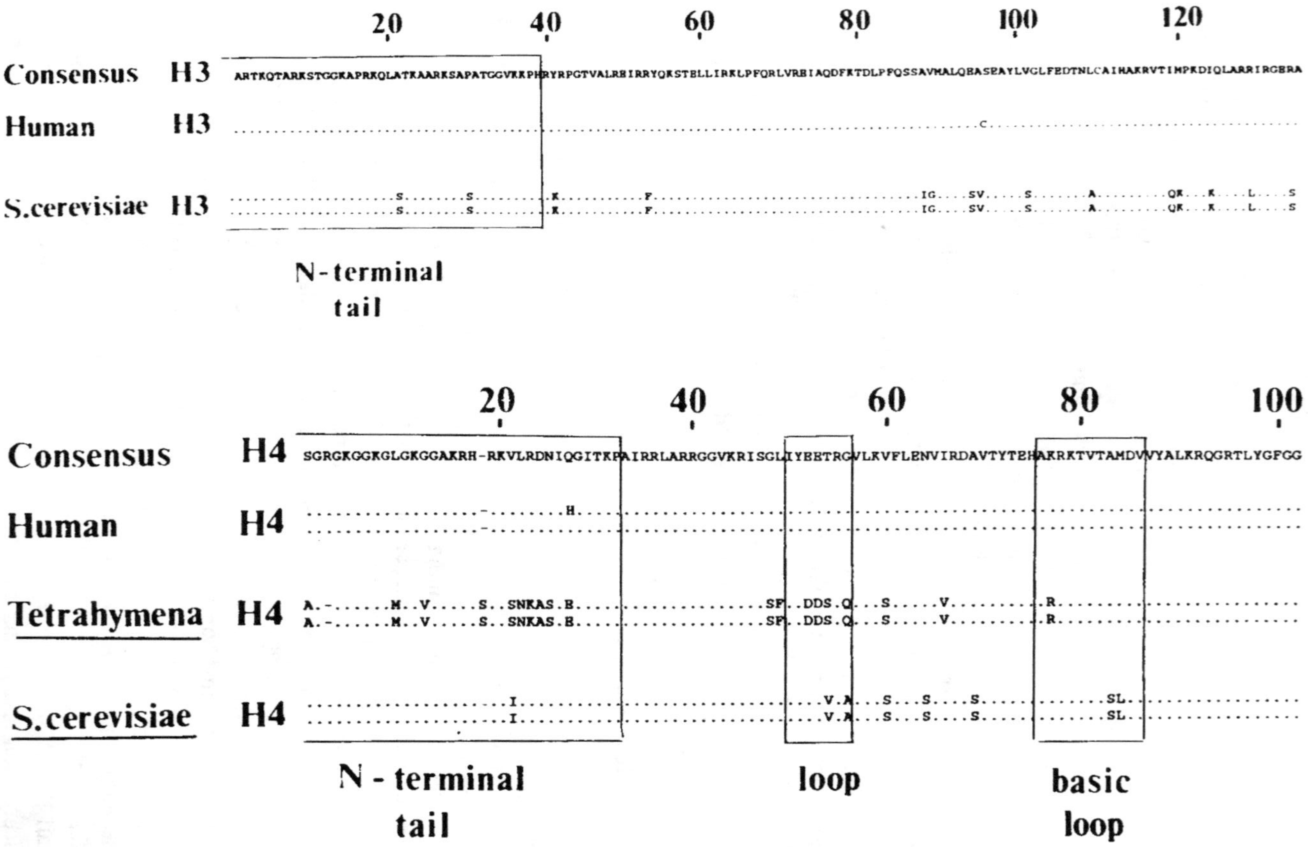

Figure 8. Comparison of different histone H3 and H4 primary sequences.

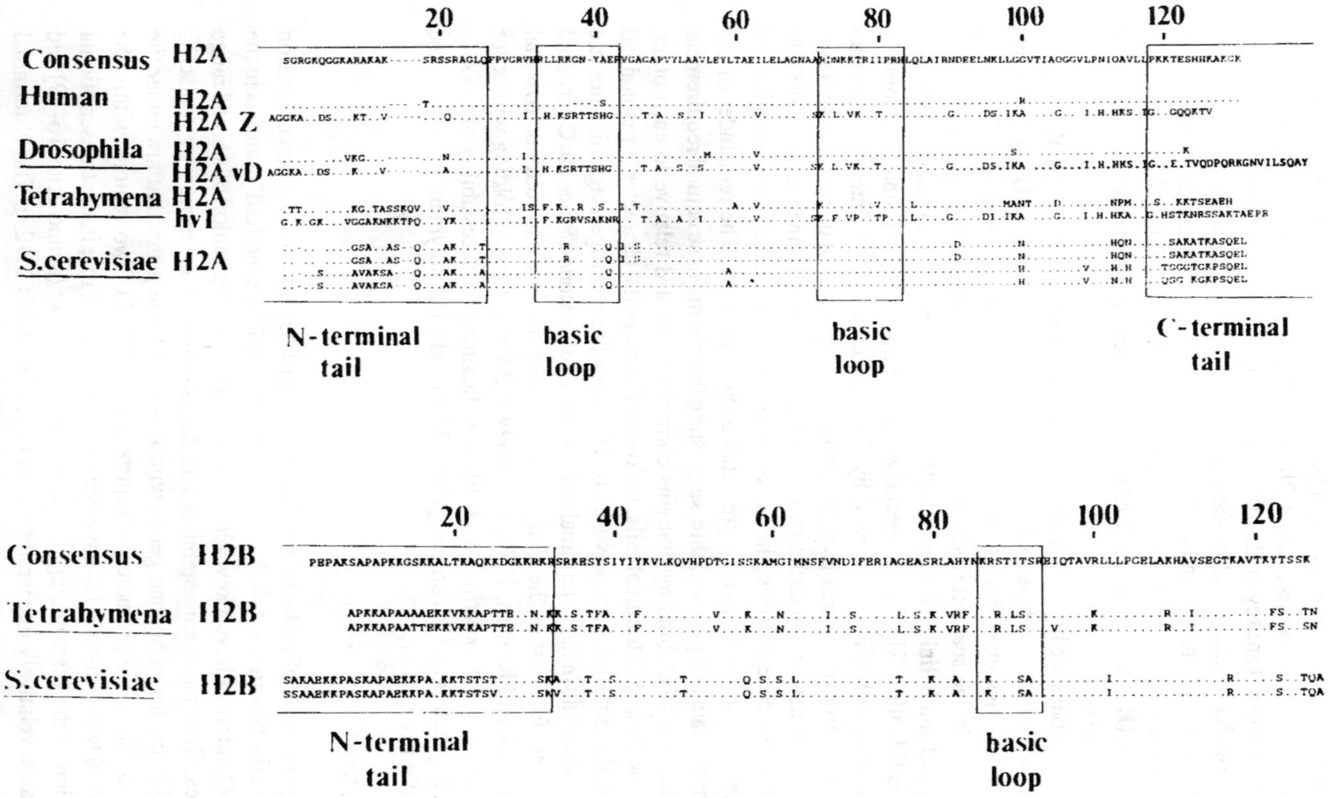

Figure 9. Comparison of different histone H2A and H2B primary sequences.

H2A.X (Mannironi et al., 1989), which has an extended C-terminal tail. Such extensions are not uncommon in variant H2A molecules, the most extreme is found in wheat H2A$_1$ where there is a nineteen amino-acid C-terminal extension which has the potential to protect about 20 bp of DNA immediately adjacent to the nucleosome core (Lindsey et al., 1991). This is again consistent with the proposed position of this domain within the nucleosome. Variant histones are especially prevalent in spermatozoa, and a H2B variant present in sea urchin sperm has an N-terminal tail with a twenty-one amino-acid extension. This tail interacts with linker DNA (Hill and Thomas, 1990) and significantly increases the stability of sperm nucleosomes.

Linker histones are highly variable in primary sequence, so much so that it is not always possible to recognize such proteins by sequence similarity. Distinct variants of linker histones exist during the development of several organisms including vertebrates. In the sea urchin (*S. purpuratus*), there are distinct sperm, cleavage (H1α), and adult linker histones (H1β and H1γ). In *Xenopus laevis* there is a cleavage B4, adult (H1) and a specialized linker histone only found in terminally differentiated cells (H1°) (reviewed by Dimitrov et al., 1993). Interesting predictions emerge from comparisons of the amino acid sequences of these different linker histones with the known structure of histone H5 (Ramakrishnan et al., 1993). In contrast to the core histones, the linker histones are most conserved in regions of the globular domain known or predicted to bind to DNA. The remainder of the globular domain is highly variable, suggesting that the major structural requirement is to have the DNA binding domains correctly orientated relative to each other. Within the N- and C-terminal tails, the overall hydrophilic character is maintained between the different histone variants. However the cleavage stage histones are much more acidic in the C-terminal tail than the adult histones. Since the C-terminal tail accounts for most of the charge neutralization necessary to fold nucleosomal arrays (Clark and Kimura, 1990) it is likely that less extensive folding will occur in cleavage stage chromatin. This is likely to facilitate the accessibility of DNA to transacting factors and the processivity of DNA and RNA polymerases through the chromatin fiber.

C. Mutants

The generation of mutant core histones in yeast has allowed the demonstration that both the N-terminal tail domains and C-terminal 'histone-fold' domains in the (H3, H4)$_2$ tetramer have a key role in the transcriptional regulation of a broad range of genes. This has been established in *Saccharomyces cerevisiae* by replacing individual copies of histone genes with mutant histone genes that are under the control of inducible promoters (Grunstein et al., 1992). Two regions within the N-terminal tail of histone H4 have been shown to have a role in the transcriptional repression of the yeast mating loci: a very basic segment (amino acids 16–19) and an adjacent relatively uncharged sequence (amino acids 21–29). The N-terminal tail of H4 also has a role in gene activation, since amino acids 4–23 are required for activation of the GAL1 promoter and several other genes (Grunstein et al., 1992). Interestingly, the N-terminal tail of H3 is required for the repression of many of the same genes that require the N-terminal tail domains of H4 for activation. These effects probably reflect specific interactions between the H3 and H4 tail domains and regulatory factors. However it is also possible that they represent different structural configurations of the histone octamer directed by post-translational modification states specific to either tail domain.

A role for the histones in the regulation of inducible gene expression has also been established through an independent line of investigation. Studies on the expression of both the HO endonuclease gene involved in the yeast mating-type switch and the SUC2 invertase gene involved in sucrose metabolism have led to the identification of genes encoding components of a multiprotein "general" activator complex (Herskowitz et al., 1992). These sucrose nonfermenting (SNF) and switch (SWI) genes influence transcription pleiotropically. The important protein components of the general activator complex (SNF5, SNF6, SWI1, SWI2, and SWI3) appear to function together as a single large complex. This complex is required for the expression of many inducible genes (e.g., PHO5, CUP1, and HIS3) and also for transcriptional induction by the glucocorticoid receptor when it is expressed in yeast (Yoshinaga et al., 1992). The link to chromatin structure follows from the isolation of suppressor mutations in switch-independent (SIN) genes that allow inducible gene transcription in yeast strains deficient for SNF or SWI gene activities. These mutations include deletion of histones H2A-H2B (Winston and Carlson, 1992) and point mutations in the C-terminal globular domain of histone H3 (Herskowitz et al., 1992).

The point mutations are especially interesting since they suggest that destabilization of the $(H3, H4)_2$ tetramer and its interaction with DNA relieves the requirement for the general activator complex. By implication in a normal cell, the general activator complex acts to destabilize chromatin. This destabilization may be through the $(H3, H4)_2$ tetramer, but it may also be through any other aspect of chromatin structure. As we have discussed most aspects of chromatin structure are at least partially dependent on the prior association of the $(H3, H4)_2$ tetramer with DNA.

VII. CONCLUSIONS

The nucleosome has an unexpectedly complex anatomy. The histone proteins undergo highly selective interactions between themselves and with DNA. Although the basis for this selectivity has not been resolved, several new features have been recently recognized. Most notable is the extended helical structure of the C-terminal domains of the histones. The C-terminal domains of the core histones do not form monomeric globules, but have extensive protein-protein and protein-DNA contacts. The interfaces between histone heterodimers are not extensive, and although specific, offer the possibility of conformational flexibility. Moreover the stability of the interaction of the histones with DNA depends upon the presence of continuous contact with DNA. This may be disrupted by *trans*-acting factors or RNA polymerase. Conformational changes in the histone octamer may follow from sequestration of linker histones into the nucleosome or from modification or mutation of the tail domains of the core histones. Since these changes are implicated in the regulation of *trans*-acting factor access to DNA in chromatin, they are likely to be an important area for future investigation.

Nucleosome structure is seen to be built upon the interaction of the $(H3, H4)_2$ tetramer with DNA. Conservation of the H3 and H4 histones through evolution is remarkable and it is not surprising that mutation or modification of these essential structural proteins has major consequences for gene regulation. Importantly the mutations that exert these effects occur not only in the N-terminal tail domains that are also the sites of acetylation and phosphorylation, but also in the C-terminal histone fold domain. This suggests that events potentially modifying the integrity of the tetramer itself and its interaction with DNA have important consequences. In contrast, histones H2A and H2B have much more flexibility in structure. Many variants and modifications exist, most of which are likely to modulate the folding of linker DNA or the interaction of linker histones within the nucleosome.

We have discussed the problems that new structural data on the histone octamer and linker histones pose for the existing paradigms for linker histone incorporation into the nucleosome and the chromatin fiber. A new model in which the linker histone is asymmetrically placed within the nucleosome is compatible with existing structural data and supports a solenoidal model of the chromatin fiber in which the linker DNA is wound in a continual superhelix with the DNA in the nucleosome. Linker histones are very variant through evolution, this structural diversity indicates that a number of independent solutions have been found for stabilizing the folding of the nucleosomal array into the chromatin fiber. Since many of these variants are developmentally regulated, we suggest that different chromatin fiber structures may have distinct functional roles during gametogenesis, embryogenesis, and within somatic cells (Bouvet et al., 1994). The future offers the exciting prospect of interrelating this particular chromatin structure with gene regulation in molecular detail.

REFERENCES

Allan, J., Hartman, P.G., Crane-Robinson, C., & Aviles, F.X. (1980). The structure of histone H1 and its location in chromatin. Nature 288, 675–679.

Arents, G., Burlingame, R.W., Wang, B.W., Love, W.E., & Moudrianakis, E.N. (1991). The nucleosomal core histone octamer at 3.1Å resolution: a tripartite protein assembly and a left-handed superhelix. Proc. Natl. Acad. Sci. USA 88, 10148–10152.

Arents, G., & Moudrianakis, E.N. (1993). Topography of the histone octamer surface: repeating structural motifs utilized in the docking of nucleosomal DNA. Proc. Natl. Acad. Sci. USA 90, 10489–10493.

Bauer, W.R., Hayes, J.J., White, J.H., & Wolffe, A.P. (1994). Nucleosome structural changes due to acetylation. J. Mol. Biol. 236, 685–690.

Bavykin, S.G., Usachenko, S.I., Zalensky, A.O., & Mirzabekov, A.D. (1990). Structure of nucleosomes and organization of internucleosomal DNA in chromatin. J. Mol. Biol. 212, 495–511.

Boulikas, T., Wiseman, J.M., & Garrard, W.T. (1980). Points of contact between histone H1 and the histone octamer. Proc. Natl. Acad. Sci. USA 77, 121–131.

Bouvet, P., Dimitrov, S., & Wolffe, A.P. (1994). Specific regulation of *Xenopus* chromosomal 5S rRNA gene transcription *in vivo* by histone H1. Genes Dev. 8, 1147–1159.

Cary, P.D., Crane-Robinson, C., Bradbury, E.M., & Dixon, G.H. (1982). Effect of acetylation on the binding of N-terminal peptides of histone H4 to DNA. Eur. J. Biochem. 127, 137–143.

Chen, H., Li, B., & Workman, J.L. (1994). A histone-binding protein nucleoplasmin stimulates transcription factor binding to nucleosomes and factor-induced nucleosome disassembly. The EMBO J. 13, 380–390.

Clark, D.J., & Kimura, T. (1990). Electrostatic mechanism of chromatin folding. J. Mol. Biol. 211, 883–896.

Clark, K.L., Halay, E.D., Lai, E., & Burley, S.K. (1993). Co crystal structure of the HNF 3/fork head DNA recognition motif resembles histone H5. Nature 364, 412–417.

Dong, F., & van Holde, K.E. (1991). Nucleosome positioning is determined by the (H3/H4)$_2$ tetramer. Proc. Natl. Acad. Sci. USA 88, 10596–10600.

Ebralidse, K.K., Grachev, S.A., & Mirzabekov, A.D. (1988). A highly basic histone H4 domain bound to the sharply bent region of nucleosomal DNA. Nature 331, 365–367.

Eickbusch, T.H., & Moudrianakis, E.N. (1978). The histone core complex: an octamer assembled by two sets of protein-protein interactions. Biochemistry 17, 4955.

Garcia-Ramirez, M., Dong, F., & Ausio, J. (1992). Role of the histone 'tails' in the folding of oligonucleosomes depleted of histone H1. J. Biol. Chem. 267, 19587–19595.

Graziano, V., Gerchman, S.E., Schneider, D.K., & Ramakrishnan, V. (1994). Histone H1 is located in the interior of the chromatin 30nm filament. Nature 368, 351–354.

Grunstein, M., Durrin, L.K., Mann, R.K., Fisher-Adams, G., & Johnson, L.M. (1992). In: Transcriptional Regulation (McKnight, S., & Yamamoto, K., eds.), pp. 1295–1315. Cold Spring Harbor Press, New York.

Gushchin, D.Y., Ebralidse, K.K., & Mirzabekov, A.D. (1991). Structure of the nucleosome: localization of the segments of the H2A histones that interact with DNA by DNA-protein crosslinking. Molek. Biol. 25, 1400–1411.

Hansen, J.C., & Wolffe, A.P. (1994). A role for histones H2A/H2B in chromatin folding and transcriptional repression. Proc. Natl. Acad. Sci. USA 91, 2339–2343.

Hayes, J.J., Tullius, T.D., & Wolffe, A.P. (1990). The structure of DNA in a nucleosome. Proc. Natl. Acad. Sci. USA 87, 7405–7409.

Hayes, J.J., Clark, D.J., & Wolffe, A.P. (1991a). Histone contributions to the structure of DNA in the nucleosome. Proc. Natl. Acad. Sci. USA 88, 6829–6833.

Hayes, J.J., Bashkin, J., Tullius, T.D., & Wolffe, A.P. (1991b). The histone core exerts a dominant constraint on the structure of DNA in a nucleosome. Biochemistry 30: 8434–8440.

Hayes, J.J., & Wolffe, A.P. (1992). Histones H2A/H2B inhibit the interactions of transcription factor IIIA with the *Xenopus borealis* somatic 5S RNA gene in a nucleosome. Proc. Natl. Acad. Sci. USA 89, 1229–1233.

Hayes, J.J., & Wolffe, A.P. (1993). Preferential and asymmetric interaction of linker histones with 5S DNA in the nucleosome. Proc. Natl. Acad. Sci. USA 90, 6415–6419.

Hayes, J.J., Pruss, D., & Wolffe, A.P. (1994). Histone domains required to assemble a chromatosome including the *Xenopus borealis* somatic 5S rRNA gene. Proc. Natl. Acad. Sci. USA (in press).

Herskowitz, I., Andrews, B., Kruger, W., Ogas, J., Sil, A., Coburn, C., & Peterson, C. (1992). Integration of multiple regulatory inputs in the control of HO expression in yeast. In: Transcriptional

Regulation (McKnight, S., & Yamamoto, K. eds.), Vol. 2, p. 949. Cold Spring Harbor Press, Ne York.

Hill, C.S., & Thomas, J.O. (1990). Core histone-DNA interactions in sea urchin sperm chromatin: t N-terminal tail of H2B interacts with linker DNA. Eur. J. Biochem. 187, 145–153.

Hill, C.S., Rimmer, J.M., Green, B.M., Finch, J.T., & Thomas, J.O. (1991). Histone-DNA interactio and their modulation by phosphorylation of Ser-Pro-X-Lys/Arg motifs. EMBO J. 10, 1939–194

Hogan, M.E., Rooney, T.F., & Austin, R.H. (1987). Evidence for kinks in DNA folding in t nucleosome. Nature 328, 554–557.

Klug, A., & Lutter, L.C. (1981). The helical periodicity of DNA on the nucleosome. Nucl. Acids R 9, 4267–4283.

Kornberg, R. (1974). Chromatin structure: a repeating unit of histones and DNA. Science 184, 868–8

Kornberg, R., & Thomas, J.O. (1974). Chromatin structure: oligomers of histones. Science 18 865–868.

Lambert, S., Muyldermans, S., Baldwin, J., Kilner, J., Ibel, K., & Wijns, L. (1991). Neutron scatteri studies of chromatosomes. Biochem. Biophys. Res. Comm. 179, 810–816.

Lee, D.Y., Hayes, J.J., Pruss, D., & Wolffe, A.P. (1993). A positive role for histone acetylation transcription factor binding to nucleosomal DNA. Cell 72, 73–84.

Lennard, A.C., & Thomas, J.O. (1985). The arrangement of H5 molecules in extended and condens chicken erythrocyte chromatin. EMBO J. 4, 3455–3462.

Levinger, L., & Varshavsky, A. (1982). Selective arrangement of ubiquitinated and D1 protein-conta ing nucleosomes within the Drosophila genome. Cell 28, 375–385.

Lindsey, G.G., Orgeig, S., Thompson, P., Davies, N., & Maeder, D.L. (1991). Extended C-terminal t of wheat histone H2A interacts with DNA of the linker region. J. Mol. Biol. 218, 805–813.

Mahadevan, L.C., Willis, A.C., & Barrah, M.J. (1991). Rapid histone H3 phosphorylation in respor to growth factors, phorbol esters, okadaic acid and protein synthesis inhibitors. Cell 65, 775–7

Mannironi, C., Bonner, W.M., & Hatch, C.L. (1989). H2A.X a histone isoprotein with a conserv C-terminal sequence is encoded by a novel mRNA with both DNA replication type and poly 3′ processing signals. Nucl. Acids Res. 17, 9113–9126.

Mirzabekov, A.D., Bavykin, S.G., Belyavsky, A.V., Karpov, V.L., Preobrazhenskaya, O.V., Schick, V. & Ebralidze, K.K. (1989). Mapping DNA-protein interactions by crosslinking. Methods Enzyn 170, 386–408.

Mirzabekov, A.D., Pruss, D.V., & Elbralidse, K.K. (1990). Chromatin super-structure depend cross-linking with DNA of the histone H5 residues Thr1, His25 and His62. J. Molec. Biol. 2 479–491.

Muyldermans, S., & Travers, A.A. (1994). DNA sequence organization in chromatosomes. J. Mol. Bi 235, 855–870.

Pruss, D., & Wolffe, A.P. (1993). Histone-DNA contacts in a nucleosome core containing a Xenopus rRNA gene. Biochemistry 32, 6810–6814.

Pruss, D., Bushman, F.D., & Wolffe, A.P. (1994). HIV integrase directs integration to sites of sev DNA distortion within the nucleosome core. Proc. Natl. Acad. Sci. USA 91, 5913–5917.

Ramakrishnan, V., Finch, J.T., Graziano, V., Lee, P.L., & Sweet, R.M. (1993). Crystal structure globular domain of histone H5 and its implications for nucleosome binding. Nature 362, 219–2

Richmond, T.J., Finch, J.T., Rushton, B., Rhodes, D., & Klug, A. (1984). Structure of the nucleoso core particle at 7Å resolution. Nature 311, 532–537.

Simpson, R.T. (1978). Structure of the chromatosome, a chromatin core particle containing 160 b pairs of DNA and all the histones. Biochemistry 17, 5524–5521.

Simpson, R.T. (1991). Nucleosome positioning: occurrence, mechanisms and functional consequenc Prog. Nucl. Acids Res. Molec. Biol. 40, 143–184.

Staynov, D.Z., & Crane-Robinson, C. (1988). Footprinting of linker histones H5 and H1 on nucleosome. EMBO J. 7, 3685–3691.

Thoma, F., Koller, T., & Klug, A. (1979). Involvement of histone H1 in the organization of the nucleosome and if the salt dependent superstructures of chromatin. J. Cell Biol. 83, 407–427.

Thomas, J.O., Rees, C., & Finch, J.T. (1992). Cooperative binding of the globular domains of histones H1 and H5 to DNA. Nucl. Acids Res. 20, 187–194.

van Daal, A., & Elgin, S.C.R. (1992). A histone variant, H2AvD, is essential in *Drosophila melanogaster*, Mol. Biol. Cell. 3, 593–602.

van Holde, K.E. (1989). Chromatin. Springer-Verlag, New York.

van Holde, K.E., Lohr, D., & Robert, C.H. (1992). What happens to nucleosomes during transcription? J. Biol. Chem. 267, 2837–2840.

Varshavsky, A.J., Bakayev, V.V., & Georgiev, G.P. (1976). Heterogeneity of chromatin subunits *in vitro* and location of histone H1. Nucl. Acids Res. 3, 477–492.

Winston, F., & Carlson, M. (1992). Yeast SNF/SWI transcriptional activators and the SPT/SIN chromatin connection. Trends Genet. 8, 387.

Yao, J., Lowary, P.T., & Widom, J. (1990). Direct detection of linker DNA bending in defined-length oligomers of chromatin. Proc. Natl. Acad. Sci. USA 87, 7603–7607.

Yao, J., Lowary, P.T., & Widom, J. (1991). Linker DNA bending by the core histones of chromatin. Biochemistry 30, 8408–8414.

Yoshinaga, S.K., Peterson, S.L., Herskowitz, I., & Yamamoto, K.R. (1992). Roles of SWI1, SWI2 and SWI3 proteins for transcriptional enhancement by steroid receptors. Science 258, 1598–1604.

Chapter 2

Histone Acetylation During Chromatin Replication and Nucleosome Assembly

ANTHONY T. ANNUNZIATO

ABSTRACT

In the following communication, the relationship between histone acetylation and nucleosome assembly is reviewed. Following an overview of the data regarding the acetylation of newly synthesized histones, and the possible enzymes involved, a description of the biochemical and genetic approaches to this topic is presented.

The Nucleus
Volume 1, pages 31–56.
Copyright © 1995 by JAI Press Inc.
All rights of reproduction in any form reserved.
ISBN: 1-55938-940-0

Histone storage complexes are then detailed, and the involvement of acetylation in histone deposition and chromatin remodeling is addressed. In addition, possible connections between acetylation and the propagation of transcriptionally active and inactive nucleoprotein structures during chromatin replication are outlined, and correlations between the phosphorylation of H2A and the generation of physiologically spaced chromatin are discussed. A brief summary of our current understanding of assembly-related histone acetylation and suggestions for future research are offered in the concluding remarks.

I. INTRODUCTION

The topics of chromatin replication and nucleosome assembly have recently been the subjects of several review articles (Annunziato, 1990; Wolffe, 1991; Gruss and Sogo, 1992; Almouzni and Wolffe, 1993a; Laskey et al., 1993), as has histone acetylation (Csordas, 1990; Turner, 1991; van Holde et al., 1992; Ausio, 1992; Bradbury, 1992; Hansen and Ausio, 1992; Lopez-Rodas et al., 1993; Tordera et al., 1993). Therefore, a comprehensive discussion of these subjects will not be attempted here. Rather, this report will focus on the involvement of histone acetylation in chromatin biosynthesis.

It is now approximately 20 years since Gordon Dixon and his colleagues first presented evidence that newly synthesized H4 is reversibly acetylated during trout spermatogenesis (Louie et al., 1974). Since then extraordinary progress has been made both in our understanding of the architecture of chromatin, and in our ability to define the pathway(s) of nucleosome assembly. Still, a definition of the role of histone post-translational modifications during chromatin replication remains elusive. The following analysis will begin with a summary of the evidence that specific histone post-translational modifications accompany histone deposition, along with a brief description of the enzymes that are likely to be involved. Experiments aimed at determining the function of histone acetylation during nucleosome assembly will then be discussed, followed by an examination of the relationship between histone acetylation and the propagation of transcriptionally competent and repressed chromatin structures.

II. ACETYLATION OF NEWLY SYNTHESIZED HISTONES

A. The Mode and Sites of Modification

The first detailed study of the reversible acetylation of newly synthesized H4 in somatic cells was performed by Ruiz-Carrillo et al. (1975), using duck erythroblasts. The principle findings of this landmark paper can be summarized as follows: 1) newly synthesized nuclear H4 (labeled with [³H]leucine for one minute, and chased for 2–4 minutes) was found in three subtypes, identified as the unmodified form, the monophosphorylated form, and a "dimodified" monoacetylated/monophosphorylated form; 2) within 20 minutes of deposition onto DNA, the modified forms of H4 were almost completely converted to the unmodified form; 3) the kinase and acetyltransferase activities responsible for modifying new H4 were located in a cytoplasmic preparation (post-mitochondrial supernatant); and 4) newly synthesized H3 was not reversibly modified, but instead entered the nucleus "acetylated in about the same proportions as the 'old' H3" (Ruiz-Carrillo et al., 1975). Further, reversible acetylation was localized to lysine residues in the H4 N-terminal "tail," while phosphorylation occurred on the N-terminal serine residue. In earlier reports, irreversible acetylation of this N-terminal serine had also been described (Phillips, 1963; Liew et al., 1970).

The next major study of the modification of nascent histones was presented by Chalkley and colleagues using hepatoma tissue culture (HTC) cells (Jackson et al., 1976). Although the overall features of histone synthesis and maturation resembled those presented by Ruiz-Carrillo and colleagues, significant differences emerged. Whereas newly synthesized H3 in duck erythroblasts was found to be essentially unmodified, in HTC cells this histone was apparently acetylated prior to deposition, then rapidly deacetylated, much like H4. The "dimodified" form of new H4 was described as predominantly diacetylated (though a modest level of phosphorylation could not be excluded); moreover, the level of H4 acetylation increased from mono- to diacetylated *following import into the nucleus.* In fact, although it is often convenient to refer to new H4 as diacetylated prior to its deposition onto DNA, the time course of the modification of newly synthesized H4 (relative to its import into the nucleus and subsequent deposition) has yet to be fully defined for somatic cells.

A number of points can be made concerning the differences outlined above. First, it must be recalled that these studies were performed prior to the introduction of sodium butyrate as a deacetylase inhibitor, and that the presence of active histone deacetylases can dramatically affect the analysis of histone acetylation. Second, Ruiz-Carillo and colleagues used sodium fluoride to block the potential action of protein phophorylases, while Jackson and colleagues did not. In this regard, it should be noted that treatment of nascent *Tetrahymena* H4 with alkaline phosphatase does not alter its "dimodified" position in acid-urea gels, consistent with a diacetylated isoform (Allis et al., 1985); similar results have been observed for newly synthesized H4 in HeLa cells (author's unpublished observations). Third, subsequent experiments performed in the presence of sodium butyrate have failed to demonstrate that new vertebrate H3 is significantly modified relative to preexisting H3 molecules (Sealy and Chalkley, 1979; Cousens and Alberts, 1982; Sobel et al., 1995), although a doublet in the region of H3 has been occasionally (but inconsistently) seen in fluorographs of acid-urea gels (see Cousens and Alberts, 1982 for a further discussion). This more slowly migrating protein has not been positively identified as H3.

The diacetylated state of newly synthesized H4 is probably a universal feature of histone metabolism: it also has been observed in trout (Louie et al., 1974), *Xenopus* oocytes (Woodland, 1979; Dimitrov et al., 1993), developing sea urchin

(Chambers and Shaw, 1984), starfish (Ikegami et al., 1993), *Drosophila* (Giancotti et al., 1984), *Tetrahymena* (Allis et al., 1985), and *Physarum* (although new H4 in this system was mostly monoacetylated) (Waterborg and Matthews, 1984). Nascent H2A and H2B appear to enter chromatin unmodified (Jackson et al., 1976; Cousens and Alberts, 1982).

The situation with H3 is more complicated. As discussed above, newly synthesized H3 is mostly unacetylated in the vertebrate systems thus far studied. In *Tetrahymena*, new H3 is present primarily in either a di- or a-monacetylated form (Allis et al., 1985). New H3 in *Physarum* is probably also mono- or diacetylated (Waterborg and Matthews, 1984). The stored H3 in *Xenopus* oocytes is acetylated (but no more so than chromatin-bound H3), and becomes deacetylated during the maturation of oocyte to egg (Woodland, 1979). Curiously, although in both *Tetrahymena* and *Drosophila* newly synthesized H3 is diacetylated, different patterns of acetylation are used in each organism: nascent *Tetrahymena* H3 is acetylated at lysines 9 and 14, *Drosophila* H3 at lysines 14 and 23 (Sobel et al., 1995). To date there has been no explanation for this variability.

Histone H4 is potentially acetylated at lysines at positions 5, 8, 12, and 16 (Bradbury, 1992). An obvious difficulty in determining the specific lysines involved in the acetylation of nascent histones is that preexisting histones are also subject to acetylation (van Holde, 1988; Wolffe, 1992a). One means of dealing with this problem has been to exploit systems in which deposition versus transcription-related acetylation can be distinguished. For example, in synchronously growing *Physarum* plasmodia, acetyl turnover on H4 is detected at positions 8, 12, and 16 during G2 phase, while turnover at position 5 is uniquely correlated with S (Pesis and Matthews, 1986). Histone acetylation sites have also been determined in *Tetrahymena*. This ciliated protozoan possesses a macronucleus that is involved in both transcription and replication, as well as a micronucleus that is transcriptionally inert. During DNA replication in both macro- and micronuclei, the deposition-related acetylation of *Tetrahymena* H4 occurs exclusively at positions 4 and 11 (Chicoine et al., 1986). Because the arginine at position 3 has been deleted in *Tetrahymena*, these sites correspond to lysines 5 and 12 in H4 of higher organisms. Thus, in both *Tetrahymena* and *Physarum*, the acetylation of lysine 5 is correlated with histone deposition.

In more recent work, the specific acetyl-lysines present in newly synthesized human H4 have been identified (Sobel et al., 1995). HeLa cells were labeled for 2 to 10 minutes in the presence of sodium butyrate. H4 was purified, deblocked, and sequenced; the radiolabeled lysines at positions 5, 8, 12, and 16 were then subjected to HPLC fractionation under conditions in which acetyl-lysine is separated from unmodified lysine (as described in Chicoine et al., 1986). Remarkably, acetylation of nascent human H4 occurs exclusively on lysines 5 and 12, in precise correspondence to the pattern found in *Tetrahymena* (allowing for the single arginine deletion). Further, the "5-12" pattern of nascent H4 acetylation was also observed in *Drosophila* cultured cells (Sobel et al., 1995). That the identical acetylation pattern of new H4 has been so rigorously conserved—at least since the divergence of multicellular organisms from the Protista—argues compellingly for a fundamental role for this modification during nucleosome assembly.

It may be useful here to compare the acetylation pattern of nascent H4 to the preferred order in which H4 acetylation sites are used in cycling cells. Using a number of means (e.g., Edman degradation, peptide mapping, specific antisera), the acetylation frequency of the lysines at positions 5, 8, 12, and 16 of H4 have been determined for several systems. In human cells (Turner et al., 1989; Thorne et al., 1990), calf thymus (Couppez et al., 1987), and yeast (Clarke et al., 1993), monoacetylated H4 (H4Ac$_1$) is acetylated almost exclusively at lysine 16, while lysines 16 and 12 are the predominant forms found associated with H4Ac$_2$ (although some lysine 8 and a lesser amount of lysine 5 are also detected). Moreover, lysine 8 is preferentially selected to generate H4Ac$_3$, and lysine 5 is typically the last to be acetylated to produce H4Ac$_4$. Similar observations have been made for plants (Waterborg, 1992). Although at first this might appear to be at odds with the "5–12" pattern exhibited by newly synthesized H4 in some species, it must be recalled that transcription-related acetylation will normally eclipse the deposition-related pattern unless specific measures are taken to avoid this (hence the advantage of systems like *Physarum* and *Tetrahymena*). It would also seem that deposition-related acetylation must be partially or entirely erased (especially at lysine 5) to establish the transcription-related pattern (e.g., diacetylation at lysines 12 and 16). Interestingly, in cuttlefish testis (where the acetylation of H4 is involved in the displacement of histones by protamines during spermiogenesis), only the lysines at positions 5 and 12 are acetylated in H4Ac$_2$ (Couppez et al., 1987). Thus, as discussed by Couppez and colleagues the acetylation of these specific lysine residues is strongly correlated with either histone deposition or histone removal.

In the interest of thoroughness, it must be added that transcription-related acetylation patterns are not always so well conserved. For example, in *Drosophila* cultured cells H4Ac$_1$ is acetylated at lysines 5, 8, or 12 with more or less equal frequency (Munks et al., 1991), and in *Tetrahymena* lysine 15 (the equivalent of lysine 16 in other systems) is the probably the *last* to be acetylated (Chicoine et al., 1986). It is not clear whether these differences are due to several acetyltransferases with distinct specificities, or to variations in the accessibility of H4 N-terminal lysines in different species (or cell type).

B. Deposition-Related Histone Acetyltransferases

The enzyme(s) responsible for acetylating newly synthesized H4 have yet to be positively identified. Nevertheless, likely candidates have been found, namely the "type B" (i.e., histone acetyltransferase B or HATB) class of enzymes, which appear in the cytosol following nuclear isolation. Among the hallmark features of canonical HATB enzymes are that they are specific for H4, acetylate only free H4 (but not H4 in chromatin), and appear to be cytoplasmic (Garcea and Alberts, 1980; Sures and Gallwitz, 1980; Weigand and Brutlag, 1981; Lopez-Rodas et al., 1991a, 1991b).

Interestingly, the *Tetrahymena* enzyme was found both in the cytoplasm and in micronuclei (but not macronuclei), and could acetylate H3 as well as H4 under certain conditions (Richman et al., 1988).

In pea (Mingarro et al., 1993) and *Tetrahymena* (Richman et al., 1988) the specific lysine residues of H4 acetylated by HATB have been identified. Perhaps not surprisingly, in both cases the deposition-related pattern of acetylation was generated: lysines 5 and 12 in pea; lysines 4 and 11 in *Tetrahymena*. In addition (and rather confusingly), in both organisms a third acetyllysine was detected following incubation of H4 with HATB. In *Tetrahymena* this was ascribed to the use of a partially acetylated substrate; however, this could not explain the result obtained with the pea acetylase. Whether or not this reflects the pea activity *in vivo* is unclear: The acetylation pattern of newly synthesized H4 in pea has not yet been determined. Recently, a HATB activity from *Drosophila* embryo extracts was characterized (Sobel et al., 1994). In this case, only the lysine at position 12 was acetylated *in vitro*, suggesting that another activity may be responsible for acetylating lysine 5 (to produce the diacetylated pattern of nascent *Drosophila* H4).

A thorough explanation of the events that generate nascent, chromatin-bound H4Ac$_2$ will necessarily hinge on an understanding of the HATB activity. As alluded to earlier, it is formally possible that in at some systems somatic new H4 enters the nucleus in a monoacetylated form, and is subsequently diacetylated by a nuclear acetylase either before or after nucleosome assembly. Indeed, in *Tetrahymena* a macronuclear acetyltransferase has been described that will diacetylate nucleosomal H4 at lysines 4 and 7, but can also acetylate *free* H4 in the deposition-related pattern (4–11) (Chicoine et al., 1987). If, however, cytoplasmic HATB is universally found to provide nascent H4 with the complete deposition-related acetylation pattern, a requirement for a nuclear activity during chromatin assembly would not be expected (at least insofar as the generation of transcriptionally inert chromatin is concerned).

III. FUNCTIONAL ANALYSIS OF HISTONE ACETYLATION DURING CHROMATIN ASSEMBLY

A. Biochemical Approaches

The biological significance of reversible acetylation during chromatin biosynthesis is still unknown. One approach to this question has involved the examination of chromatin assembled in the presence of the deacetylase inhibitor sodium butyrate. During normal replication and nucleosome assembly, new H4 is deacetylated within 20–30 minutes of its deposition (Ruiz-Carrillo et al., 1975; Jackson et al., 1976), concomitant with the maturation of nascent chromatin (reviewed in Annunziato, 1990). The presence of butyrate during this period prevents the deacetylation of new H4, preserving this usually transient modification. Rather surprisingly, the maintenance of "deposition-type" histone acetylation on newly replicated/newly assembled chromatin does not impede nucleosome assembly per se, nor does it alter the repeat-length of the assembled region (Annunziato and

Seale, 1983; Shimamura and Worcel, 1989). Thus, histone deacetylation and nucleosome assembly are not obligatorily coordinated. However, chromatin replicated and assembled in the presence of butyrate (bu-chromatin) retains an increased sensitivity to DNaseI (Annunziato and Seale, 1983), a feature that is diagnostic of newly replicated chromatin as well as of acetylated chromatin in general (reviewed in Csordas, 1990; Turner, 1991). Notably, Shimamura and Worcel (1989) found that non-replicating plasmids assembled by a *Xenopus* oocyte extract in the presence of sodium butyrate were *not* preferentially DNaseI sensitive. As suggested by Shimamura and Worcel (1989), the inability of small DNA plasmids to form higher order chromatin structures may have precluded detecting the influence of butyrate on chromatin folding.

Subsequent studies of chromatin replicated in butyrate have shown that: 1) mononucleosomes prepared from bu-chromatin do not exhibit heightened DNaseI sensitivity, in agreement with results obtained for hyperacetylated mononucleosomes (Simpson, 1978); 2) mono- and oligonucleosomes of bu-chromatin are preferentially soluble in magnesium (relative to control chromatin), which is related to a reduced H1 content *in vitro*; 3) reconstitution with excess histone H1 abolishes the preferential Mg^{2+}-solubility of chromatin replicated in butyrate, in a concentration-dependent manner; and 4) trypsinization of the histone N-terminal tails (the sites of acetylation) also eliminates differential Mg^{2+}-solubility (Perry and Annunziato, 1989, 1991).

These findings suggest that the properties of acetylated nascent chromatin are mediated at an organizational level above the nucleosome, involving a reduction in H1-dependent nucleosomal interactions, and perhaps in H1 content. Consistent with this are observations that hyperacetylated chromatin also has an increased Mg^{2+}-solubility (e.g., Perry and Chalkley, 1982; Ridsdale et al., 1990) and a partially relaxed higher order structure (McGhee et al., 1983; Annunziato et al., 1988). In light of the proposed involvement of the histone N-termini in chromatin condensation (Chahal et al., 1980; Allan et al., 1982; Garcia-Ramirez et al., 1992), and the observation that the acetylation of H3 and H4 reduces DNA supercoiling around the nucleosome core (Norton et al., 1989, 1990; Thomsen et al., 1991), it has been suggested that acetylation may modify the binding of H1 to newly assembled nucleosomes (Perry and Annunziato, 1989, 1991). This could in turn help to regulate the folding of nascent chromatin during assembly.

The effects of acetylation on the binding of linker histones to chromatin is not entirely clear. For example, Reeves and colleagues (1985) observed that when plasmids were transfected into mammalian cells in the presence of butyrate, the (acetylated) assembled minichromosomes were preferentially sensitive to DNaseI and H1-depleted. In other experiments, however, SV40 assembled in butyrate-treated cells appeared to have a normal H1-complement, despite high levels of core histone acetylation (Lutter et al., 1992). Moreover, the presence of butyrate during *Xenopus* embryogenesis did not prevent the transition from B4 linker histone to histone H1 which occurs during development, or significantly lower the total level of nuclear H1 (Dimitrov et al., 1993).

It is possible that some of the inconsistencies in these results can be ascribed to differences in experimental design. In cases in which the H1 content of whole nuclei was examined (Dimitrov et al., 1993), a global picture of H1 deposition was provided within a system containing significant levels of un- and monoacetylated H4 (in addition to the more highly acetylated isoforms). Analyses of nucleosomes replicated in the presence of butyrate, in contrast, focus on regions containing diacetylated newly synthesized H4, and undergoing dynamic chromatin assembly. Here, too, complications may arise because isolated chromatin fibers (prepared by nuclease digestion) almost certainly have properties that differ from chromatin *in situ* (Giannasca et al., 1993). *In vitro* assays will also be highly sensitive to ionic conditions, especially if acetylation has only a subtle effect on DNA topology, or on the affinity of H1 for newly assembled nucleosomes. Indeed, in experiments in which Mg^{2+}-solubility was used to monitor nucleosomal interactions, most acetylated nascent chromatin was found to be H1-*depleted* (not H1-*free*), with ~30 percent of new chromatin apparently containing a normal complement of linker histone (Perry and Annunziato, 1989, 1991). Moreover, it has now been shown that core histone hyperacetylation has little effect on the reconstitution of histone H5 onto a mononucleosome containing a 5 S rRNA gene (Ura et al., 1994), indicating that there is no simple relationship between acetylation and linker histone content (at least for H5).

Histone-DNA interactions are particularly labile during the earliest stages of nucleosome assembly, especially for histone H1 (Jackson et al., 1981; Seale, 1981; Bavykin et al., 1993). As discussed by Bavykin et al. (1993), linker histones might be deposited onto newly replicated chromatin within minutes of DNA synthesis (Annunziato et al., 1981; Bavykin et al., 1993), but in a metastable manner that does not promote formation of the 30-nm fiber. Similarly, the maintenance of histone acetylation during chromatin assembly may prolong this (normally short-lived) instability, resulting in demonstrable H1-depletion under certain conditions *in vitro*. It cannot be too strongly emphasized, however, that even if the preservation of acetylation weakens or retards the deposition of H1 onto new DNA, the actual function of acetylation during chromatin assembly may be completely different. A more informative approach would be to block the initial acetylation of newly synthesized H4 (rather than its deacetylation), so as to uncover possible adverse effects. Unfortunately, a reasonably specific inhibitor of HATB that will act *in vivo* has not been identified. With a biochemical reagent unavailable, researchers have turned to genetically manipulating the histones themselves. This work will be discussed in the following section.

B. Genetic Analysis of Histone Acetylation

Possessing a low histone gene copy number and a high frequency of homologous recombination, the budding yeast *Saccharomyces cerevisiae* is perhaps an ideal organism for the genetic analysis of histone function. In a series of watershed experiments, Grunstein and colleagues first explored replacing the native histone genes of yeast with mutant copies under inducible control. The predominant thrust of this type of analysis, now performed in several laboratories, has been to determine the contribution of individual histones (or histone domains) in cell viability and gene regulation. Most of this work has recently been detailed in a number of review articles (Bradbury, 1992; Tordera et al., 1993; Turner, 1993; Wolffe and Dimitrov, 1993). The following discussion will be limited to investigations that bear upon nucleosome assembly and cell replication.

Cells completely lacking newly synthesized H2B or H4 can successfully complete a full round of DNA replication, producing nucleosome-depleted chromatin; however, repression of histone synthesis results in complete disruption of mitosis, synchronous arrest in G2, and cell death (Han et al., 1987; Smith and Stirling, 1988; Kim et al., 1988). Although nucleosomes can be assembled on non-replicating DNA following the resumption of histone synthesis in G2, this does not reverse the lethal phenotype. A likely explanation for this is that delayed nucleosome assembly is inefficient, resulting in chromatin that remains nucleosome-depleted as measured by nuclease sensitivity and DNA supercoiling (Han et al., 1987; Kim et al., 1988; Norris et al., 1988).

The N-termini of H3 and H4 of course contain the sites of reversible post-translational modifications (including acetylation and phosphorylation), and have been evolutionarily conserved. In direct contrast to intuitive predictions, it was found that yeast cells with deletions of the N-terminus of either H3 or H4 were fully viable (Kayne et al., 1988; Morgan et al., 1991; Mann and Grunstein, 1992), in agreement with previous results obtained for terminal deletions of H2A or H2B (Schuster et al., 1986). Nevertheless, progression through the G2/M phase of the cell cycle was slowed following the mutation of H3 or H4, resulting in an approximately 1.5- to 2-fold increase in doubling time, depending on the size of deletion and the histone affected; a double mutant, lacking both H3 *and* H4 N-termini, was inviable (Morgan et al., 1991). Although the chromatin structure of the double mutant was not directly analyzed, aspects of the lethal phenotype indicated that cell death was not due to nucleosome depletion (i.e., a block in nucleosome assembly), but rather to a functionally defective chromatin structure. The data further suggested that the N-termini of H3 and H4 can operationally substitute for one another *in vivo*, at least to some degree.

To more precisely define the function of histone modifications, the acetylatable lysines of the H4 N-terminus have been replaced with other amino acids, either singly or in combination. By substituting lysines 5, 8, 12, or 16 with positively charged arginines, which cannot be acetylated, processes that require acetylation can be uncovered; conversely, replacing these lysines with uncharged residues (e.g., glycine, glutamine, or alanine) should detect events that require *deacetylation*. When each of the four lysines was individually replaced by arginine, no adverse effects on cell viability or doubling time were detected (Megee et al., 1990; Durrin et al., 1991); however, changing lysine 16 to an uncharged amino acid uniquely prevented repression of the mating type loci (Johnson et al., 1990; Park and Szostak, 1990), in accord with an earlier deletion analysis of the H4 N-terminal domain

(Kayne et al., 1988). Substituting selected lysine pairs (e.g., 5–8; 5–12; 8–12) with arginine increased cell doubling time slightly, and converting three lysines at once (5–8–12) yielded an even more pronounced effect (Megee et al., 1990; Park and Szostak, 1990; Durrin et al., 1991).

When all four acetylatable lysines of H4 were simultaneously replaced by arginine, the mutants either grew very slowly (~3–fold increase in doubling time) (Park and Szostak, 1990; Durrin et al., 1991), or were inviable (Megee et al., 1990); the differing results may be due to the strain of yeast used, or to the method of scoring viability (Durrin et al., 1991). When the same four lysines were changed to uncharged amino acids—to mimic irreversible acetylation—doubling time increased about 2-fold, with the delay occurring predominantly in G2/M; also, mating efficiency fell precipitously, as expected (see above) (Megee et al., 1990; Park and Szostak, 1990). In contrast, similar mutations of H3 had only minor effects on mating or generation time (Mann and Grunstein, 1992).

What do the above data tell us about histone acetylation? Before discussing possible interpretations, it is well to recall that yeast chromatin is unique in some respects. Certainly one example of this is the absence of any identifiable linker histone. In addition, with some noteworthy exceptions (Braunstein et al., 1993), the vast majority of yeast histones are hyperacetylated during log phase (unlike the situation in multicellular organisms) (Nelson, 1982). As a consequence, most of yeast chromatin is highly sensitive to digestion with DNaseI (Lohr and Hereford, 1979), and is probably maintained in a transcriptionally competent structure. Recent work has shown that the N-termini of H3 and H4 can either activate or repress certain genes, but that other genes are unaffected by N-terminal deletions (Kayne et al., 1988; Durrin et al., 1991; Mann and Grunstein, 1992). If acetylation greatly weakens or abolishes the binding of histone N-termini to DNA (Cary et al., 1982; Hong et al., 1993), then it is perhaps not surprising (in light of the general hyperacetylation of yeast chromatin) that complete removal of the H4 N-terminus has no effect on yeast viability, but instead interferes with specific cell functions (such as mating). Evidently, the genes regulated by single histone tails are inessential for cell viability under normal laboratory conditions. Significantly, the lethality of the H3–H4 double mutants (with both tails deleted) occurs randomly in the cell cycle, after a variable number of cell divisions. As discussed by Megee and colleagues (1990), this may reflect the stochastic assembly of defective H3–H4 tetramers onto essential regions of chromatin—regions that require at least one H3 or H4 N-terminus for functionality.

The histones of mammalian metaphase chromosomes are for the most part underacetylated, as documented by loss of the more highly acetylated isoforms of histone H4 (Chahal et al., 1980; D'Anna et al., 1983; Turner, 1989; Jeppesen et al., 1992). Nevertheless, some acetylation of H4 persists throughout the M phase, apparently concentrated in transcriptionally competent loci (Jeppesen et al., 1992; Jeppesen and Turner, 1993). The reverse is true for transcriptionally inert regions: Both pericentriolar heterochromatin and the inactivated X chromosome of female mammals lack H4 acetylation (Jeppesen and Turner, 1993). It has therefore been suggested that the maintenance of acetylation during mitosis identifies chromatin to be transcribed in the succeeding G1 phase (Jeppesen and Turner, 1993), while a more global deacetylation facilitates chromosome condensation (Chahal et al., 1980; Roth and Allis, 1992). If a similar pattern operates during mitosis in yeast, then the comparatively minor (cell cycle) effects of simultaneously substituting the acetylatable lysines of H4 with uncharged residues might be expected: under normal conditions, much of the yeast genome may be tagged by acetylation during M for transcription in the following G1.

To the author's knowledge, the increase in yeast doubling time observed when all the acetylatable lysines of H4 were concurrently changed to arginine has not been ascribed to a particular stage of the cell cycle. In any event, the data indicate that acetylation of H4, although not essential in yeast, at least greatly facilitates chromatin function. As suggested by Allis and colleagues (1985), one possibly fruitful area of research would be to combine the multiple lysine>arginine substitutions of H3 *and* H4 within one cell, to see if acetylation of these key histones is functionally redundant. Still, the fact that the $R_{(5,8,12,16)}$-mutant of H4 is viable, albeit slow-growing, argues against this mutation causing severe nucleosome depletion (which causes cell cycle arrest). It therefore follows that H4 molecules that cannot be acetylated *are still able to be assembled into nucleosomes*. Similarly, the viability of the lys>(gly/glut/ala) mutants, and of cells lacking either the H3 or the H4 N-terminus, genetically demonstrates that deacetylation is not required for chromatin assembly, in complete accord with the biochemical studies discussed above.

Do the data then rule out a significant role for histone acetylation during chromatin assembly? Although the reversible acetylation of H4 is apparently not required for nucleosome assembly in yeast, it would seem premature to assert that acetylation has no function whatsoever in chromatin biosynthesis. The slow growth of the lys>arg mutants, which in the case of $R_{5,12}$ is attributable to a 50 percent increase in the length of S phase (Megee et al., 1990), suggests that chromatin replication and/or nucleosome assembly may be relatively inefficient in the absence of acetylation. Strains of yeast with such slow growth rates are obviously at a selective disadvantage with respect to wild type cells, and will eventually be eliminated under natural conditions. Perhaps dynamic acetylation should therefore be considered as an adaptation that facilitates the assembly and transcription of chromatin (thereby accelerating doubling time), while remaining dispensable in most instances. Of course, dispensability need not imply lack of function, and it remains to be seen how essential the acetylation of nascent H4 is in *human* cells, where new H3 is unacetylated (Sobel et al., 1995). Still, the significance of the acetylation of nascent H4 remains enigmatic.

IV. HISTONE STORAGE COMPLEXES

Many species accumulate large histone pools during oogenesis which are later used to satisfy the assembly requirements of early development. With sufficient core histones to assemble approximately 15,000–20,000 nuclei (Woodland and Adamson, 1977), the oocyte of *Xenopus laevis* is by far the most thoroughly studied of these systems. Histones in the oocyte are sequestered in well-characterized storage complexes: H3/H4 are associated with the 110kDa proteins N1/N2, while H2A/H2B are complexed with a pentamer of nucleoplasmin (reviewed in Laskey et al., 1993). Extracts prepared from *Xenopus* oocytes (or eggs), as well as the purified complexes, have been demonstrated to assemble chromatin *in vitro*, and it is thought that both complexes become actively engaged in nucleosome assembly after fertilization (reviewed in Annunziato, 1990; Almouzni and Wolffe, 1993a; Laskey et al., 1993). A discussion of the large body of data obtained by exploiting the *Xenopus* oocyte system is clearly far beyond the scope of this report; instead, the emphasis in the following section will be on modifications to histones accompanying their storage *in vivo*.

As previously noted, stored H4 is diacetylated in *Xenopus* oocytes (Woodland, 1979). Unlike chromatin-bound nascent H4 in somatic cells (e.g., Cousens and Alberts, 1982), stored diacetylated H4 is not further modified to tri- and tetraacetylated forms when oocytes are incubated with the deacetylase inhibitor sodium butyrate (Dimitrov et al., 1993). Similar results have been obtained using butyrate-treated cytosolic extracts from HeLa cells, and exogenous H4 (according to L. Chang and A.T. Annunziato's unpublished observations). It would thus appear that the HATB activity in these systems is highly site-specific and ignores remaining acetylatable lysines once diacetylation is achieved. Moreover, the data argue against the diacetylated state normally being maintained through a dynamic equilibrium between acetylation of all four acetylatable lysines and ongoing deacetylation.

It has been postulated that acetylation may facilitate the interaction of H4 with the molecular chaperones required for replication-coupled nucleosome assembly (Smith and Stillman, 1991b; Dimitrov et al., 1993). This would account for the diacetylation of stored H4 in *Xenopus* oocytes[1], and of nascent H4 in somatic cells. Because the *Xenopus* storage protein N1 (and the H3/H4·N1 complex) can assemble both replicating and non-replicating DNA (e.g., Kleinschmidt et al., 1985, 1990; Dilworth et al., 1987; Zucker and Worcel, 1990; Almouzni et al., 1990a), the only factor known to be uniquely involved in replication-coupled nucleosome assembly (using native histones) is mammalian CAF-I. First isolated and characterized by Smith and Stillman (1989, 1991a, 1991b), CAF-I is a multisubunit protein found in the nucleus of human cells. CAF-I is required to assemble replicating SV40 into chromatin *in vitro*, but ordinarily does not promote the assembly of non-replicating viral DNA (Smith and Stillman, 1989, 1991b). The replication-coupled assembly reaction relies entirely on the small pool of soluble histones found in the cytoplasm of growing cells (Smith and Stillman, 1989, 1991b); it is very likely that these histones are newly synthesized in transit to the nucleus (Oliver et al., 1974; Bonner et al., 1988; Jackson, 1990). Intriguingly, only the cytoplasmic histones (and thus presumably diacetylated H4), can be exclusively targeted to replicating DNA by CAF-I: The addition of even small amounts of purified nuclear H3/H4 promotes the assembly of both replicating and non-replicating plasmids alike (Smith and Stillman, 1991b). Whether this difference is due solely to the modification state of the native soluble histones, or to additional (unidentified) assembly factors associated with them, is unknown.

The histone pool in dividing somatic cells is extremely small, and it may be that storage proteins such as nucleoplasmin and N1 have no direct counterparts in non-embryonic systems.[2] On the other hand, as yet unknown factors may escort H3 and H4 into the nucleus, where they are then guided to newly replicated DNA by CAF-I, an apparently strictly nuclear protein (Smith and Stillman, 1991a). The rapidity with which histones are typically translocated to the nucleus makes analysis of the import process difficult, and, therefore, one approach to this problem has involved the study of histones synthesized in the absence of DNA replication. It has long been known that when replication is inhibited histone synthesis continues at a reduced rate for up to several hours (Nadeau et al., 1978). Both the replication-linked histone subtypes, as well as the basally synthesized non-allelic variants of H2A, H2B, and H3 are produced during DNA synthesis inhibition, although synthesis of the replication-linked subtypes declines to a greater degree (Zweidler, 1980; Wu and Bonner, 1981; Urban and Zweidler, 1983). Newly synthesized H2A and H2B continue to enter chromatin in the absence of replication (through a process of histone exchange), but most new H3 and H4 remain undeposited (Louters and Chalkley, 1985; Senshu and Ohashi, 1979; Bonner et al., 1988), reflecting their usual targeting to newly synthesized DNA (e.g., Jackson and Chalkley, 1981a, 1981b; Annunziato et al., 1982). This "unbound" H3/H4 can be detected in the cytoplasm following nuclear isolation (Louters and Chalkley, 1985; Bonner et al., 1988), but there is evidence that these histones may form a nuclear pool, perhaps interacting weakly with chromatin (Jackson, 1990).

The diacetylated form of new H4 accumulates when replication is blocked (Bonner et al., 1988), presumably in preparation for nucleosome assembly. Recently it was found that antibodies that recognize acetylated H4 not only immunoprecipitate new cytoplasmic H4, but newly synthesized H3 as well (Perry et al., 1993b). This raises the possibility that new H3 and H4 are associated in a somatic nucleosome assembly complex, formed prior to deposition. Evidence that a molecular chaperone may also be involved comes from sucrose gradient analyses which indicate that the complex sediments as a 5–6S particle (Perry et al., 1994), in agreement with an earlier analysis of soluble H3/H4 (Senshu and Yamada, 1980). This is larger than expected for the ~3S H3/H4 tetramer (Kornberg and Thomas, 1974), but smaller than an 11S mononucleosome (Thomas and Kornberg, 1975). Notably, the complete histone octamer (~6.5S or 108kDa) is unstable at physiological ionic strength in the absence of DNA (Eickbush and Moudrianakis, 1978).

Cytoplasmic H3 and H4 have rather anomalous characteristics, including the inability to bind to phosphocellulose (Smith and Stillman, 1991b), or to chromatin *in vivo* (Jackson, 1990). An escort protein that shielded histone positive charges could account for this behavior, as well as for the rapid sedimentation of the H3/H4 complex in sucrose gradients. However, it is unlikely that the complete CAF-I complex is associated with the cytoplasmic histones, as indicated both by the biochemistry of *in vitro* nucleosome assembly (Smith and Stillman, 1989), and by immunohistochemical analysis (Smith and Stillman, 1991a). Although a number of other "assembly factors" from somatic cells have been inferred or described (Ishimi et al., 1987; Banerjee and Cantor, 1990; Gruss et al., 1990; Lassle et al., 1992), their ability to assemble non-replicating DNA distinguishes them sharply from the CAF-I activity, and thus they may be more analogous to nucleoplasmin and/or N1. The degree to which any of these other factors actually participate in chromatin assembly *in vivo* (and to which the diacetylation of H4 aids in their function) has not yet been determined.

V. ACETYLATION, DNA REPLICATION, AND HISTONE DEPOSITION

A powerful approach to the study of acetylation involves the use of antisera that specifically recognize the acetylated histone isoforms. As described above, Turner and colleagues have used antisera specific for each monoacetylated species of H4 (e.g., $H4Ac_1$ at lysine 5; $H4Ac_1$ at lysine 8, etc.) to determine the frequency of use of each acetylatable lysine, and to map the distribution of $H4Ac_1$ subtypes within chromosomal domains (Turner et al., 1989, 1992; Munks et al., 1991; Jeppesen et al., 1992; Jeppesen and Turner, 1993; Sommerville et al., 1993; Clarke et al., 1993). Similarly, Crane-Robinson and colleagues have established a strong correlation between histone acetylation and chromatin transcriptional activity/competence, by immunoprecipitating nucleosomes with an antibody that recognizes acetylated core histones (Hebbes et al., 1988, 1992; Clayton et al., 1993). Allis and colleagues have also developed antisera that specifically recognize acetylated H4 (Lin et al., 1989); these have now been used to document several aspects of chromatin metabolism, including: 1) deposition and transcription-related acetylation in *Tetrahymena* (Lin et al., 1989), and *Euplotes* (Olins et al., 1991); 2) displacement-related acetylation in rat spermatids (Meistrich et al., 1992); and 3) the inverse relationship between transcriptional silencing and histone acetylation in both *Tetrahymena* (Lin et al., 1991) and yeast (Braunstein et al., 1993).

The acetylation-specific antibodies developed by Lin and colleagues (1989) have also been used to analyze nucleosome assembly and histone exchange in HeLa cells (Perry et al., 1993b). Not surprisingly, it was found that chromatin replicated in the presence of sodium butyrate was immunoprecipitated with high efficiency by antibodies that recognize acetylated H4, consistent with the selective deposition of diacetylated nascent H4 onto newly replicated DNA. Moreover, the enrichment for new DNA in the immunopellet was paralleled by a similar enrichment for all four newly synthesized histones. Because the antibodies that were used in this study recognize both new and old acetylated H4, this experiment did not specifically address the question of conservative versus non-conservative assembly mechanisms; in this regard, it is now generally held that (at least) a significant proportion of new H3/H4 tetramers are deposited separately from new H2A/H2B (reviewed in Annunziato, 1990). However, the immunoprecipitation results underscore that the acetylation of H4 is central to the deposition of newly synthesized H2A, H2B, and H3, *whether or not* these histones are conservatively assembled, suggesting that acetylation may facilitate assembly. In the same study (Perry et al., 1993b) it was further demonstrated that the new H2A and H2B which enter chromatin during the inhibition of DNA replication are also preferentially targeted to acetylated chromatin regions. The acetylation of H4 can therefore be correlated with histone deposition, histone replacement, chromatin remodeling, and transcriptional competence. One feature that these processes may share in common is a localized unfolding of chromatin higher order structures.

Another process that almost certainly requires chromatin decondensation is DNA replication. To the extent that transcription can be used as a model for replication, it might be predicted that parental histones are acetylated in advance of the replication fork, thereby aiding in unfolding the 30-nm fiber. If this were the case, then segregated nucleosomes should be efficiently precipitated by antibodies that recognize acetylated H4, much like transcriptionally active nucleosomes, or nucleosomes assembled *de novo*. However, when the acetylation status of parental histones was examined (by immunoprecipitating chromatin replicated in the absence of concurrent protein synthesis), it was found that the occurrence of acetylated H4 among segregated nucleosomes was approximately equal to that found in bulk chromatin (Perry et al., 1993a); thus, replication-competent chromatin is not obligatorily acetylated. Similar conclusions were also drawn from an immunohistochemical study of replication in the protozoan *Euplotes*, in which macronuclear DNA synthesis is confined to a migrating nuclear organelle, the replication band: chromatin in the rear zone of the band, where histone deposition and nucleosome assembly occur, was strongly stained with antibodies that recognize acetylated H4; chromatin in the forward zone of the replication band (immediately preceding replication) stained very weakly (Olins et al., 1991).

If parental nucleosomes need not be acetylated for chromatin replication, then the acetylation status of prereplicative chromatin may be preserved during DNA synthesis and histone segregation. This could provide an epigenetic mechanism for differentiating potentially active from inactive chromatin structures (Figure 1). It is now well established that nucleosome assembly occurs in a stepwise fashion, with the deposition of H3/H4 preceding that of H2A/H2B (Dilworth et al., 1987; Fotedar and Roberts, 1989; Kleinschmidt et al., 1990; Sapp and Worcel, 1990; Zucker and Worcel, 1990; Almouzni et al., 1990a; Smith and Stillman, 1991b). The staged assembly of chromatin facilitates the binding of transcription factors to DNA (Almouzni et al., 1991; Hayes and Wolffe, 1992; Crippa et al., 1993), as does histone acetylation (Lee et al., 1993). Acetylation may also modulate chromatin higher order structures (Annunziato et al., 1988), which can in turn regulate gene activity in an H1-dependent manner (Weintraub, 1984, 1985; Wolffe and Brown, 1988; Wolffe, 1989, 1991; Shimamura et al., 1989; Garrard, 1991; Laybourn and Kadonaga, 1991; Hansen and Wolffe, 1992; Zlatanova and van Holde, 1992). Because of the dispersive segregation of parental histones (Annunziato and Seale, 1984; Cusick et al., 1984; Jackson and Chalkley, 1985; Bonne-Andrea et al., 1990; Burhans et al., 1991; Krude and Knippers, 1991; Randall and Kelly, 1992;

A) Replication of Unacetylated (Inactive) Chromatin

● Nucleosomes with old, unacetylated H3/H4
○ Nucleosomes with new, diacetylated H4

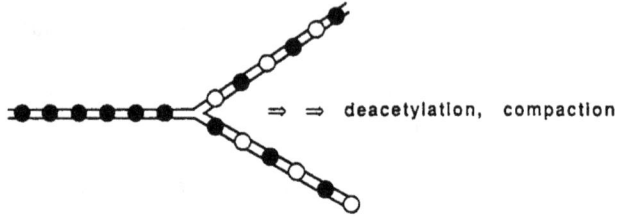

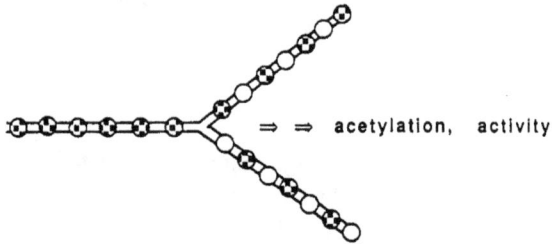

⇒ ⇒ deacetylation, compaction

B) Replication of Acetylated (Active) Chromatin

⊗ Nucleosomes with old, acetylated H3/H4
○ Nucleosomes with new, diacetylated H4

⇒ ⇒ acetylation, activity

Figure 1. Contribution of the acetylation status of segregated histones to the maintenance of active versus inactive structures during chromatin replication. Because there appears to be no requirement for the acetylation of prereplicative nucleosomes, the degree of acetylation of segregated histones may aid in the propagation of chromatin organization. **A)** Replication of inactive, unacetylated chromatin: Histones of prereplicative nucleosomes are unacetylated, and remain unacetylated as they segregate, causing diacetylated new H4 to be deposited in an "unacetylated domain"; unacetylated segregated nucleosomes (possessing H1 [Annunziato and Seale, 1982]) promote the deacetylation of new H3/H4, and compaction of this chromatin region. **B)** Replication of active, acetylated chromatin: Prereplicative nucleosomes are acetylated, and remain so as they segregate to new DNA; diacetylated new H4 is thus deposited amid acetylated old histones; the resulting "open" structure fosters maintained acetylation and gene activity (as mediated by the binding of specific transactivating factors, HMG proteins, etc.). *Note:* Dispersive nucleosome segregation has been depicted as an alternating mechanism for illustrative purposes; parental nucleosomes probably segregate in groups (reviewed in Annunziato, 1990); the cooperative interaction of (segregated) unacetylated nucleosome clusters may potentiate the folding of inactive chromatin, in an H1-dependent fashion (see Weintraub, 1985; Wolffe, 1991; Garrard, 1991; Zlatanova and van Holde, 1992 for discussions of the role of H1 in gene regulation).

Sugasawa et al., 1992), both arms of the replication fork are involved in *de novo* nucleosome assembly, and therefore become complexed with diacetylated nascent H4. As has been suggested (Perry et al., 1993a), if the assembly and subsequent deacetylation of newly replicated chromatin yields transcriptionally inactive chromatin structures (Almouzni et al., 1990b; Lee et al., 1993), then the segregation of unacetylated parental histones may identify chromatin regions that are to be fully deacetylated, condensed into higher order structures, and transcriptionally inert. Indeed, replication-coupled chromatin assembly may be the primary mechanism for the repression of basal transcription (Almouzni and Wolffe, 1993b). In contrast, the segregation of acetylated parental histones may help to generate transcriptionally active chromatin, by promoting further (or continued) acetylation (Cousens and Alberts, 1979, 1982), and thus H1-depletion (Kamakaka and Thomas, 1990; Bresnick et al., 1992). Consistent with this, it has been found that parental histones become readily accessible to histone acetylases once replication has occurred (Perry et al., 1993a). However, if the acetylation status of prereplicative chromatin influences the propagation of transcriptionally competent chromatin, it probably acts in a secondary role, subsequent to the binding of specific *trans*-activating factors (see Elgin, 1990; Felsenfeld, 1992; Wolffe, 1992b).

VI. NUCLEOSOME SPACING AND THE PHOSPHORYLATION OF HISTONE H2A

It is now well established that ATP is required to assemble chromatin possessing a physiological repeat-length (i.e., 180 bp) in extracts prepared from *Xenopus* oocytes or eggs (Glikin et al., 1884; Ruberti and Worcel, 1986; Almouzni and Méchali, 1988; Sessa and Ruberti, 1990; Almouzni et al., 1991). Although the reason for this is not understood, there is some evidence that phosphorylation, perhaps of histone H2A, may be involved. The major species of H2A in *Xenopus* oocytes is H2A.X (Dilworth et al., 1987), a minor nonallelic sequence variant of the primary H2A subtype(s) (West and Bonner, 1980). H2A.X is rapidly phosphorylated during chromatin assembly *in vitro*, as well as during the *in vivo* assembly of plasmids microinjected into *Xenopus* germinal vesicles (Kleinschmidt and Steinbeisser, 1991). Phosphorylation of H2A.X is DNA-dependent, does not occur in the absence of nucleosome assembly, and turns over as the proper spacing is achieved. Moreover, premature dephosphorylation of H2A.X (as well as of other phosphoproteins) with alkaline phosphatase during nucleosome assembly prevents the establishment of the physiological spacing, yielding a 160 bp repeat-length instead (Kleinschmidt and Steinbeisser, 1991).

A phosphorylation-dephosphorylation cycle appears to be a widespread element in chromatin assembly: the phosphorylation of H2A has also been correlated with an increase in chromatin spacing during nucleosome assembly in HeLa cell extracts (Banerjee and Cantor, 1990; Banerjee et al., 1991). Tremethick and Frommer (1992) have partially purified an ATP-dependent activity from *Xenopus* oocytes that aids in the generation of regularly spaced chromatin (with a repeat length of 165 bp); however, as this activity neither phosphorylates H2A.X *in vitro*, nor effects a 180 bp repeat during assembly, its identity with the DNA-dependent kinase described by Kleinschmidt and Steinbeisser would seem unlikely.

VII. CONCLUSIONS

It is doubtful that the diacetylation of newly synthesized H4 would be so highly conserved had it no function during chromatin biosynthesis. Nevertheless, the biochemical and genetic data indicate that the reversible acetylation of nascent H4 is dispensable for histone deposition and physiological nucleosome alignment. Of course, the genetic evidence obtained from yeast may in part reflect the peculiarities of this system, while the ability of acetylated H3 to substitute for H4Ac$_2$ remains untested. These matters should be fertile areas for future research. Other questions worth examining include whether acetylation is involved in the translocation of new H4 into the nucleus, and if H4 must be diacetylated to interact with the molecular chaperones that link histone deposition to DNA replication. Approaches for addressing these problems might include a conditional mutant for HATB, or a specific inhibitor of this enzyme. *In vitro* assembly systems in which the acetylation of H4 can be manipulated should also prove valuable.

Given that heightened DNaseI sensitivity is a feature of acetylated chromatin in general, it is perhaps ironic that the only clearly identified assembly-related effect of histone acetylation is the prolonged DNaseI sensitivity (and H1-depletion?) of chromatin replicated in the presence of butyrate. Whether or not this represents an actual role for acetylation during nucleosome assembly *in vivo*, it does underscore the importance of timely deacetylation in the proper and complete formation of chromatin. Indeed, the transience of H4Ac$_2$ during nucleosome assembly can be taken as evidence that histone deacetylase is targeted to (or associated with) replicating chromatin *in vivo*, as well as to plasmids assembled in *Xenopus* extracts; presumably the requisite deacetylase only recognizes H4Ac$_2$ that is nucleosome-associated. If, in fact, the acetylation status of segregated parental histones (and the transition from deposition to transcription-related acetylation patterns) identifies potentially active chromatin during nucleosome assembly, then histone deacetylation emerges as a key regulatory mechanism (see Lopez-Rodas et al., 1993). It would therefore seem that an explanation of the relationship between reversible acetylation and chromatin assembly will ultimately entail an understanding of the acetyltransferases and deacetylases involved, as well as of the steps that selectively deposit core and linker histones onto replicating DNA.

ACKNOWLEDGMENT

The author was supported by a grant from the National Institutes of Health (GM 35837). This communication is dedicated to the memory of Dr. Abraham Worcel, who more than 15 years ago insightfully proposed the stepwise assembly of newly replicated chromatin (Worcel et al., 1978).

NOTES

1. Although it is uncertain whether N1 is directly involved in chromatin assembly *in vivo*, it is probable that either N1 or another escort protein (see Sapp and Worcel, 1990) couples the deposition of diacetylated H4 to DNA replication during *Xenopus* development (Dimitrov et al., 1993).

2. Cotton and Chalkley (1987) have presented evidence for a nucleoplasmin-like protein (termed nucleoplasmin S) in a somatic cell line derived from *Xenopus* kidney. Nucleoplasmin S was shown to mediate nucleosome assembly *in vitro*, much like oocyte nucleoplasmin, but the presence of this protein in other cell types, or its action as an assembly factor *in vivo*, has not been demonstrated.

REFERENCES

Allan, J., Harborne, N., Rau, D.C., & Gould, H. (1982). Participation of the core histone "tails" in the stabilization of the chromatin solenoid. J. Cell Biol. 93, 285–297.

Allis, C.D., Chicoine, L.G., Richman, R., & Schulman, I.G. (1985). Deposition-related histone acetylation in micronuclei of conjugating *Tetrahymena*. Proc. Natl. Acad. Sci., USA 82, 8048–8052.

Almouzni, G., & Méchali, M. (1988). Assembly of spaced chromatin involvement of ATP and DNA Topoisomerase Activity. EMBO J. 7, 4355–4365.

Almouzni, G., & Wolffe, A.P. (1993a). Nuclear assembly, structure, and function—The use of *Xenopus in vitro* systems. Exp. Cell Res. 205, 1–15.

Almouzni, G., & Wolffe, A.P. (1993b). Replication-coupled chromatin assembly is required for the repression of basal transcription *in vivo*. Genes & Dev. 7, 2033–2047.

Almouzni, G., Clark, D.J., Méchali, M., & Wolffe, A.P. (1990a). Chromatin assembly on replicating DNA *in vitro*. Nucleic Acids Res. 18, 5767–5774.

Almouzni, G., Méchali, M., & Wolffe, A.P. (1990b). Competition between transcription complex assembly and chromatin assembly on replicating DNA. EMBO J. 9, 573–582.

Almouzni, G., Méchali, M., & Wolffe, A.P. (1991). Transcription complex disruption caused by a transition in chromatin structure. Mol. Cell. Biol. 11, 655–665.

Annunziato, A.T. (1990). Chromatin replication and nucleosome assembly. The eukaryotic nucleus: Molecular biochemistry and macromolecular assemblies (Strauss, P., & Wilson, S.H., Eds.) Vol. 2. The Telford Press, Caldwell, NJ, 687–712.

Annunziato, A.T., & Seale, R.L. (1982). Maturation of nucleosomal and nonnucleosomal components of nascent chromatin, differential requirements for concurrent protein synthesis. Biochemistry 21, 5431–5438.

Annunziato, A.T., & Seale, R.L. (1983). Histone acetylation is required for the maturation of newly replicated chromatin. J. Biol. Chem. 258, 12675–12684.

Annunziato, A.T., & Seale, R.L. (1984). Presence of nucleosomes within irregularly cleaved fragments of newly replicated chromatin. Nucleic Acids Res. 12, 6179–6196.

Annunziato, A.T., Schindler, R.K., Thomas, C.A., Jr., & Seale, R.L. (1981). Dual nature of newly replicated chromatin, evidence for nucleosomal and non-nucleosomal DNA at the site of native replication forks. J. Biol. Chem. 256, 11880–11886.

Annunziato, A.T., Schindler, R.K., Riggs, M.G., & Seale, R.L. (1982). Association of newly synthesized histones with replicating and nonreplicating regions of chromatin. J. Biol. Chem. 257, 8507–8515.

Annunziato, A.T., Frado, L.-L.Y., Seale, R.L., & Woodcock, C.L.F. (1988). Treatment with sodium butyrate inhibits the complete condensation of interphase chromatin. Chromosoma (Berl). 96, 132–138.

Ausio, J. (1992). Structure and dynamics of transcriptionally active chromatin. J. Cell Sci. 102, 1–5.

Banerjee, S., & Cantor, C.R. (1990). Nucleosome assembly of simian virus 40 DNA in a mammalian cell extract. Mol. Cell. Biol. 10, 2863–2873.

Banerjee, S., Bennion, G.R., Goldberg, M.W., & Allen, T.D. (1991). ATP dependent histone phosphorylation and nucleosome assembly in a human cell free extract. Nucleic Acids Res. 19, 5999–6006.

Bavykin, S., Srebreva, L., Banchev, T., Tsanev, R., Zlatanova, J., & Mirzabekov, A. (1993). Histone H1 deposition and histone DNA interactions in replicating chromatin. Proc. Natl. Acad. Sci. USA 90, 3918–3922.

Bonne-Andrea, C., Wong, M.L., & Alberts, B.M. (1990). *In vitro* replication through nucleosomes without histone displacement. Nature (London) 343, 719–726.

Bonner, W.M., Wu, R.S., Panusz, H.T., & Muneses, C. (1988). Kinetics of accumulation and depletion of soluble newly synthesized histone in the reciprocal regulation of histone and DNA synthesis. Biochemistry 27, 6542–6550.

Bradbury, E.M. (1992). Reversible histone modifications and the chromosome cell cycle. Bioessays 14, 9–16.

Braunstein, M., Rose, A.B., Holmes, S.G., Allis, C.D., & Broach, J.R. (1993). Transcriptional silencing in yeast is associated with reduced nucleosome acetylation. Genes & Dev. 7, 592–604.

Bresnick, E.H., Bustin, M., Marsaud, V., Richard-Foy, H., & Hager, G.L. (1992). The transcriptionally-active MMTV promoter Is depleted of histone H1. Nucleic Acids Res. 273–278.

Burhans, W.C., Vassilev, L.T., Wu, J., Sogo, J.M., Nallaseth, F.S., & DePamphilis, M.L. (1991). Emetine allows identification of origins of mammalian DNA replication by imbalanced DNA synthesis, not through conservative nucleosome segregation. EMBO J. 10, 4351–4360.

Cary, P.D., Crane-Robinson, C., Bradbury, E.M., & Dixon, G.H. (1982). Effect of acetylation on the binding of N-terminal peptides of histone H4 to DNA. J. Biol. Chem. 127, 137–143.

Chahal, S.S., Matthews, H.R., & Bradbury, E.M. (1980). Acetylation of histone H4 and its role in chromatin structure and function. Nature (London) 287, 76–79.

Chambers, S.A.M., & Shaw, B.R. (1984). Levels of histone H4 diacetylation decrease dramatically during sea urchin embryonic development and correlate with cell doubling rate. J. Biol. Chem. 259, 13458–13463.

Chicoine, L.G., Schulman, I.G., Richman, R., Cook, R., & Allis, C.D. (1986). Nonrandom utilization of acetylation sites in histones isolated from *Tetrahymena*. Evidence for functionally distinct H4 acetylation sites. J. Biol. Chem. 261, 1071–1076.

Chicoine, L.G., Richman, R., Cook, R.G., Gorovsky, M.A., & Allis, C.D. (1987). A single histone acetyltransferase from *Tetrahymena* macronuclei catalyzes deposition-related acetylation of free histones and transcription-related acetylation of nucleosomal histones. J. Cell Biol. 105, 127–135.

Clarke, D.J., O'Neill, L.P., & Turner, B.M. (1993). Selective Use of H4 Acetylation Sites in the Yeast *Saccharomyces cerevisiae*. Biochem. J. 294, 557–561.

Clayton, A.L., Hebbes, T.R., Thorne, A.W., & Crane-Robinson, C. (1993). Histone acetylation and gene induction in human cells. FEBS Lett. 336, 23–26.

Cotten, M., & Chalkley, R. (1987). Purification of a novel, nucleoplasmin-like protein from somatic nuclei. EMBO J. 6, 3945–3954.

Couppez, M., Martin-Ponthieu, A., & Sautiere, P. (1987). Histone H4 from cuttlefish testis is sequentially acetylated: Comparison with acetylation of calf thymus histone H4. J. Biol. Chem. 262, 2854–2860.

Cousens, L.S., & Alberts, B.M. (1979). Different accessibilities in chromatin to histone acetylase. J. Biol. Chem. 254, 1716–1723.

Cousens, L.S., & Alberts, B.M. (1982). Accessibility of newly synthesized chromatin to histone acetylase. J. Biol. Chem. 257, 3945–3949.

Crippa, M.P., Trieschmann, L., Alfonso, P.J., Wolffe, A.P., & Bustin, M. (1993). Deposition of chromosomal protein HMG-17 during replication affects the nucleosomal ladder and transcriptional potential of nascent chromatin. EMBO J. 12, 3855–3864.

Csordas, A. (1990). On the biological role of histone acetylation. Biochem. J. 265, 23–38.

Cusick, M.F., DePamphilis, M.L., & Wassarman, P.M. (1984). Dispersive segregation of nucleosomes during replication of simian virus 40 chromosomes. J. Mol. Biol. 178, 249–271.

D'Anna, J.A., Gurley, L.R., & Tobey, R.A. (1983). Extent of histone modification and H1^{0} content during cell cycle progression in the presence of butyrate. Exp. Cell Res. 147, 407–417.

Dilworth, S.M., Black, S.J., & Laskey, R.L. (1987). Two complexes that contain histones are required for nucleosome assembly *in vitro*: Role of nucleoplasmin and N1 in *Xenopus* oocytes. Cell 51, 1009–1018.

Dimitrov, S., Almouzni, G., Dasso, M., & Wolffe, A.P. (1993). Chromatin transitions during early *Xenopus* embryogenesis—changes in histone-H4 acetylation and in linker histone type. Dev. Biol. 160, 214–227.

Durrin, L.K., Mann, R.K., Kayne, P.S., & Grunstein, M. (1991). Yeast histone H4 N-terminal sequence is required for promoter activation *in vivo*. Cell 65, 1023–1031.

Eickbush, T.H., & Moudrianakis, E.N. (1978). The histone core complex: An octamer assembled by two sets of protein-protein interactions. Biochemistry 17, 4956–4964.

Elgin, S.C.R. (1990). Chromatin structure and gene activity. Current Opin. Cell Biol. 2, 437–445.

Felsenfeld, G. (1992). Chromatin as an essential part of the transcriptional mechanism. Nature (London) 355, 219–224.

Fotedar, R., & Roberts, J.M. (1989). Multistep pathway for replication-dependent nucleosome assembly. Proc. Natl. Acad. Sci. USA 86, 6459–6463.

Garcea, R.L., & Alberts, B.M. (1980). Comparative studies of histone acetylation in nucleosomes, nuclei, and intact cells: Evidence for special factors which modify acetylase action. J. Biol. Chem. 255, 11454–11463.

Garcia-Ramirez, M., Dong, F., & Ausio, J. (1992). Role of the histone tails in the folding of oligonucleosomes depleted of histone-H1. J. Biol. Chem. 267, 19587–19595.

Garrard, W.T. (1991). Histone H1 and the conformation of transcriptionally active chromatin. Bioessays 13, 87–88.

Giancotti, V., Russo, E., Cristini, F.D., Grazio si, G., Micali, F., & Crane-Robinson, C. (1984). Histone modification in early and late *Drosophila* embryos. Biochem. J. 218, 321–329.

Giannasca, P.J., Horowitz, R.A., & Woodcock, C.L. (1993). Transitions between *in situ* and isolated chromatin. J. Cell Sci. 105, 551–561.

Glikin, G.C., Ruberti, I., & Worcel, A. (1884). Chromatin assembly in *Xenopus* oocytes: *In vitro* studies. Cell 37, 33–41.

Gruss, C., & Sogo, J.M. (1992). Chromatin replication. Bioessays 14, 1–8.

Gruss, C., Gutierrez, C., Burhans, W.C., DePamphilis, M.L., Koller, T., & Sogo, J.M. (1990). Nucleosome assembly in mammalian extracts before and after DNA replication. EMBO J. 9, 2911–2922.

Han, M., Chang, M., Kim, U.-J., & Grunstein, M. (1987). Histone H2B repression causes cell-cycle-specific arrest in yeast: Effects on chromosomal segregation, replication, and transcription. Cell 48, 589–597.

Hansen, J.C., & Ausio, J. (1992). Chromatin dynamics and the modulation of genetic activity. Trends Biochem. Sci. 17, 187–191.

Hansen, J.C., & Wolffe, A.P. (1992). Influence of chromatin folding on transcription initiation and elongation by RNA polymerase-III. Biochemistry 31, 7977–7988.

Hayes, J.J., & Wolffe, A.P. (1992). Histones H2A/H2B inhibit the interaction of transcription factor-IIIA with the *Xenopus borealis* somatic 5S RNA gene in a nucleosome. Proc. Natl. Acad. Sci. USA 89, 1229–1233.

Hebbes, T.R., Thorne, A.W., & Crane-Robinson, C. (1988). A direct link between core histone acetylation and transcriptionally active chromatin. EMBO J. 7, 1395–1402.

Hebbes, T.R., Thorne, A.W., Clayton, A.L., & Crane-Robinson, C. (1992). Histone acetylation and globin gene switching. Nucleic Acids Res. 20, 1017–1022.

Hong, L., Schroth, G.P., Matthews, H.R., Yau, P., & Bradbury, E.M. (1993). Studies of the DNA binding properties of histone H4 amino terminus—thermal denaturation studies reveal that acetylation markedly reduces the binding constant of the H4 tail to DNA. J. Biol. Chem. 268, 305–314.

Ikegami, S., Ooe, Y., Shimizu, T., Kasahara, T., Tsuruta, T., Kijima, M., Yoshida, M., & Beppu, T. (1993). Accumulation of multiacetylated forms of histones by trichostatin-A and its developmental consequences in early starfish embryos. Roux Arch. Dev. Biol. 202, 144–151.

Ishimi, Y., Kojima, M., Yamada, M., & Hanaoka, F. (1987). Binding mode of nucleosome-assembly protein (AP-I) and histones. Eur. J. Biochem. 162, 19–24.

Jackson, V. (1990). *In vivo* studies on the dynamics of histone-DNA interaction: Evidence for nucleosome dissolution during replication and transcription and a low level of dissolution independent of both. Biochemistry 29, 719–731.

Jackson, V., & Chalkley, R. (1981a). A new method for the isolation of replicative chromatin: Selective deposition of histone on both new and old DNA. Cell 23, 121–134.

Jackson, V., & Chalkley, R. (1981b). A reevaluation of new histone deposition on replicating chromatin. J. Biol. Chem. 256, 5095–5103.

Jackson, V., & Chalkley, R. (1985). Histone segregation on replicating chromatin. Biochemistry 24, 6930–6938.

Jackson, V., Shires, A., Tanphaichitr, N., & Chalkley, R. (1976). Modifications to histones immediately after synthesis. J. Mol. Biol. 104, 471–483.

Jackson, V., Marshall, S., & Chalkley, R. (1981). The sites of deposition of newly synthesized histone. Nucleic Acids Res. 9, 4563–4581.

Jeppesen, P., & Turner, B.M. (1993). The inactive X-chromosome in female mammals is distinguished by a lack of histone H4 acetylation, a cytogenetic marker for gene expression. Cell 74, 281–289.

Jeppesen, P., Mitchell, A., Turner, B., & Perry, P. (1992). Antibodies to defined histone epitopes reveal variations in chromatin conformation and underacetylation of centric heterochromatin in human metaphase chromosomes. Chromosoma 101, 322–332.

Johnson, L.M., Kayne, P.S., Kahn, E.S., & Grunstein, M. (1990). Genetic evidence for an interaction between SIR3 and Histone H4 in the repression of the silent mating loci in *Saccharomyces cerevisiae*. Proc. Natl. Acad. Sci. USA. 87, 6286–6290.

Kamakaka, R.T., & Thomas, J.O. (1990). Chromatin structure of transcriptionally competent and repressed genes. EMBO J. 9, 3997–4006.

Kayne, P.S., Kim, U.-J., Han, M., Mullen, J.R., Yoshizaki, F., & Grunstein, M. (1988). Extremely conserved histone H4 N-Terminus is dispensable for growth but essential for repressing the silent mating loci in yeast. Cell 55, 27–39.

Kim, U.-J., Han, M., Kayne, P., & Grunstein, M. (1988). Effects of histone H4 depletion on the cell cycle and transcription of *Saccharomyces cerevisiae*. EMBO J. 7, 2211–2219.

Kleinschmidt, J.A., & Steinbeisser, H. (1991). DNA-dependent phosphorylation of histone H2A.X during nucleosome assembly in *Xenopus laevis* oocytes—involvement of protein phosphorylation in nucleosome spacing. EMBO J. 10, 3043–3050.

Kleinschmidt, J.A., Fortkamp, E., Krohne, G., Zentgraf, H., & Franke, W.W. (1985). Coexistence of two different types of soluble histone complexes in nuclei of *Xenopus laevis* oocytes. J. Biol. Chem. 260, 1166–1176.

Kleinschmidt, J.A., Seiter, A., & Zentgraf, H. (1990). Nucleosome assembly *in vitro*: Separate transfer and synergistic interaction of native histone complexes purified from nuclei of *Xenopus laevis* oocytes. EMBO J. 9, 1309–1318.

Kornberg, R.D., & Thomas, J.O. (1974). Chromatin structure: oligomers of histones. Science 184, 865–868.

Krude, T., & Knippers, R. (1991). Transfer of nucleosomes from parental to replicated chromatin. Mol. Cell. Biol. 11, 6257–6267.

Laskey, R.A., Mills, A.D., Philpot, A., Leno, G.H., Dilworth, S.M., & Dingwall, C. (1993). The role of nucleoplasmin in chromatin assembly and disassembly. Phil. Trans. R. Soc. Lond. B. 339, 263–269.

Lassle, M., Richter, A., & Knippers, R. (1992). Comparison of replicative and nonreplicative chromatin assembly pathways in HeLa cell extracts. Biochim. Biophys. Acta 1132, 1–10.

Laybourn, P.J., & Kadonaga, J.T. (1991). Role of nucleosomal cores and histone-H1 in regulation of transcription by RNA Polymerase-II. Science 254, 238–245.

Lee, D.Y., Hayes, J.J., Pruss, D., & Wolffe, A.P. (1993). A positive role for histone acetylation in transcription factor access to nucleosomal DNA. Cell 72, 73–84.

Liew, C.C., Haslett, G.W., & Allfrey, V. (1970). N-acetyl-seryl-tRNA and polypeptide chain initiation during histone biosynthesis. Nature (London) 226, 414–417.

Lin, R., Leone, J.W., Cook, R.G., & Allis, C.D. (1989). Antibodies specific to acetylated histones document the existence of deposition and transcription-related histone acetylation in *Tetrahymena*. J. Cell Biol. 108, 1577–1588.

Lin, R.L., Cook, R.G., & Allis, C.D. (1991). Proteolytic removal of core histone amino termini and dephosphorylation of Histone H1 correlate with the formation of condensed chromatin and transcriptional silencing during *Tetrahymena* macronuclear development. Genes & Dev. 5, 1601–1610.

Lohr, D., & Hereford, L. (1979). Yeast chromatin is uniformly digested by DNaseI. Proc. Natl. Acad. Sci. USA 76, 4285–4288.

Lopez-Rodas, G., Georgieva, E.I., Sendra, R., & Loidl, P. (1991a). Histone acetylation in *Zea mays* 1. Activities of histone acetyltransferases and histone deacetylases. J. Biol. Chem. 266, 18745–18750.

Lopez-Rodas, G., Tordera, V., Delpino, M.M.S., & Franco, L. (1991b). Subcellular localization and nucleosome specificity of yeast histone acetyltransferases. Biochemistry 30, 3728–3732.

Lopez-Rodas, G., Brosch, G., Georgieva, E.I., Sendra, R., Franco, L., & Loidl, P. (1993). Histone deacetylase—a key enzyme for the binding of regulatory proteins to chromatin. FEBS Lett. 317, 175–180.

Louie, A.J., Candido, E.P.M., & Dixon, G. (1974). Enzymatic modifications and their possible roles in regulating the binding of basic proteins to DNA and in controlling chromosomal structure. Cold Spring Harb. Symp. Quant. Biol. 38, 808–819.

Louters, L., & Chalkley, R. (1985). Exchange of histones H1, H2A, and H2B *in vivo*. Biochemistry 24, 3080–3085.

Lutter, L.C., Judis, L., & Paretti, R.F. (1992). Effects of histone acetylation on chromatin topology *in vivo*. Mol. Cell. Biol. 12, 5004–5014.

Mann, R.K., & Grunstein, M. (1992). Histone H3 N-terminal mutations allow hyperactivation of the yeast GAL1 gene *in vivo*. EMBO J. 11, 3297–3306.

McGhee, J.D., Nickol, J.M., Felsenfeld, G., & Rau, D.C. (1983). Histone hyperacetylation has little effect on the higher order folding of chromatin. Nucleic Acids Res. 11, 4065–4074.

Megee, P.C., Morgan, B.A., Mittman, B.A., & Smith, M.M. (1990). Genetic analysis of histone H4: Essential role of lysines subject to reversible acetylation. Science 247, 841–845.

Meistrich, M.L., Trostleweige, P.K., Lin, R.L., Bhatnagar, Y.M., & Allis, C.D. (1992). Highly acetylated H4 Is associated with histone displacement in rat spermatids. Mol. Reprod. Dev. 31, 170–181.

Mingarro, I., Sendra, R., Salvador, M.L., & Franco, L. (1993). Site specificity of pea histone acetyltransferase-B *in vitro*. J. Biol. Chem. 268, 13248–13252.

Morgan, B.A., Mittman, B.A., & Smith, M.M. (1991). The highly conserved N-terminal domains of histone H3 and histone H4 are required for normal cell cycle progression. Mol. Cell. Biol. 11, 4111–4120.

Munks, R.J.L., Moore, J., O'Neill, L.P., & Turner, B.M. (1991). Histone H4 acetylation in *Drosophila*— frequency of acetylation at different sites defined by immunolabelling with site-specific antibodies. FEBS Lett. 284, 245–248.

Nadeau, P., Oliver, D.R., & Chalkley, R. (1978). Effect of inhibition of DNA synthesis on histone synthesis and deposition. Biochemistry 17, 4885–4893.

Nelson, D.A. (1982). Histone acetylation in bakers yeast, maintenance of the hyperacetylated configuration in log phase protoplasts. J. Biol. Chem. 257, 1565–1568.

Norris, D., Dunn, B., & Osley, M.A. (1988). The effect of histone gene deletions on chromatin structure in *Saccharomyces cerevisiae*. Science 242, 759–760.

Norton, V.G., Imai, B.S., Yau, P., & Bradbury, E.M. (1989). Histone acetylation reduces nucleosomal core particle linking number change. Cell 57, 449–457.

Norton, V.G., Marvin, K.W., Yau, P., & Bradbury, E.M. (1990). Nucleosome linking number change controlled by acetylation of histones H3 and H4. J. Biol. Chem. 265, 19848–19852.

Olins, D.E., Olins, A.L., Herrmann, A., Lin, R.L., Allis, C.D., & Robert-Nicoud, M. (1991). Localization of acetylated histone H4 in the macronucleus of *Euplotes*. Chromosoma 100, 377–385.

Oliver, D., Granner, D., & Chalkley, R. (1974). Identification of a distinction between cytoplasmic histone synthesis and subsequent histone deposition. Biochemistry 13, 746–749.

Park, E.-C., & Szostak, J.W. (1990). Point mutations in the yeast histone H4 gene prevent silencing of the silent mating type locus *HML*. Mol. Cell. Biol. 10, 4932–4934.

Perry, C.A., Allis, C.D., & Annunziato, A.T. (1993a). Parental nucleosomes segregated to newly replicated chromatin are underacetylated relative to those assembled *de novo*. Biochemistry. 32, 13615–13623.

Perry, C.A., & Annunziato, A.T. (1989). Influence of histone acetylation on the solubility, H1 content, and DNase I sensitivity of newly assembled chromatin. Nucleic Acids Res. 17, 4275–4291.

Perry, C.A., & Annunziato, A.T. (1991). Histone acetylation reduces H1-mediated nucleosome interactions during chromatin assembly. Exp. Cell. Res. 196, 337–345.

Perry, C.A., Chang, L., Allis, C.D., & Annunziato, A.T. (1994). Native non-nucleosomal histone complexes in somatic cells. Mol. Biol. Cell 5, 211a.

Perry, C.A., Dadd, C.A., Allis, C.D., & Annunziato, A.T. (1993b). Analysis of nucleosome assembly and histone exchange using antibodies specific for acetylated H4. Biochemistry 32, 13605–13614.

Perry, M., & Chalkley, R. (1982). Histone acetylation increases the solubility of chromatin and occurs sequentially over most of the chromatin. J. Biol. Chem. 257, 7336–7347.

Pesis, K.H., & Matthews, H.R. (1986). Histone acetylation in replication and transcription: Turnover at specific acetylation sites in histone H4 from *Physarum polycephalum*. Arch. Biochem. Biophys. 251, 665–673.

Phillips, D.M.P. (1963). The presence of acetyl groups in histones. Biochem. J. 87, 258–263.

Randall, S.K., & Kelly, T.J. (1992). The fate of parental nucleosomes during SV40 DNA replication. J. Biol. Chem. 267, 14259–14265.

Reeves, R., Gorman, C.M., & Howard, B. (1985). Minichromosome assembly of nonintegrated plasmid DNA transfected into mammalian cells. Nucleic Acids Res. 13, 3599–3615.

Richman, R., Chicoine, L.G., Collini, C.M.P., Cook, R.G., & Allis, C.D. (1988). Micronuclei and the cytoplasm of growing *Tetrahymena* contain a histone acetylase activity which is highly specific for free histone H4. J. Cell Biol. 106, 1017–1026.

Ridsdale, J.A., Hendzel, M.J., Delcuve, G.P., & Davie, J.R. (1990). Histone acetylation reduces the capacity of H1 histones to condense transcriptionally active/competent chromatin. J. Biol. Chem. 265, 5150–5156.

Roth, S.Y., & Allis, C.D. (1992). Chromatin condensation—does histone H1 dephosphorylation play a role. Trends Biochem. Sci. 17, 93–98.

Ruberti, I., & Worcel, A. (1986). Mechanism of chromatin assembly in *Xenopus* oocytes. J. Mol. Biol. 189, 457–476.

Ruiz-Carrillo, A., Wangh, L.J., & Allfrey, V.G. (1975). Processing of newly synthesized histone molecules. Science 190, 117–128.

Sapp, M., & Worcel, A. (1990). Purification and mechanism of action of a nucleosome assembly factor from *Xenopus* oocytes. J. Biol. Chem. 265, 9357–9365.

Schuster, T., Han, M., & Grunstein, M. (1986). Yeast histone H2A and H2B amino termini have interchangeable functions. Cell 45, 445–451.

Seale, R.L. (1981). *In vivo* assembly of newly synthesized histones. Biochemistry 20, 6432–6437.

Sealy, L., & Chalkley, R. (1979). Modification of histones immediately following synthesis. Arch. Biochem. Biophys. 197, 78–82.

Senshu, T., & Ohashi, M. (1979). The fate of newly synthesized histones shortly after interruption of DNA replication. J. Biochem. 86, 1259–1267.

Senshu, T., & Yamada, F. (1980). Involvement of cytoplasmic soluble fraction in the assembly of nucleosome-like materials under near physiological conditions. J. Biochem. 87, 1659–1668.

Sessa, G., & Ruberti, I. (1990). Assembly of correctly spaced chromatin in a nuclear extract from *Xenopus laevis* oocytes. Nucleic Acids Res. 18, 5449–5445.

Shimamura, A., & Worcel, A. (1989). The assembly of regularly spaced nucleosomes in the *Xenopus* oocyte S-150 extract is accompanied by deacetylation of histone H4. J. Biol. Chem. 264, 14524–14530.

Shimamura, A., Sapp, M., Rodriguez-Campos, A., & Worcel, A. (1989). Histone H1 represses transcription from minichromosomes assembled *in vitro*. Mol. Cell. Biol. 9, 5573–5584.

Simpson, R.T. (1978). Structure of chromatin containing extensively acetylated H3 and H4. Cell 13, 691–699.

Smith, M.M., & Stirling, V.B. (1988). Histone H3 and H4 gene deletions in *Saccharomyces cerevisiae*. J. Cell Biol. 106, 557–566.

Smith, S., & Stillman, B. (1989). Purification and characterization of CAF-I, a human cell factor required for chromatin assembly during DNA replication *in vitro*. Cell 58, 15–25.

Smith, S., & Stillman, B. (1991a). Immunological characterization of chromatin assembly factor-I, a human cell factor required for chromatin assembly during DNA replication *in vitro*. J. Biol. Chem. 266, 12041–12047.

Smith, S., & Stillman, B. (1991b). Stepwise assembly of chromatin during DNA replication *in vitro*. EMBO J. 10, 971–980.

Sobel, R.E., Cook, R.G., & Allis, C.D. (1994). Non-random acetylation of histone H4 by a cytoplasmic histone acetyltransferase as determined by novel methodology. J. Biol. Chem. 269, 18576–18582.

Sobel, R.E., Cook, R.G., Perry, C.A., Annunziato, A.T., & Allis, C.D. (1995). Conservation of deposition-related acetylation sites in newly synthesized histones H3 and H4. Proc. Natl. Acad. Sci. USA 92, 1237–1241.

Sommerville, J., Baird, J., & Turner, B.M. (1993). Histone H4 acetylation and transcription in amphibian chromatin. J. Cell Biol. 120, 277–290.

Sugasawa, K., Ishimi, Y., Eki, T., Hurwitz, J., Kikuchi, A., & Hanaoka, F. (1992). Nonconservative segregation of parental nucleosomes during Simian Virus-40 chromosome replication *in vitro*. Proc. Natl. Acad. Sci. USA 89, 1055–1059.

Sures, I., & Gallwitz, D. (1980). Histone specific acetyltransferases from calf thymus. Isolation, properties, and substrate specificity of three different enzymes. Biochemistry 19, 943–951.

Thomas, J.O., & Kornberg, R.D. (1975). An octamer of histones in chromatin and in solution. Proc. Natl. Acad. Sci. USA. 72, 2626–2630.

Thomsen, B., Bendixen, C., & Westergaard, O. (1991). Histone hyperacetylation is accompanied by changes in DNA topology *in vivo*. Eur. J. Biochem. 201, 107–111.

Thorne, A.W., Kmiciek, D., Mitchelson, K., Sautiere, P., & Crane-Robinson, C. (1990). Patterns of histone acetylation. Eur. J. Biochem. 193, 701–713.

Tordera, V., Sendra, R., & Pérez-Ortín, J.E. (1993). The role of histones and their modifications in the informative content of chromatin. Experientia 49, 780–788.

Tremethick, D.J., & Frommer, M. (1992). Partial purification, from *Xenopus-laevis* oocytes, of an ATP-dependent activity required for nucleosome spacing *in vitro*. J. Biol. Chem. 267, 15041–15048.

Turner, B.M. (1989). Acetylation and deacetylation of histone H4 continue through metaphase with depletion of more-acetylated isoforms and altered site usage. Exp. Cell Res. 182, 206–214.

Turner, B.M. (1991). Histone acetylation and control of gene expression. J. Cell Sci. 99, 13–20.

Turner, B.M. (1993). Decoding the nucleosome. Cell 75, 5–8.

Turner, B.M., O'Neill, L.P., & Allan, I.M. (1989). Histone H4 acetylation at different sites in human cells: Frequency of acetylation at different sites defined by immunolabeling with site-specific antibodies. FEBS Lett. 253, 141–145.

Turner, B.M., Birley, A.J., & Lavender, J. (1992). Histone H4 isoforms acetylated at specific lysine residues define individual chromosomes and chromatin domains in *Drosophila* polytene nuclei. Cell 69, 375–384.

Ura, K., Wolffe, A.P., & Hayes, J.J. (1994). Core histone acetylation does not block linker histone binding to a nucleosome including a *Xenopus borealis* 5 S rRNA gene. J. Biol. Chem. 269, 27171–27174,

Urban, M.K., & Zweidler, A. (1983). Changes in nucleosomal core histone variants during chicken development and maturation. Dev. Biol. 95, 421–428.

van Holde, K.E. (1988) Chromatin. Springer Series in Molecular Biology, (Rich, A., Ed.) Springer-Verlag, New York.

van Holde, K.E., Lohr, D.E., & Robert, C. (1992). What happens to nucleosomes during transcription? J. Biol. Chem. 267, 2837–2840.

Waterborg, J.H. (1992). Identification of five sites of acetylation in alfalfa histone H4. Biochemistry 31, 6211–6219.

Waterborg, J.H., & Matthews, H.R. (1984). Patterns of histone acetylation in *Physarum polycephalum*. H2A and H2B acetylation is functionally distinct from H3 and H4 acetylation. Eur. J. Biochem. 142, 329–335.

Weigand, R.C., & Brutlag, D.L. (1981). Histone acetylase from *Drosophila melanogaster* specific for H4. J. Biol. Chem. 256, 4578–4583.

Weintraub, H. (1985). Assembly and propagation of repressed and derepressed chromosomal states. Cell 42, 705–711.

Weintraub, H. (1984). Histone-H1-dependent chromatin superstructures and the suppression of gene activity. Cell 38, 17–27.

West, H.H.P., & Bonner, W.M. (1980). Histone 2A, a heteromorphous family of eight protein species. Biochemistry 19, 3238–3245.

Wolffe, A.P. (1989). Dominant and specific repression of *Xenopus* oocyte 5S RNA genes and satellite I DNA by histone H1. EMBO J. 8, 527–537.

Wolffe, A.P. (1991). Implications of DNA replication for eukaryotic gene expression. J. Cell Sci. 99, 201–206.

Wolffe, A.P. (1992a).Chromatin: structure and function. Academic Press, San Diego, CA.

Wolffe, A.P. (1992b). New Insights into Chromatin Function in Transcriptional Control. FASEB J. 6, 3354–3361.

Wolffe, A.P., & Brown, D.D. (1988). Developmental regulation of two 5S ribosomal RNA genes. Science 241, 1626–1632.

Wolffe, A.P., & Dimitrov, S. (1993). Histone-modulated gene activity: Developmental implications. Critic. Rev. Euk. Gene Expr. 3, 167–191.

Woodland, H.R. (1979). The modification of stored histones H3 and H4 during the oogenesis and early development of *Xenopus laevis*. Dev. Biol. 68, 360–370.

Woodland, H.R., & Adamson, E.D. (1977). The synthesis and storage of histones during the oogenesis of *Xenopus laevis*. Dev. Biol. 57, 118–135.

Worcel, A., Han, S., & Wong, M.L. (1978). Assembly of newly replicated chromatin. Cell 15, 969–977.

Wu, R.S., & Bonner, W.M. (1981). Separation of basal histone synthesis from S-phase histone synthesis in dividing cells. Cell 27, 321–330.

Zlatanova, J., & van Holde, K. (1992). Histone H1 and transcription—still an enigma. J. Cell Sci. 103, 889–895.

Zucker, K., & Worcel, A. (1990). The histone H3/H4-N1 complex supplemented with histone H2A-H2B dimers and DNA topoisomerase I Forms nucleosomes on circular DNA under physiological conditions. J. Biol. Chem. 265, 14487–14496.

Zweidler, A. (1980). Nonallelic histone variants in development and differentiation. Dev. Biochem. 15, 47–56.

Part II

Transcription Factor Interaction with Nucleosomes

Chapter 3

Manipulating Chromatin Structure to Study the Interaction of Transcription Factors with Nucleosomes *In Vitro* and *In Vivo*

RANDALL H. MORSE

ABSTRACT

The packaging of DNA into nucleosomes in eukaryotic cells raises the question of how proteins such as transcription factors gain access to their sites in chromatin. Various experimental approaches to this problem *in vitro* and *in vivo* are discussed.

The Nucleus
Volume 1, pages 59–78.
Copyright © 1995 by JAI Press Inc.
ISBN: 1-55938-940-0

A central difficulty is obtaining templates which are homogeneous with respect to having a transcription factor binding site within a nucleosome; this dilemma has been resolved by manipulating chromatin structure, either by removing templates having nonnucleosomal sites or by using nucleosome positioning signals to make "designer chromatin." Experimental results are presented which indicate a hierarchy with respect to transcription factors and chromatin. Parameters which may govern whether a transacting factor will be occluded from its site by histones or will successfully compete against the histones for its binding site include the strength and stability with which a factor binds to its site, and the type of DNA-binding domain which it uses. Finally, some implications of what we know about transcription factor binding in chromatin for the mechanism of enhancer action are presented.

A eukaryotic cell must squeeze up to several billion base pairs (bp) of DNA into a nucleus whose diameter is 1–10 microns. The four to five orders of magnitude of compaction required to fit this DNA into the cell nucleus is accomplished with the aid of the histone proteins.

The first level of packaging of DNA in eukaryotes is the nucleosome (see Chapter 1). Although wrapping DNA around the histone octamer core only achieves about a fivefold compaction, already a central problem arises: How do processes requiring DNA as template take place when the DNA is already closely engaged with proteins? How is chromatin replicated? How is it repaired? And how is it transcribed?

I. *IN VITRO* STUDIES ON TRANSCRIPTION AND CHROMATIN

Early studies focused on the last question by reconstituting chromatin *in vitro* from histones and DNA and adding bacterial or eukaryotic RNA polymerase preparations. A general inhibition of transcriptional initiation and elongation was observed, whether the polymerase was of eukaryotic or prokaryotic origin (Williamson and Felsenfeld, 1978; Wasylyk and Chambon, 1979; Wasylyk et al., 1979). So the problem remained: Chromatin inhibited transcription, and eukaryotic polymerases were not specially equipped to overcome this inhibition.

These early studies were limited by the nonspecific nature of the transcription measured: Transcription initiated from multiple undetermined sites, and was measured as gross incorporation of mononucleotides into polymers. Subsequent years saw the unraveling of the design of the eukaryotic promoter, and the problem could be redefined: What prevented packaging of promoter elements into chromatin from repressing their function *in vivo*?

The first aspect of this question that needed to be addressed was whether such packaging would indeed inhibit function. It seemed likely that it would: The close apposition of histones and DNA seemed certain to impose steric constraints, and the severe bending of DNA and altered ionic environment also were good bets to impair DNA-protein recognition. Knezetic and Luse (1986) examined this issue by using a plasmid containing the adenovirus major late promoter as a template for specific transcription by RNA polymerase II. When the plasmid was packaged into nucleosomes using an extract prepared from frog oocytes, transcriptional initiation was inhibited.

This work was an improvement over earlier studies, but was still deficient in that it examined a heterogeneously packaged template of which only an unknown, probably small fraction was transcribed. Since the structure of the transcribed fraction was perforce unknown, it could not be determined whether inhibition resulted from a general compaction of the DNA or whether it was caused by specific histone-DNA interactions. Lorch and colleagues (1987) addressed this issue by examining small linear DNA fragments packaged into nucleosomes using a protocol which yielded templates in which promoter sequences were incorporated into nucleosomes or were not. These templates could be transcribed by SP6 RNA polymerase and RNA polymerase II only when the promoter sequences were nonnucleosomal. Transcription by RNA polymerase II in these experiments arose from single-stranded poly C tails, and so what was measured in both cases was polymerase accessibility, rather than accessibility to other transacting factors required for transcription by RNA polymerase II *in vivo*.

The next logical step was to examine the ability of transacting factors required for polymerase recruitment to bind DNA when their recognition sequences were incorporated into nucleosomes. Two early studies approached this problem by comparing the transcription of templates assembled into nucleosomes before or after incubation with extracts containing promoter-binding factors (Matsui, 1987; Workman and Roeder, 1987). In both cases, nucleosome assembly was strongly inhibitory to transcription, but the inhibition could be largely overcome by preincubation with transacting factors; in one case, prebinding of the TFIID fraction required for basal transcription by RNA polymerase II to a TATA site was sufficient to overcome nucleosomal inhibition (Workman and Roeder, 1987; see Chapter 4).

In spite of these advances, because individual plasmid molecules differed in which sequences were incorporated into nucleosomes, it still remained uncertain whether nucleosomal inhibition resulted from a general compaction or specific histone-DNA interactions. To overcome this difficulty, a protocol was devised which eliminated those molecules in which a promoter sequence was not packaged into nucleosomes by restriction endonuclease cleavage (see Figure 1) (Morse, 1989). This protocol was first used to investigate whether packaging specific sequences of the 5S RNA gene from *Xenopus borealis* into a nucleosome would inhibit its transcription by RNA polymerase III. The 5S RNA gene depends on an internal promoter for its transcription. This promoter is first recognized by the transcription factor TFIIIA; TFIIIB and TFIIIC then bind and recruit RNA polymerase III (reviewed in Wolffe and Brown, 1988).

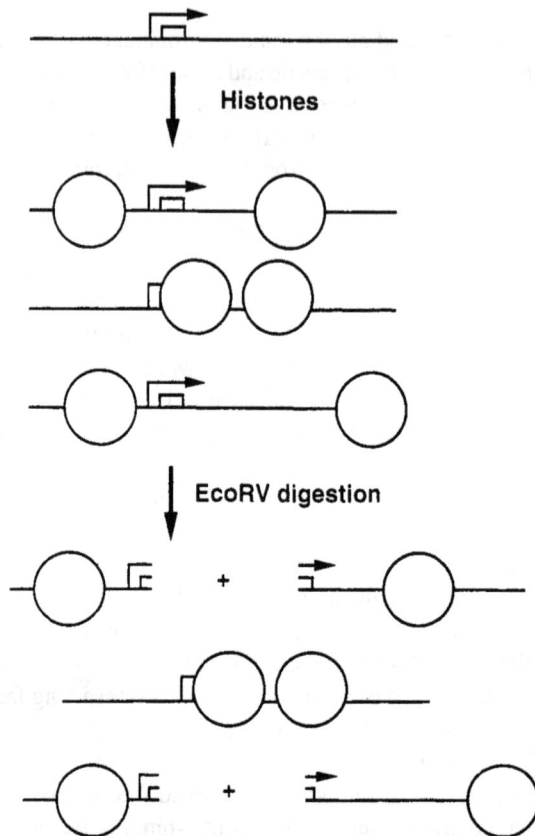

Figure 1. Restriction enzyme protection assay. DNA containing a transcribable sequence such as the 5S RNA gene (arrow; the box is the TFIIIA binding site/internal promoter), is assembled into nucleosomes using purified histones. Subsequent restriction at a site in the promoter eliminates as transcribable templates those molecules not having the site assembled into a nucleosome.

When a plasmid bearing the 5S RNA gene is packaged into nucleosomes *in vitro* with purified histones, a heterogeneous assemblage of nucleosomal templates results, in which the promoter sequence for the 5S RNA gene is nucleosomal in some molecules and nonnucleosomal in others (see Figure 1). Digestion of the assembled molecules with EcoRV cleaves the internal promoter of those molecules whose promoter is nonnucleosomal, rendering them incapable of being transcribed. The remaining templates are uncut, but their EcoRV sites and hence their internal promoters are packaged into nucleosomes. Incubation of these templates in a highly efficient transcriptional extract from *Xenopus* oocytes (Birkenmeier et al., 1978) resulted in no detectable transcripts, indicating that packaging of this promoter into nucleosomes completely inhibited transcription. This result was confirmed and extended by others (Clark and Wolffe, 1991; Hansen and Wolffe, 1992). Demonstration that packaging of the 5S RNA gene into a nucleosome prevents binding of TFIIIA *in vitro* provided a mechanistic explanation for the observed inhibition (Hayes and Wolffe, 1992). Later, the same protocol was used to demonstrate that packaging the start site of a gene transcribed by RNA polymerase II into a nucleosome prevented its transcription (Laybourn and Kadonaga, 1991).

II. *IN VIVO* STUDIES ON TRANSCRIPTION AND CHROMATIN

The studies described above and others too numerous to detail established that nucleosomes could inhibit transcription *in vitro* by interfering with protein-DNA interactions at promoters. What, then, operates *in vivo* to counter this repressive potential of chromatin? Genetic experiments with yeast showed that altered histone stoichiometry could affect promoter utilization *in vivo*, strongly suggesting that the repressive potential exerted by nucleosomes *in vitro* also existed *in vivo* (Clark-Adams et al., 1988; Han and Grunstein, 1988). Nuclease hypersensitive sites were mapped at many promoters, suggesting an incompatibility of nucleosomes and transcriptional initiation (Gross and Garrard, 1988; Elgin, 1988). Positioned nucleosomes present on promoters of inactive genes have been found to be lost upon activation, supporting this view (Benezra et al., 1986; Almer et al., 1986; Richard-Foy and Hager, 1987; see Chapter 6 for a discussion of the extensive work done in this area with the mouse mammary tumor virus promoter).

Studies on extant genes have taught us much about the complex interplay between regulatory factors and chromatin in gene activation. However, because these promoters invariably contain multiple binding sites for transcription factors, it has been difficult to glean from such studies what allows access of transacting factors to DNA in chromatin. An alternative approach has followed the line used for *in vitro* studies, in which chromatin structure is manipulated to place binding sites for transacting factors in nucleosomes and to ask what effect on function results.

A. Competition between tRNA Transcription and Nucleosome Positioning in Yeast

The TRP1ARS1 yeast minichromosome has provided a useful vehicle for manipulating chromatin structure *in vivo*. This circular minichromosome (see Figure 2), 1453 bp in length, contains the *TRP1* gene and a functional ARS (autonomously replicating sequence); together these elements allow this plasmid to be maintained as a multicopy episome in *trp*-yeast cells. The plasmid is packaged into seven positioned nucleosomes—that is, the same DNA sequences are packaged into nucleosomes with the same orientation relative to the protein core in the great majority of plasmid molecules in all cells. This has been demonstrated first by

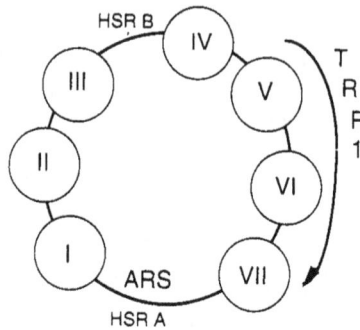

Figure 2. The TRP1ARS1 minichromosome. The numbered circles represent positioned nucleosomes. HSR A and HSR B are nuclease hypersensitive sites.

indirect end-label analysis following micrococcal nuclease digestion (Thoma et al., 1984) and confirmed in numerous later studies (reviewed in Simpson, 1991 and Thoma, 1992).

The A element, an 11 bp sequence essential for ARS function, is just outside the border of nucleosome I of TRP1ARS1 (Figures 2 and 3). Thus, histones would not normally be expected to interfere with access of transacting factors to this sequence. However, deletion of 60–80 bp of DNA from the side of this nucleosome flanking the A element caused the A element to be incorporated into a nucleosome, and copy number of the resulting plasmid was reduced by about 60-fold (Simpson, 1990). This suggested that the function of a cis-acting element could be inhibited by incorporation into a nucleosome *in vivo*, probably by interfering with access of a transacting factor or factors.

Would incorporation of transcriptional promoter elements into a nucleosome similarly impede function? To address this question a tRNA gene was inserted into a TRP1ARS1 derivative such that its start site was incorporated into a predicted positioned nucleosome (see Figure 3) (Morse et al., 1992). The tRNA gene was marked by insertion of an oligonucleotide so that its transcript would be larger than the endogenous yeast tRNA transcripts, which might otherwise have obscured detection of the introduced gene by cross-hybridization. The tRNA gene on the plasmid *TAt-100* was transcribed efficiently, and the positioned nucleosome was found not to be present. Mutations that inactivated the tRNA gene restored the positioned nucleosome, demonstrating that the sequences were not inhibitory toward nucleosome formation. Rather, the transcriptionally competent tRNA gene overrode the nucleosome positioning information.

These results indicated the existence of a hierarchy, in which factors binding to a tRNA gene promoter could override a nucleosome positioning signal, whereas the same signal could inhibit function of the ARS A element. This conclusion was buttressed by a second experiment, in which a different nucleosome positioning

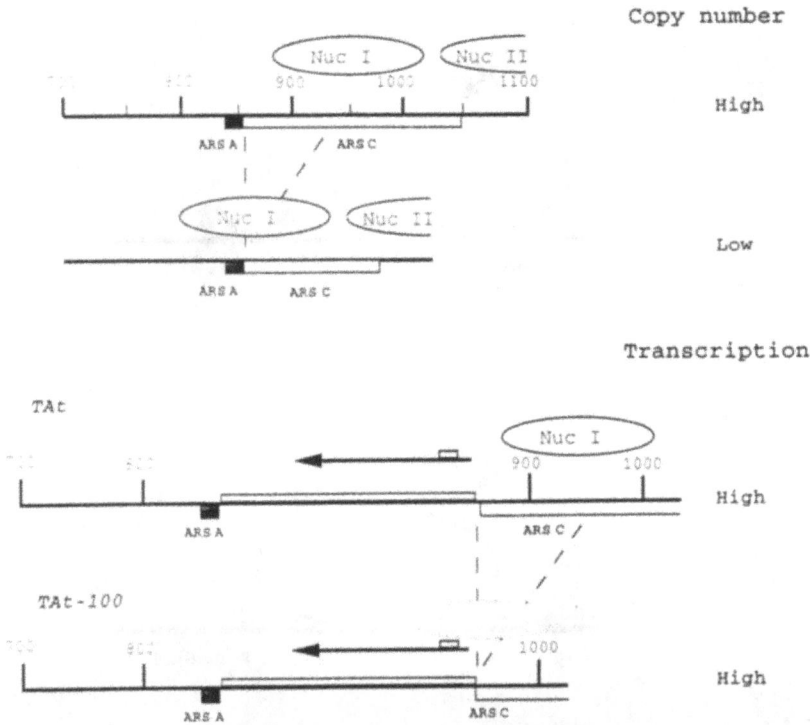

Figure 3. Effect of the ARS C nucleosome positioning element on replication of TRP1ARS1 and on transcription of a tRNA gene. The region of the TRP1ARS1 plasmid near the ARS A element, from 700–1,100 map units, is depicted. (Adapted from Morse et al., 1992; with permission from the American Society for Microbiology).

signal was juxtaposed to the same tRNA gene (Morse et al., 1992). This signal comprised the α2/MCM1 operator, which positions a nucleosome in yeast haploid α cells and diploids (a/α cells) (Roth et al., 1990; Chapter 5). The normal function of this operator is to repress a-specific genes in diploids (a/α cells) and α cells, and part of the mechanism of repression appears to be incorporation of TATA elements into the nucleosome positioned by the operator (Shimizu et al., 1991; Roth et al., 1992; Chapter 5). However, in contrast to this repression of genes transcribed by RNA polymerase II, the tRNA gene was not repressed by the α2/MCM1 operator. Once again, the positioned nucleosome was not present unless the tRNA gene was rendered inactive. Consistent with this result, the α2/MCM1 complex was also shown not to repress transcription of *SNR6*, which is transcribed by RNA polymerase III, as well as another tRNA gene (Herschbach and Johnson, 1993). Interestingly, this latter study also showed that the α2/MCM1 repressor could repress transcription by RNA polymerase I. Chromatin structure was not examined, but we

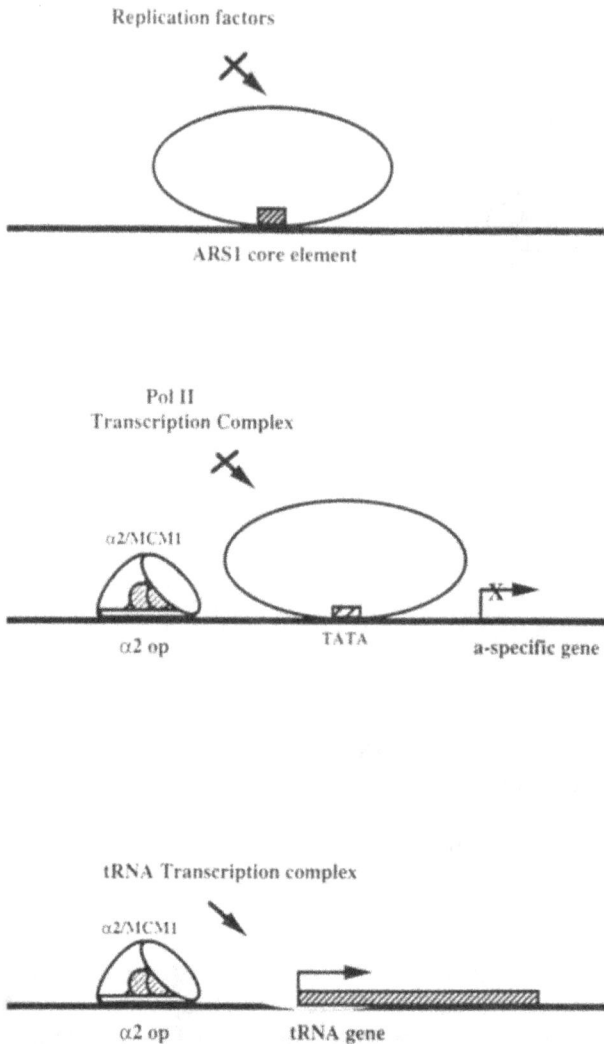

Figure 4. Hierarchies in nucleosome positioning and function.

would expect that a positioned nucleosome would be found adjacent to the α2/MCM1 operator over the repressed pol I promoter.

Taken together, the results discussed above indicate the existence of a hierarchy between transcription factors (possibly including RNA polymerases) and chromatin, in which unknown parameters determine which party wins in competing for sites on DNA *in vivo* (see Figure 4).

B. Class III Genes and Chromatin

The strength and stability with which a factor binds to a particular site may be a strong determinant of its ability to compete with histones *in vivo*. As already discussed, a 5S RNA gene from *Xenopus* can be repressed *in vitro* by incorporation of the TFIIIA binding site into a nucleosome (Morse, 1989). *In vivo*, active *Xenopus* 5S RNA genes are nucleosome-free, as one might expect; inactive oocyte 5S RNA genes present in somatic tissues are organized into nucleosomes, and histone H1 appears to play an important role in their repression (Gottesfeld and Bloomer, 1980; Young and Carroll, 1983; Schlissel and Brown, 1984; Wolffe, 1989; Chipev and Wolffe, 1992; Bouvet et al., 1994). TFIIIA alone binds to the 5S promoter with relatively low affinity [K_d about 10^{-8} (Hanas et al., 1983)], and can be displaced by a nucleosome to yield a repressed template *in vitro* (Tremethick et al., 1990). Binding of TFIIIC and TFIIIB to the TFIIIA-DNA complex results in a much more stable complex (reviewed by Wolffe and Brown, 1988) which is resistant to nucleosomal repression (Tremethick et al., 1990). Moreover, the complete transcription complex is more stable in the case of the *Xenopus* somatic 5S RNA gene than for the oocyte gene (Wolffe and Brown, 1987), consistent with the repression of the latter but not the former by chromatin. Thus, the pattern of repression of 5S RNA genes, in which nucleosomes can repress *in vitro* if TFIIIA alone is bound but not when the stable complex including TFIIIB and C has been formed, and the less stable oocyte complex is repressed by nucleosomes under conditions where the more stable somatic complex is not, fits the idea that the stability and strength of interaction of transcription factors with DNA affects their ability to be repressed by chromatin.

What about tRNA genes? Transcription of these genes is initiated by binding of TFIIIC to the internal promoter; this binding is about two orders of magnitude stronger than that of TFIIIA to the 5S RNA gene promoter (reviewed in Geiduschek and Tocchini-Valentini, 1988). This stronger binding is paralleled by the finding that tRNA genes remain transcriptionally active when juxtaposed to a nucleosome positioning signal, as already discussed (Morse et al., 1992). Furthermore, under conditions where chromatin assembly in *Xenopus* oocytes represses 5S RNA genes, tRNA genes remain active (Almouzni et al., 1991), and a tRNA gene isolated as chromatin on a yeast episome was transcribable *in vitro*, whereas a 5S RNA gene was not (Lassar et al., 1985). Although these latter results could be due to different nucleosome positioning on tRNA and 5S RNA genes, the concordance among these varied experiments is striking. A direct investigation of nucleosomal repression of tRNA genes *in vitro* has not been reported to date, nor has the effect of juxtaposing a nucleosome positioning signal with a 5S RNA gene *in vivo*.

A tRNA gene with a TFIIIC binding site mutated such that it has a weakened interaction with TFIIIC but can still be transcribed *in vitro* was found not to be transcribed in wild type yeast; a suppressor designated *tap1-1* was found that rescued the crippled tRNA gene (DiSegni et al., 1993). Transcription in the *tap1-1* mutant strain was context dependent, and it was suggested that the crippled tRNA gene could not be transcribed if flanking sequences imparted a strongly repressive chromatin structure on the gene (Aldrich et al., 1993). Preliminary experiments indicate that the chromatin structure is indeed better defined (i.e., less random) in the context in which the gene cannot be transcribed in the *tap1-1* mutant than in the context in which it can be transcribed according to unpublished results by Morse and Hall. This would suggest that weakening the interaction of TFIIIC with the tRNA gene promoter makes it more prone to being repressed by chromatin.

Finally, the yeast U6 snRNA gene is worth briefly discussing. This gene has a 5′ promoter which is sufficient for its transcription *in vitro*, catalyzed by TFIIIB; TFIIIC is not required. *In vivo*, however, a TFIIIC binding site 3′ to the gene is required for its function (Brow and Guthrie, 1990). When the gene is assembled into chromatin *in vitro*, TFIIIC and its binding site becomes necessary for transcriptional activation (Burnol et al., 1993). Thus, the strong TFIIIC-DNA interaction appears capable of overcoming nucleosomal repression of the U6 snRNA gene *in vitro* and possibly *in vivo*.

C. Interaction of GAL4 with Chromatin

An approach to learning more about what determines a given factor's ability to compete with histones for sites on DNA *in vivo* is illustrated in Figure 5. A site for a given factor, such as GAL4, is introduced into a plasmid or into the yeast genome in a location predicted to be in a positioned nucleosome or outside its border. In the latter case, the factor should be able to bind and if situated appropriately with respect to a reporter gene, it should activate transcription. In the former case, the question is open, and variables such as the amount of factor, type of factor, and when in the cell cycle the factor is expressed can be experimentally manipulated.

Some experiments along these lines have been carried out with the transcription factor GAL4 (Morse, 1993). GAL4 is a zinc-binding transcription factor which is responsible for induction by galactose of a number of genes in yeast (Johnston, 1987; Trumbly, 1992). In the presence of glucose, the *GAL4* gene is itself repressed and GAL4 binding sites on DNA are not occupied (Figure 6). If glucose is not present, GAL4 is produced and binds to its upstream activating sequences (UASs) upstream of the genes which it activates, but if galactose is not present, a repressor protein, GAL80, binds to GAL4 and prevents transcriptional activation. In the presence of galactose, the activation domain of GAL4 is unmasked and transcription is induced (Johnston et al., 1987; Ma and Ptashne, 1987b).

GAL4 binding to its sites in genomic chromatin in yeast has been investigated at the *GAL1-10* and *GAL7* loci (Lohr, 1984; Fedor and Kornberg, 1989; Fedor, Lue, and Kornberg, 1988; Axelrod et al., 1993; Cavalli and Thoma, 1993). In both cases, the binding sites are accessible to nucleases even in glucose, suggesting that they may be constitutively accessible to GAL4. Strongly positioned nucleosomes surrounding the GAL4 binding sites in the *GAL1-10* promoter depend on the protein

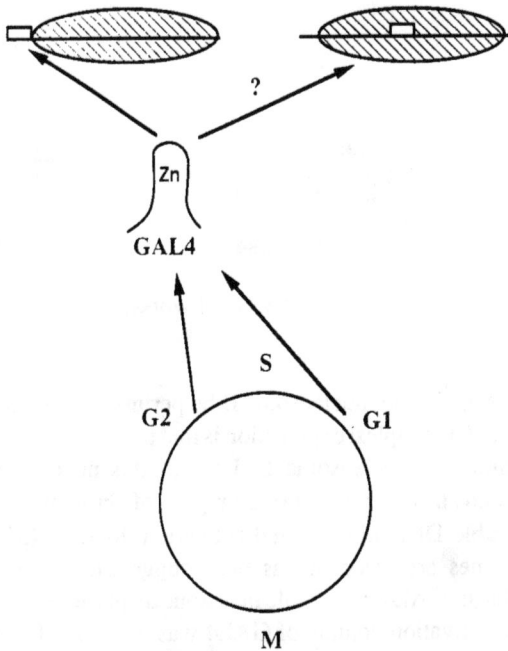

Figure 5. Strategy for investigating binding of transcription factors (for example, GAL4) to nucleosomal DNA in yeast.

GRF2 and its binding site, and it has been hypothesized that this protein may function to prevent nucleosomes from obscuring the GAL4 binding site (Fedor et al., 1988; Chasman et al., 1990).

To test whether a nucleosome could impede GAL4 binding *in vivo*, the TRP1ARS1 minichromosome was again used as a starting vehicle. A single 17 bp strong binding site for GAL4 was introduced into nucleosome I of a TRP1ARS1 derivative, and the resulting plasmid, called TA17Δ80, was introduced into yeast cells. Chromatin was then prepared and analyzed by indirect end-labeling.

When chromatin was prepared from cells containing TA17Δ80 grown in glucose, nucleosomes I and II were positioned as predicted (Figure 7). In galactose medium, however, a different picture was seen: new nuclease cleavage sites were seen in the regions of both of these nucleosomes, and the strong cleavages at the borders were reduced in intensity, similar to the naked DNA controls. This suggested that GAL4 was binding and preventing formation of the two positioned nucleosomes. Experiments in which the DNA fragment encompassing nucleosomes I and II of TA17Δ80 was fused to a *lacZ* reporter gene confirmed this by showing strongly induced expression in the presence of galactose.

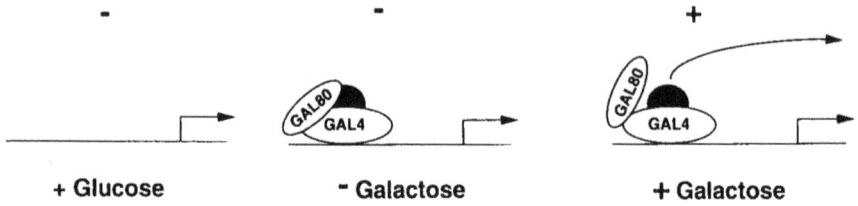

Figure 6. How GAL4 works.

Why should the neighboring nucleosome II be perturbed by GAL4 binding to a site in nucleosome I? The simplest explanation is that positioned nucleosome I helps determine the position of nucleosome II. Loss of this nucleosome could then randomize nucleosome position over a small region of chromatin.

GAL4 has separable DNA-binding and activation domains (Ma and Ptashne, 1987a). Since histones are basic, it has been suggested that acidic activation domains, such as that of GAL4, play a role in histone displacement from chromatin. To see whether the activation domain of GAL4 was important for its binding to a site in chromatin, chromatin was analyzed from cells harboring TA17Δ80 grown in raffinose, conditions under which GAL4 should bind but with its activation domain masked by GAL80 (see Figure 6). Again, nucleosomes I and II were perturbed, indicating that GAL4 could bind without aid from its acidic activation domain (see Figure 7). This was further confirmed by experiments showing that a derivative of GAL4 lacking its activation domain could also bind and perturb nucleosomes I and II of TA17Δ80 (Morse, 1993).

The nucleosome positioning signal of nucleosome I in TRP1ARS1 is weaker than that of the α2/MCM1 repressor (Simpson, 1990; Chapter 5). Binding sites for GAL4 adjacent to an α2/MCM1 operator can be bound by GAL4 with concomitant perturbation of nucleosome positioning, according to unpublished data, indicating that a GAL4 binding site, like a tRNA gene, can dominate this nucleosome positioning signal as well as that of nucleosome I of TRP1ARS1.

In vitro experiments have shown that GAL4 can bind to a nucleosome containing five binding sites, leading to a metastable structure in which both histones and GAL4 are bound more weakly than when either is bound in the absence of the other (Workman and Kingston, 1992; Chapter 4). Is this "non-canonical" nucleosome structure also generated by binding of GAL4 to TA17Δ80 in yeast? The *in vitro* binding, like that seen *in vivo*, does not depend on the activation domain of GAL4. Little can be said, however, about the structure of either entity. It is clear from the mode of binding of GAL4 to DNA that some perturbation of the nucleosome must accompany GAL4 binding (Marmorstein et al., 1992), as indicated by the micro-coccal nuclease cleavage sites in the region of nucleosome I of TA17Δ80 (Figure

7). However, whether histones are still bound in close proximity to GAL4 has not been determined. Neither is it known whether GAL4 can invade a nucleosome already formed in yeast, as seen *in vitro* (Workman and Kingston, 1992; Vettese-Dadey et al., 1994; Chapter 4), or whether a round of replication is required for binding to occur (Brown, 1984).

The acidic activation domain of GAL4 is not required for binding to nucleosomal templates *in vitro* or *in vivo*. However, a chromatin related effect distinct from GAL4 binding may be mediated by the activation domain *in vitro*. As mentioned earlier, a transcriptional reporter gene rendered refractory to transcription by assembly into nucleosomes *in vitro* could be rescued by preincubation with an extract containing the TATA-binding protein (Workman and Roeder, 1987). When a similar template with GAL4 binding sites upstream of the TATA site was packaged into nucleosomes, it too was transcriptionally inert (Workman et al., 1991; Chapter 4). Binding of GAL4-VP16 (in which the activation domain of VP16 replaces that of GAL4) before assembly allows transcription even when TFIID is added after nucleosome assembly; this "bypass" effect requires the presence of an activation domain. Similarly, the "antirepressive" effect by which GAL4-VP16 stimulates transcription from H1-containing chromatin templates *in vitro* depends on the presence of an activation domain (Croston et al., 1992). In both cases, it was suggested that the activation domain functioned to "open" chromatin and allow access by the basal transcription apparatus.

Does the activation domain perform a similar function *in vivo*? Recent work indicates that enhancers exert a chromatin-related antirepressive effect in early mouse embryos (Majumder et al., 1993). Similarly, chromatin-mediated repression of basal transcription in *Xenopus* oocytes can be relieved by binding of GAL4-VP16 to an upstream site (Almouzni and Wolffe, 1993). A hint that the activation domain of GAL4 may affect chromatin structure apart from the effect due to GAL4 binding comes from comparing the nuclease cleavage pattern of TA17Δ80 in glucose, galactose and raffinose (see Figure 7). The strong cleavage sites at the outer borders of nucleosomes I and II seen in glucose are cut much more weakly in galactose, when the positioned nucleosomes I and II are no longer present. In raffinose, however, when GAL4 also binds but with its activation domain masked, these sites are still strongly cleaved (see Figure 7); they are also strongly cleaved when a GAL4 derivative lacking an activation domain binds (Morse, 1993). This suggests that the activation domain may indeed be exerting a specific effect on chromatin. This question is currently being pursued.

D. Enhancer/UAS's and Chromatin

To the hierarchy of transcription factors and chromatin depicted in Figure 4 can be added GAL4, which can bind and perturb a positioned nucleosome. The glucocorticoid receptor can also bind to a positioned nucleosome, but without displacing it, to form a ternary complex *in vitro* (Perlmann and Wrange, 1988). *In*

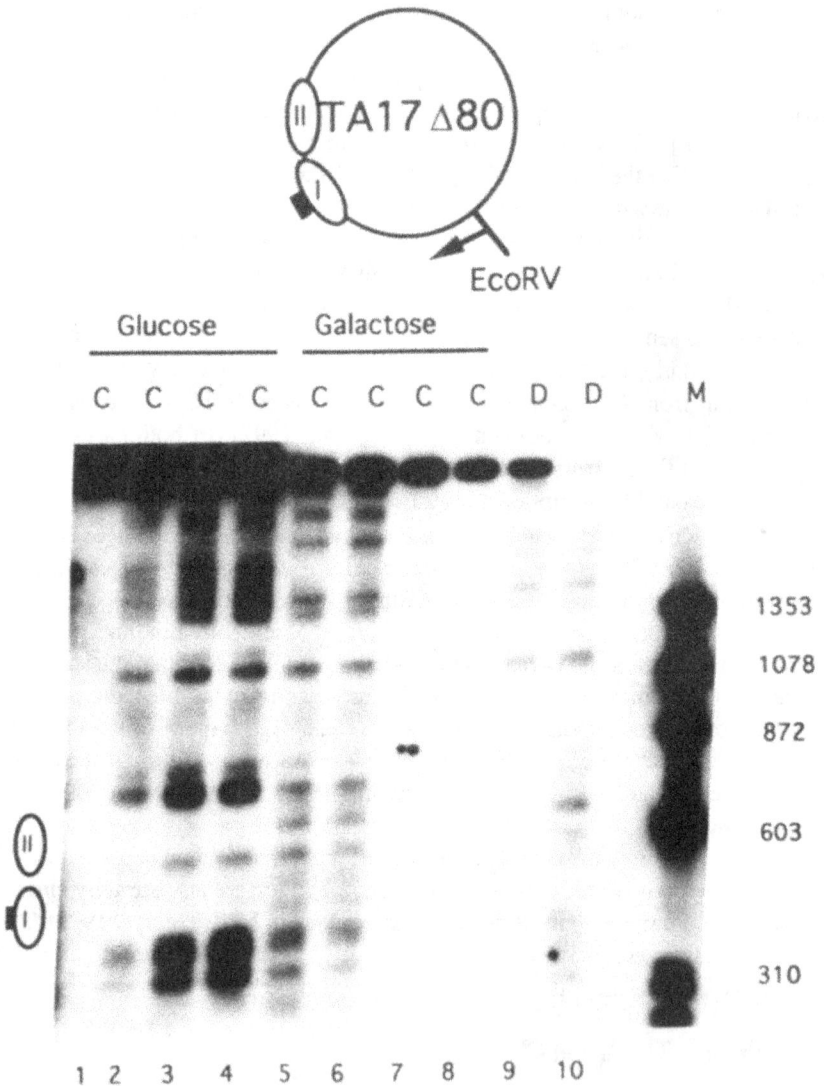

Figure 7. Disruption of positioned nucleosomes in the yeast episome TA17Δ80 by GAL4 binding. Micrococcal nuclease cleavage sites in chromatin (C lanes) and DNA (D lanes) were mapped relative to the unique EcoRV site of TA17Δ80 in cells grown in glucose, galactose, or raffinose medium, as indicated. Increasing amounts of micrococcal nuclease were used in lanes 1–4 and 8–5 (left panel), and 6–4 (right

(continued)

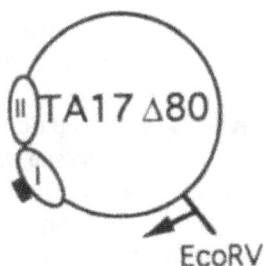

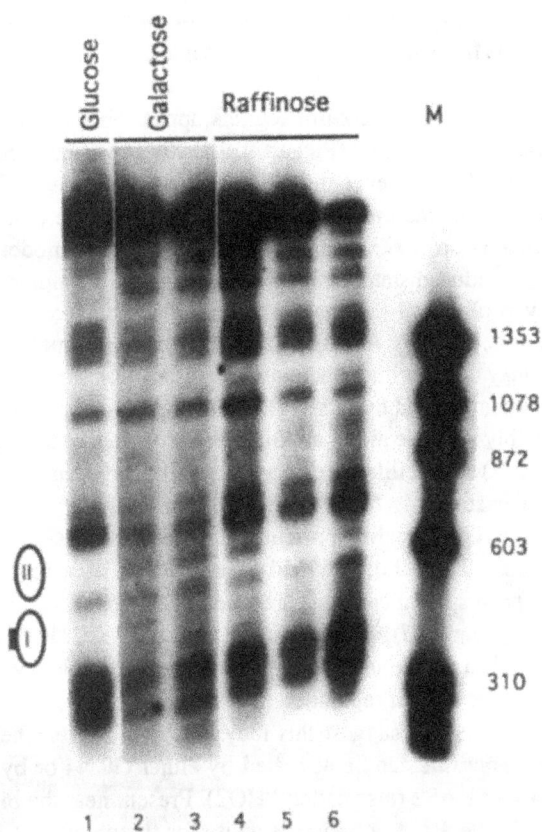

Figure 7. (continued)panel). The positions of nucleosomes I and II present in glucose are indicated on the left of each panel; the small box represents the single GAL4 binding site in nucleosome I. (From "Nucleosome disruption by transcription factor binding in yeast," by R.H. Morse in *Science*, Volume 262, pp. 1563–1566, December 3, 1993; ©AAAS; with permission.)

vivo, binding of GR to its site in a nucleosome is followed by perturbation of chromatin structure, binding of other factors, and transcription (Chapter 6). Another transcriptional activator, yeast heat shock factor (HSF), was found to bind to a nucleosome and activate transcription without histone displacement *in vivo* in one study (Pederson and Fidrych, 1994); in another study, overexpression of yeast HSF resulted in loss of a positioned nucleosome containing weak HSF binding sites (Gross et al., 1993; Chapter 7). A likely source of the different fates of nucleosomes in these two studies is the location of the binding site with respect to the positioned nucleosome; binding to more accessible peripheral sites (Simpson, 1990; Kladde and Simpson, 1994; Vettese-Dadey et al., 1994) may be less disruptive than binding to sites nearer the center of the nucleosome. This notion remains to be directly tested. Interestingly, *Drosophila* and mammalian heat shock factors do not bind to nucleosomal sites *in vitro* (Becker et al., 1991; Taylor et al., 1991); more work will be required to determine whether the interaction between metazoan HSFs and nucleosomes differs from that of yeast HSF with nucleosomes in some fundamental way.

Enhancers/UASs and locus control regions appear able to "open" chromatin, thereby potentiating transcription (Pikaart et al., 1992; Almouzni and Wolffe, 1993; Georgel et al., 1993; Majumder et al., 1993; Reitman et al., 1993). This opening of chromatin may also depend on the presence of a functional promoter (Georgel et al., 1993; Reitman et al., 1993). This suggests a three-state model as depicted in Figure 8. In the uninduced state, nothing is bound to the promoter and no DNaseI hypersensitivity is observed (Elgin, 1988); positioned nucleosomes may occupy the entire promoter. Binding of a transcriptional activator such as GAL4 to the UAS/enhancer may occur without activating transcription (for example, when the activation domain is masked by GAL80). An exposed activation domain "opens" chromatin (possibly with the aid of auxiliary proteins such as the SWI complex in yeast; see Chapter 8) and facilitates transcription complex assembly at the promoter, and transcription ensues.

Can all enhancer/UAS binding proteins therefore bind to sites in chromatin *in vivo*? Such a general capability would imply a lack of a regulatory role for chromatin, except in special cases such as the genes repressed by the α2/MCM1 complex in yeast (Chapter 5); in this case, the activating MCM1 function could be masked by α2 while a positioned nucleosome prevented any leaky basal transcription in order to preserve tight regulation over mating type.

Some evidence exists to suggest this may not be the case. The *HIS4* gene in *Saccharomyces cerevisiae* can be activated by either GCN4 or by the combined action of BAS1 and BAS2 (also called PHO2). Present near the binding sites for these activators in the *HIS4* promoter is a site for the ubiquitous protein RAP1. RAP1 itself cannot activate transcription of *HIS4*, but if its binding site is removed from the promoter, neither GCN4 nor BAS1/BAS2 activation can take its place (Devlin et al., 1991). Increased nuclease accessibility of the locus when the RAP1 binding site is present led to the suggestion that RAP1 binding to the *HIS4* promoter

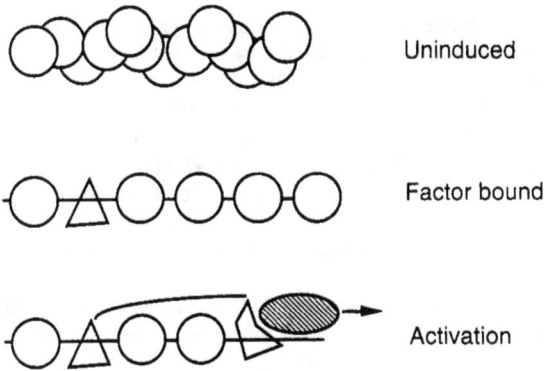

Figure 8. A simple three-state model for transcriptional activation of a promoter packaged into nucleosomes.

clears chromatin, allowing access by GCN4 or BAS1/BAS2. GCN4 is a member of the bZIP class of transcription factors, and BAS2 is a homeodomain containing protein. Perhaps transcription factors differ in their ability to bind to sites in chromatin depending on their class. Studies using engineered chromatin templates in yeast are sure to shed light on this and other pertinent issues in the near future.

ACKNOWLEDGMENTS

Work by the author described in this paper was carried out in the laboratory of Dr. Robert Simpson at the National Institutes of Health (NIH); once again, I thank Bob for his continued and generous support and encouragement. I also thank Sharon Roth for permission to use Figure 4, which she created, and Sharon Roth, Alan Wolffe, and numerous other colleagues from the NIH for discussions which have helped me think about transcription and chromatin.

REFERENCES

Aldrich, T.L., DiSegni, G., McConaughy, B.L., Keen, N.J., Whelen, S., & Hall, B.D. (1993). Structure of the yeast TAP1 protein: Dependence of transcription activation on the DNA context of the target gene. Mol. Cell. Biol. 13, 3434–3444.

Almer, A., Rudolph, H., Hinnen, A., & Horz, W. (1986). Removal of positioned nucleosomes from the yeast *PHO5* promoter upon *PHO5* induction releases additional upstream activating DNA elements. EMBO J. 5, 2689–2696.

Almouzni, G., & Wolffe, A.P. (1993). Replication-coupled chromatin assembly is required for the repression of basal transcription *in vivo*. Genes Dev. 7, 2033–2047.

Almouzni, G., Mechali, M., & Wolffe, A.P. (1991). Transcription complex disruption caused by a transition in chromatin structure. Mol. Cell. Biol. 11, 655–665.

Axelrod, J.D., Reagan, M.S., & Majors, J. (1993). GAL4 disrupts a repressing nucleosome during activation of *GAL1* transcription *in vivo*. Genes Dev. 7, 857–869.

Becker, P.B., Rabindran, S.K., & Wu, C. (1991). Heat shock-regulated transcription *in vitro* from a reconstituted chromatin template. Proc. Natl. Acad. Sci. USA 88, 4109–4113.

Benezra, R., Cantor, C.R., & Axel, R. (1986). Nucleosomes are phased along the mouse β-major globin gene in erythroid and non-erythroid cells. Cell 44, 697–704.

Birkenmeier, E.H., Brown, D.D., & Jordan, E. (1978). A nuclear extract of *Xenopus laevis* oocytes that accurately transcribes 5S RNA genes. Cell 15, 1077–1086.

Bouvet, P., Dimitrov, S., & Wolffe, A.P. (1994). Specific regulation of *Xenopus* chromosomal 5S rRNA gene transcription *in vivo* by histone H1. Genes 8, 1147–1159.

Brow, D.A., & Guthrie, C. (1990). Transcription of a yeast U6 snRNA gene requires a polymerase III promoter element in a novel position. Genes Dev. 4, 1345–1356.

Brown, D.D. (1984). The role of stable complexes that repress and activate eukaryotic genes. Cell 37, 359–365.

Burnol, A.-F., Margottin, F., Huet, J., Almouzni, G., Prioleau, M.-N., Mechali, M., & Sentenac, A. (1993). TFIIIC relieves repression of U6 snRNA transcription by chromatin. Nature 362, 475–477.

Cavalli, G., & Thoma, F. (1993). Chromatin transitions during activation and repression of galactose-regulated genes in yeast. EMBO J. 12, 4603–4613.

Chasman, D.I., Lue, N.F., Buchman, A.R., LaPointe, J.W., Lorch, Y., & Kornberg, R.D. (1990). A yeast protein that influences the chromatin structure of UAS_G and functions as a powerful auxiliary activator. Genes Dev. 4, 503–514.

Chipev, C.C., & Wolffe, A.P. (1992). Chromosomal organization of *Xenopus laevis* oocyte and somatic 5S rRNA genes *in vivo*. Mol. Cell. Biol. 12, 45–55.

Clark, D.J., & Wolffe, A.P. (1991). Superhelical stress and nucleosome-mediated repression of 5S RNA gene transcription *in vitro*. EMBO J. 10, 3419–3428.

Clark-Adams, C.D., Norris, D., Osley, M.A., Fassler, J.S., & Winston, F. (1988). Changes in histone gene dosage alter transcription in yeast. Genes Dev. 2, 150–159.

Croston, G.E., Laybourn, P.J., Paranjape, P.M., & Kadonaga, J.T. (1992). Mechanism of transcriptional antirepression by GAL4-VP16. Genes Dev. 6, 2270–2281.

Devlin, C., Tice-Baldwin, K., Shore, D. & Arndt, K.T. (1991). RAP1 is required for BAS1/BAS2- and GCN4-dependent transcription of the yeast *HIS4* gene. Mol. Cell. Biol. 11, 3642–3651.

DiSegni, G., McConaughy, B.L., Shapiro, R.A., Aldrich, T.L., & Hall, B.D. (1993). *TAP1*, a yeast gene that activates the expression of a tRNA gene with a defective internal promoter. Mol. Cell. Biol. 13, 3424–3433.

Elgin, S.C.R. (1988). The formation and function of DNase I hypersensitive sites in the process of gene activity. J. Biol. Chem. 263, 19259–19262.

Fedor, M.J., & Kornberg, R.D. (1989). Upstream activation sequence-dependent alteration of chromatin structure and transcription activation of the yeast *GAL1-GAL10* genes. Mol. Cell. Biol. 9, 1721–1732.

Fedor, M.J., Lue, N.F., & Kornberg, R.D. (1988). Statistical positioning of nucleosomes by specific protein binding to upstream activating sequence in yeast. J. Mol. Biol., 204, 109–127.

Geiduschek, E.P., & Tocchini-Valentini, G.P. (1988). Transcription by RNA polymerase III. Annu. Rev. Biochem. 57, 873–914.

Georgel, P., Dretzen, G., Jagla, K., Bellard, F., Dubrovsky, E., Calco, V. & Bellard, M. (1993). GEBF-I activates the *Drosophila* SGS3 gene enhancer by altering a positioned nucleosomal core particle. J. Mol. Biol. 234, 319–330.

Gottesfeld, J.M., & Bloomer, L.S. (1980). Non random alignment of nucleosomes on 5S RNA genes of *Xenopus laevis*. Cell 21, 751–760.

Gross, D.S., & Garrard, W.T. (1988). Nuclease hypersensitive sites in chromatin. Annu. Rev. Biochem. 57, 159–197.

Gross, D.S., Adams, C.C., Lee, S., & Stentz, B. (1993). A critical role for heat shock transcription factor in establishing a nucleosome-free region over the TATA-initiation site of the yeast *HSP82* gene. EMBO J. 12, 3931–3945.

Han, M., & M. Grunstein (1988). Nucleosome loss activates yeast downstream promoters *in vivo*. Cell 55: 1137–1145.

Hanas, J.S., Bogenhagen, D.F., & Wu, C.-W. (1983). Cooperative model for the binding of *Xenopus* transcription factor A to the 5S RNA gene. Proc. Natl. Acad. Sci. U.S.A. 80, 2142–2145.

Hansen, J.C., & Wolffe, A.P. (1992). Influence of chromatin folding on transcription initiation and elongation by RNA polymerase III. Biochemistry 31, 7977–7988.

Hayes, J.J., & Wolffe, A.P. (1992). Histones H2A/H2B inhibit the interaction of transcription factor IIIA with the *Xenopus borealis* somatic 5S RNA gene in a nucleosome. Proc. Natl. Acad. Sci. USA 89, 1229–1233.

Herschbach, B.M., & Johnson, A.D. (1993). The yeast $\alpha2$ protein can repress transcription by RNA polymerases I and II but not III. Mol. Cell. Biol. 13, 4029–4038.

Johnston, M. (1987). A model fungal gene regulatory mechanism: the GAL genes of Saccharomyces cerevisiae. Microbiol. Rev. 51, 458–476.

Johnston, S.A., Salmeron, J.M. Jr., & Dincher, S.S. (1987). Interaction of positive and negative regulatory proteins in the galactose regulon of yeast. Cell 50, 143–146.

Kladde, M.P., & Simpson, R.J. (1994). Positioned nucleosomes inhibit methylation *in vitro*. Proc. Natl. Acad. Sci. USA 91, 1361–1365,

Knezetic, J.A., & Luse, D.S. (1986). The presence of nucleosomes on a DNA template prevents initiation by RNA polymerase II in vitro. Cell 45, 95–104.

Lassar, A.B., Hamer, D.H., & Roeder, R.G. (1985). Stable transcription complex on a class III gene in a minichromosome. Mol. Cell. Biol. 5, 40–45.

Laybourn, P.J., & Kadonaga, J.T. (1991). Role of nucleosomal cores and histone H1 in regulation of transcription by RNA polymerase II. Science 254, 238–245.

Lohr, D. (1984). Organization of the *GAL1-GAL10* intergenic control region chromatin. Nucl. Acids Res. 12, 8457–8474.

Lorch, Y., LaPointe, J.W., & Kornberg, R.D. (1987). Nucleosomes inhibit the initiation of transcription but allow chain elongation with the displacement of histones. Cell 49, 203–210.

Ma, J., & Ptashne, M. (1987a). Deletion analysis of GAL4 defines two transcriptional activating segments. Cell 48, 847–853.

Ma, J., & Ptashne, M. (1987b). The carboxy-terminal 30 amino acids of GAL4 are recognized by GAL80. Cell 50, 137–142.

Majumder, S., Miranda, M., & DePamphilis, M.L. (1993). Analysis of gene expression in mouse preimplantation embryos demonstrates that the primary role of enhancers is to relieve repression of promoters. EMBO J. 12, 1131–1140.

Marmorstein, R., Carey, M., Ptashne, M., & Harrison, S.C. (1992). DNA recognition by GAL4: structure of a protein-DNA complex. Nature 356, 408–414.

Matsui, T. (1987). Transcription of adenovirus 2 major late and peptide IX genes under conditions of *in vitro* nucleosome assembly. Mol. Cell. Biol. 7, 1401–1408.

Morse, R.H. (1989). Nucleosomes inhibit both transcriptional initiation and elongation by RNA polymerase III *in vitro*. EMBO J. 8, 2343–2351.

Morse, R.H. (1993). Nucleosome disruption by transcription factor binding in yeast. Science 262, 1563–1566.

Morse, R.H., Roth, S.Y., & Simpson, R.T. (1992). A transcriptionally act e tRNA gene interferes with nucleosome positioning *in vivo*. Mol. Cell Biol. 12, 4015–4025.

Perlmann, T., & Wrange, O. (1988). Specific glucocorticoid receptor binding to DNA reconstituted in a nucleosome. EMBO J. 7, 3073–3079.

Pederson, D.S., & Fidrych, T. (1994). Heat shock factor can activate transcription while bound to nucleosomal DNA in *Saccharomyces cerevisiae*. Mol. Cell. Biol. 14, 189–199.

Pikaart, M., Feng, J., & Villeponteau, B. (1992). The polyomavirus enhancer activates chromatin accessibility on integration into the *HPRT* gene. Mol. Cell. Biol. 12, 5785–5792.

Reitman, M., Lee, E., Westphal, H., & Felsenfeld, G. (1993). An enhancer/locus control region is not sufficient to open chromatin. Mol. Cell. Biol. 13, 3990–3998.

Richard-Foy, H., & Hager, G.L. (1987). Sequence-specific positioning of nucleosomes over the steroid-inducible MMTV promoter. EMBO J. 6, 2321–2328.

Roth, S.Y., Dean, A., & Simpson, R.T. (1990). Yeast α2 repressor positions nucleosomes in TRP1/ARS1 chromatin. Mol. Cell. Biol. 10, 2247–2260.

Schlissel, M.S., & Brown, D.D. (1984). The transcriptional regulation of *Xenopus* 5S RNA genes in chromatin: the roles of active stable complexes and histone H1. Cell 37, 903–913.

Shimizu, M., Roth, S.Y., Szent-Gyorgyi, C., & Simpson, R.T. (1991). Nucleosomes are positioned with base pair precision adjacent to the α2 operator in *Saccharomyces cerevisiae*. EMBO J. 10, 3033–3041.

Simpson, R.T. (1990). Nucleosome positioning can affect the function of a cis-acting element *in vivo*. Nature 343, 387–389.

Simpson, R.T. (1991). Nucleosome positioning: Occurrence, mechanisms, and functional consequences. Prog. Nucleic Acids Res. Mol. Biol. 40, 143–184.

Taylor, I.C.A., Workman, J.L., Schuetz, T.J., & Kingston, R.E. (1991). Facilitated binding of GAL4 and heat shock factor to nucleosomal templates: Differential function of DNA-binding domains. Genes Dev. 5, 1285–1298.

Thoma, F. (1992). Nucleosome positioning. Biochim. Biophys. Acta 1130, 1–19.

Thoma, F., Bergman, L.W., & Simpson, R.T. (1984). Nuclease digestion of circular TRP1ARS1 chromatin reveals positioned nucleosomes separated by nuclease-sensitive regions. J. Mol. Biol. 177, 715–733.

Tremethick, D., Zucker, K., & Worcel, A. (1990). The transcription complex of the 5S RNA gene, but not transcription factor TFIIIA alone, prevents nucleosomal repression of transcription. J. Biol. Chem. 265, 5014–5023.

Trumbly, R.J. (1992). Glucose Repression in the yeast *saccharomyces cerevisiae*. Mol. Microbiol. 6, 15–21.

Vettese-Dadey, M., Walter, P., Chen, H., Juan, L.J., & Workman, J.L. (1994). Role of the histone amino termini in facilitated binding of a transcription factor, GAL4-AH, to nucleosome cores. Mol. Cell. Biol. 14, 970–981.

Wasylyk, B., & Chambon, P. (1979). Transcription by eukaryotic RNA polymerases A and B of chromatin assembled *in vitro*. Eur. J. Biochem. 98, 317–327.

Wasylyk, B., Thevenin, G., Oudet, P., & Chambon, P. (1979). Transcription of *in vitro* assembled chromatin by *E. coli* RNA polymerase. J. Mol. Biol. 128, 411–440.

Williamson, P., & Felsenfeld, G. (1978). Transcription of histone-covered T7 DNA by *E. coli* RNA polymerase. Biochemistry 17, 5695–5705.

Wolffe, A.P. (1989). Dominant and specific repression of *Xenopus* oocyte 5S RNA genes and satellite I DNA by histone H1. EMBO J. 8, 527–537.

Wolffe, A.P., & Brown, D.D. (1987). Differential 5S RNA gene expression *in vitro*. Cell 51, 733–740.

Wolffe, A.P., & Brown, D.D. (1988). Developmental regulation of two 5S ribosomal RNA genes. Science 241, 1626–1632.

Workman, J.L., & Kingston, R.E. (1992). Nucleosome core displacement in vitro via a metastable transcription factor-nucleosome complex. Science 258, 1780–1784.

Workman, J.L., & Roeder, R.G. (1987). Binding of transcription factor TFIID to the major late promoter during *in vitro* nucleosome assembly potentiates subsequent initiation by RNA polymerase II. Cell 51, 613–622.

Workman, J.L., Taylor, I.C., & Kingston, R.E. (1991). Activation domains of stably bound GAL4 derivatives alleviate repression of promoters by nucleosomes. Cell 64, 533–544.

Young, D., & Carroll, D. (1983). Regular arrangement of nucleosomes on 5S rRNA genes in *Xenopus laevis*. Mol. Cell. Biol. 3, 720–730.

Chapter 4

Mechanisms and Consequences of Transcription Factor Binding to Nucleosomes

PHILLIP P. WALTER, MICHELLE VETTESE-DADEY,

JACQUES COTE, CHRISTOPHER C. ADAMS,

LI–JUNG JUAN, RHEA UTLEY, and

JERRY L. WORKMAN

The Nucleus
Volume 1, pages 79–100.
Copyright © 1995 by JAI Press Inc.
All rights of reproduction in any form reserved.
ISBN: 1-55938-940-0

ABSTRACT

Biochemical studies of the interactions of regulatory transcription factors with nucleosome cores have illustrated several parameters which restrict the binding of individual transcription factors to nucleosomal DNA. These include a differential intrinsic affinity of different factors for their recognition sites on nucleosomes, the location of the binding sites within the nucleosome core, and inhibition from the core histone amino termini. However, nucleosome-mediated repression of factor binding can be overcome by the facilitated binding of multiple factors, relief of inhibition from the core histone amino termini (i.e., by histone acetylation), and through the function of accessory proteins which stimulate factor binding (i.e., by interacting with histones). The binding of regulatory factors to nucleosomes results in the formation of factor/nucleosome ternary complexes which contain bound factors, core histones, and DNA. The histones in the ternary complexes, however, are destabilized relative to those in unbound nucleosome cores and are susceptible to dissociation by histone-binding proteins and histone competitors. Thus transcription factor binding can initiate nucleosome displacement in the presence of histone chaperones.

I. INTRODUCTION

Biochemical analysis of the proteins involved in transcription by RNA polymerase II has revealed two classes of transcription factors. These include the general initiation factors, which are required in addition to RNA polymerase II for accurate initiation, and regulatory factors, which bind upstream promoter and/or enhancer elements and activate transcription initiation (reviewed in Mitchell and Tjian, 1989; Saltzman and Weinmann, 1989; Sawadogo and Sentenac, 1990; Zawel and Reinberg, 1993). A separate line of investigation characterized changes in chromatin structure at protein-coding genes (transcribed by RNA polymerase II) and revealed the formation of apparent nucleosome-free regions (nuclease hypersensitive sites) over promoter and enhancer elements which accompany or precede the activation of cellular genes (reviewed in Yaniv and Cereghini, 1986; Pederson, et al., 1986; Gross and Garrard, 1988; Elgin, 1988). Together, these studies led to the expectation that transcription factors would play a role in the remodeling of chromatin structures and that chromatin structures would in turn participate in the regulation of class II gene transcription.

A. Transcription Analysis of Nucleosome Function *In Vitro*

The initial attempts to biochemically address the function of nucleosomes in the transcription of class II genes utilized cell-free transcription systems coupled with *in vitro* nucleosome assembly. The strategy of these early studies was similar to those utilized in polymerase III systems (reviewed in Brown, 1984; Wolffe and Brown, 1988) and illustrated that nucleosomes repressed transcription initiation from class II promoters (Knezetic and Luse, 1986, Lorch et al., 1987). However, formation of stable preinitiation complexes which minimally contained the TATA binding complex, TFIID, prevented subsequent repression by nucleosome assembly (Matsui, 1987; Workman and Roeder, 1987; Knezetic et al., 1988). These functional studies suggested mutually exclusive binding of nucleosomes and transcription complexes to template DNA which resulted in nucleosome-mediated suppression of basal transcription.

The suppression of basal transcription by nucleosome assembly increased the dependence of transcription on the action of regulatory factors and thus increased the fold stimulation over basal levels. A variety of transcription activators were found to prevent and/or reverse nucleosome mediated repression as part of their activation functions. These included the Pseudorabies virus immediate early protein, USF, and derivatives of the yeast GAL4 protein (Workman et al., 1988, 1990, 1991). This enhanced activation function of the GAL4 derivatives was dependent on the derivative containing an activation domain (i.e., GAL4–VP16) and was able to prevent or reverse repression from the assembly of nucleosome cores alone (Workman et al., 1991; Lorch et al., 1992; Kamakaka et al., 1993; Juan et al., 1994; see also Almouzni and Wolffe, 1993). In addition, GAL4–VP16, Sp1, GAGA factor, NF1, and the glucocorticoid receptor have all been shown to prevent repression of transcription by the preternatural binding of purified H1 to naked DNA templates (termed "antirepression") (Croston et al., 1991, 1992; Kerrigan et al., 1991; Dusserre and Mermod, 1992; Eriksson and Wrange, 1994). While an analysis of the function of H1 versus nucleosome cores in accurately-spaced reconstituted chromatin indicates that the nucleosome core is the dominant repressor (Sandoltzopoulos et al., 1994), GAL4–VP16 and Sp1 have also been shown to prevent additional repression from the binding of histone H1 to templates which also contained nucleosome cores (Laybourn and Kadonaga, 1991; Kamakaka et al., 1993). Therefore, transcription activators have been shown to overcome repression from nucleosome cores, H1 binding to templates containing nucleosome cores, and H1 binding to naked DNA. These studies directly illustrate a function of histones in the suppression of basal transcription and thereby an enhanced degree of regulation by activators. Together they also illustrate the limitations of functional transcription assays alone in establishing the role of specific histones and chromatin structures. Similar transcription effects have been attributed to different histones (i.e., H1 versus core histones) and different histone/DNA complexes (nucleosomes versus H1/DNA complexes).

B. Direct Analysis of Transcription Factor Interactions with Nucleosomes

A complementary line of study to transcriptional analysis of nucleosome-assem-bled template DNA is the direct analysis of transcription factor interactions with nucleosomal DNA. This approach has included analysis of transcription factor binding to nucleosome-assembled plasmid templates bearing the cognate binding sequences and, more often, the binding of transcription factors to individual nucleosome cores. Nucleosome core reconstitution onto nucleosome-length frag-ments of DNA (146–180 bp) is usually achieved by the direct transfer of histone octamers from purified cellular nucleosome cores at high salt concentrations or by directly mixing purified core histones and DNA at high salt followed in either case by a slow reduction of ionic strength (reviewed in Rhodes and Laskey, 1989; Wolffe and Hayes, 1993). The products of these reconstitutions are individual nucleosome cores bearing the DNA fragment and binding sites(s) of interest. The analysis of factor interactions with reconstituted nucleosomal DNA can then proceed by footprinting and mobility shift assays.

By directly analyzing the interactions (i.e., binding) of transcription factors with nucleosomal DNA it is possible to investigate specific steps in the activation of nucleosomal templates. For example, if an activator reverses repression of a nucleosome-assembled template *in vitro* several steps need to occur prior to the detection of activity by transcription assays; these include 1) binding of the activator to its cognate site on the nucleosomal template, 2) any necessary disruption of the activator bound nucleosome, 3) formation of a transcription complex by the general factors and RNA polymerase II at the TATA-box and initiation site, 4) displacement of the nucleosome at the site of transcription complex formation, and 5) transcrip-tion initiation. By contrast, direct binding assays and analysis of the formed complexes allow a more detailed picture of the interactions between nucleosomes and transcription factors at each of these steps.

In this review, we describe general principles of the interactions between regu-latory transcription factors and nucleosome cores. Two important points relevant to this discussion are 1) that most eukaryotic enhancers and promoters contain binding elements for multiple factors, and 2) that there is a stunning diversity of configurations in which binding sites for these transcription factors are arranged in regulatory elements. Thus, there is either an equally large and diverse number of specific mechanisms by which regulatory factors deal with underlying nu-cleosomes, or there is a smaller number of more general mechanisms which can be accessed through different combinations of factor binding sites. We favor the latter possibility and explore this topic in the following sections. The concepts and supporting data are derived from work in several systems, however, the specific examples presented in the figures are from our studies using the GAL4 system. For further details of biochemical studies analyzing nucleosome function on the mouse mammary tumor virus promoter and the 5S RNA genes see chapters 3 and 6.

II. PARAMETERS GOVERNING THE INTERACTION OF TRANSCRIPTION FACTORS WITH NUCLEOSOMES

In principle, the complexing of DNA with histone octamers to form nucleosome cores generates several potential structural hindrances to the subsequent association of sequence-specific DNA binding proteins. These include steric occlusions from the surface of the histone octamer against one face of the DNA helix and from the close proximity of the adjacent DNA helix on the nucleosome core. In addition, distortion of the DNA structure from its bending around the nucleosome which alters the helical repeat (reviewed in Wolffe, 1992) might also inhibit factor binding. Experiments conducted thus far suggest that the interactions of transcription factors with nucleosomal DNA is largely modulated by specific protein domains, including the particular DNA-binding domains of individual transcription factors and the amino terminal "tails" of the core histones (see below).

A. Differential Affinities of Transcription Factors for Nucleosomal DNA

An important consideration in the potential mechanisms by which a nucleosome-imbedded promoter or enhancer could become occupied by transcription factors is the relative affinities different factors have for nucleosomal DNA. Those most readily able to access their binding sites would be expected to initiate binding and disruption of the underlying nucleosomes leading to the binding of additional factors (reviewed in Adams and Workman, 1993).

The concept of differential affinities of factors for nucleosomal DNA was first suggested by studies of the mouse mammary tumor virus (MMTV) promoter. Perlmann and Wrange (1988) demonstrated that the glucocorticoid receptor (GR) was readily able to bind a subset of its sites, glucocorticoid response elements (GREs), on a positioned nucleosome core reconstituted onto a fragment from the MMTV promoter. GR and progesterone receptor binding to nucleosomes on the MMTV promoter was also observed by the Beato and Hager groups (Pina et al., 1990; Archer et al. 1991). In addition, these latter studies also demonstrated that, by contrast, nuclear factor 1 (NF1) was unable to bind its recognition site on the same nucleosome even though its binding site was located nearer to the edge of the nucleosome (Pina et al., 1990; Archer et al., 1991). It is important to note that the position of the reconstituted nucleosomes utilized by Pina and colleagues (1990) was apparently shifted 30 bp from that used by Perlmann and Wrange (1988) or Archer and colleagues (1991). Thus, the difference in GR and NF1 binding is not solely due to precise nucleosome-positioning since similar results were observed by all groups. This is consistent with the notion that in addition to nucleosome position effects (see in section B) GR also has a greater inherent ability to access its sites on nucleosome cores than does NF1 (discussed in Archer et al., 1991). A recent analysis indicates that GR binding to GREs at positions −176, −120, and −84 (i.e., at different positions in the nucleosome) is only inhibited 2–9 fold relative to naked DNA (Perlmann, 1992). By contrast, NF1 binding to the nucleosome was not observed at 100 fold higher concentrations of factor than that required to completely bind an equivalent amount of naked DNA (Pina et al., 1990).

A direct comparison of the unrelated factors (derivatives of the yeast factor, GAL4, and the human heat shock factor) has revealed a differential effect on their binding affinities when their sites are occupied by nucleosome cores (Taylor et al., 1991). The binding of a single GAL4 dimer to nucleosome cores was inhibited from 10–100 fold, relative to naked DNA dependent on the location of the GAL4 binding site (Taylor et al., 1991; Juan et al., 1994; Vettese-Dadey et al., 1994; also see next section). By contrast, HSF binding to the nucleosome was not observed even at a 1000-fold higher concentration than that required to observe binding to naked DNA (Taylor et al., 1991). In the case of GAL4, nucleosome-binding is independent of the presence or absence of an activation domain indicating that nucleosome-binding is a property of the factor's DNA binding domain (Taylor et al., 1991; Workman and Kingston, 1992).

Differential affinities of transcription factors for nucleosomal DNA has also been observed with additional regulatory factors. The mammalian Zn-finger protein, Sp1, demonstrates nucleosome-binding properties that are similar to GAL4, including activation domain independence (Li et al., 1994), while the binding of the basic/helix-loop-helix protein, USF and NF-ICB is inhibited by nucleosomes to a much greater degree (Adams and Workman, 1995).

B. Nucleosome Positioning

The fact that specifically positioned nucleosomes are often observed on promoter and enhancer elements *in vivo* suggests that nucleosome positioning may play an important role in transcription regulation (reviewed in Simpson, 1991). To establish the function of nucleosome positioning in determining factor access it is necessary to experimentally manipulate positions of nucleosomes relative to binding elements. Experimental manipulation of nucleosome positioning in yeast has revealed functions of nucleosome positioning in both transcription and replication. These studies have been reviewed in detail elsewhere (Simpson, 1991; also see Chapters 3 and 5). Here we focus on recent *in vitro* experiments which address the role of nucleosome positioning in the binding of regulatory factors.

There are two important aspects of nucleosome positioning which might influence transcription factor binding. The first is rotational phasing which represents the orientation of the DNA helix (and thus the binding sites) on the surface of the histone octamer. The best evidence for a rotational phasing effect on the binding of factors comes from the aforementioned studies of the MMTV promoter. The precise rotational phasing of the DNA sequence on nucleosomes reconstituted on this promoter presents some GREs at positions where the major groove faces outward and other GREs where the major groove faces the surface of the histone octamer. When presented with this choice, it is clear that GR initially binds the GREs facing outward (Perlmann and Wrange, 1988; Pina et al., 1990). An attempt to measure the affinity of these sites on nucleosomes by Perlmann (1992), indicates that one of the inward facing sites (at –120) has an affinity for GR which is reduced approximately eightfold, relative to naked DNA, while the highest affinity outward facing site (–176) is reduced only 2.3-fold. However, this small difference in the affinity of these sites in spite of rotational orientation might result in part from "cooperative" binding effects of multiple receptors to this nucleosome (see the next section). The binding of the receptor to the outward facing GRE may increase affinity for the inward facing GRE. A more clear understanding of the role of rotational phasing on GR binding will result from studies which experimentally rotate an individual GRE with regard to the histone octamer surface. Such experiments have not been reported to date.

The only other attempt to measure rotational phasing effects yielded negative results (Taylor et al., 1991). In this study the affinity of GAL4 derivatives was unaffected by altering the rotational phasing of its binding element and human heat shock factor was never observed to bind its sequence in nucleosomes regardless of the rotational orientation.

The second important aspect of nucleosome positioning is translational phasing. This refers to the location of the nucleosome along the DNA sequence which then determines whether a factor's binding site will be situated in the center of a nucleosome, near the edge, or between nucleosomes. Earlier *in vivo* studies of ARS function in yeast and *in vitro* studies with prokaryotic RNA polymerases had implicated translational phasing in regulating accessibility of cis-acting elements (Simpson, 1990; Wolffe and Drew, 1989). Recent *in vitro* studies have also illustrated the effects of translational phasing of nucleosomes on regulatory factor binding (see next section).

To determine the effect of translational phasing on the binding of the GR, Li and Wrange (1994) analyzed the binding of GR to a single GRE at different translational positions within a nucleosome core while maintaining the rotational phasing. This study revealed two important points. First, translational phasing of the nucleosome over the GRE does affect GR affinity. At 40 bp from the nucleosome dyad (33 bp from the end of the nucleosome) GR affinity is only reduced 2.5 fold relative to naked DNA. However, 10.8-fold inhibition of GR binding (relative to DNA) was observed at 20 bp from the dyad (53 bp from the end) and 3.5-fold inhibition (relative to naked DNA) was observed at the dyad (center) of the nucleosome. The second point is again the striking affinity of the GR for the nucleosomal DNA at all of these translational positions.

Analysis of GAL4–AH, and Sp1 binding to nucleosome cores has illustrated a potent translational positioning effect on the binding of these factors (Vettese-Dadey et al., 1994; Li et al., 1994). Binding of GAL4–AH to a site located between seven and 21 bp from the end of the nucleosome was only inhibited about tenfold relative to naked DNA. An additional 12–15 fold inhibition of GAL4–AH binding was observed at a site located between 60 and 74 bp from the end. The differential inhibition of GAL4–AH binding by the translational positioning of nucleosomes

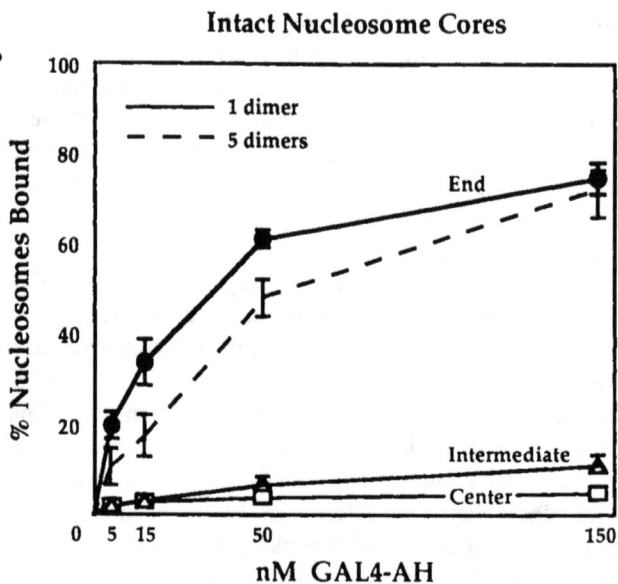

A. Intact Nucleosome Cores

% Nucleosomes Bound

1 dimer
5 dimers

End

Intermediate

Center

nM GAL4-AH

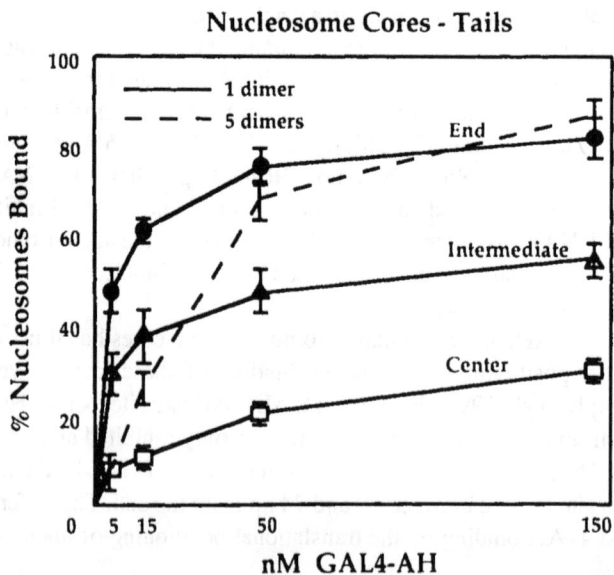

B. Nucleosome Cores - Tails

% Nucleosomes Bound

1 dimer
5 dimers

End

Intermediate

Center

nM GAL4-AH

Figure 1. The binding of GAL4–AH to reconstituted nucleosome cores. The percentage of nucleosomes bound by GAL4–AH at increasing concentrations of factor for three independent experiments is shown. Part A shows binding to intact nucleosome cores and Part B shows binding to nucleosome cores where the amino-terminal tails have been removed with trypsin. The error bars represent two standard deviations and are shown when they fall outside of the character. The probe DNA reconstituted into nucleosomes was 160 bp in all cases and contained a single GAL4 site 21 bp from the end (end position, solid circles), 40 bp from the end (intermediate position, triangles), or 74 bp from the end (center position, squares). The dashed line represents the binding of five dimers of GAL4 to a nucleosome bearing five centered sites. Experimental details of nucleosome reconstitution and GAL4–AH binding can be found in Vettese-Dadey and colleagues (1994) and Chen and colleagues (1994).

is illustrated in Part A of Figure 1. The binding of peptides containing the Zn fingers of Sp1 to nucleosomes have shown a greater than tenfold difference in affinity for GC boxes (Sp1 recognition sites) dependent on the location within a nucleosome core. Again binding was most inhibited near the center of the nucleosome (Li et al., 1994).

III. MECHANISMS TO OVERCOME NUCLEOSOME-INHIBITION OF FACTOR BINDING

It is clear from *in vivo* studies of some inducible genes that mechanisms exist for the disruption of the primary level of chromatin structure, arrays of nucleosomes, at regulatory elements in the absence of DNA replication and chromatin reassembly (reviewed in Felsenfeld, 1992; Adams and Workman, 1993). While it is not clear how general these mechanisms are, it seems reasonable to expect that similar mechanisms will apply to at least some constitutively active genes via the function of constitutive factors. Such pathways require that regulatory factors gain access to their recognition sequences when they are contained in nucleosomes. Below we describe three mechanisms to increase the affinity of regulatory transcription factors for nucleosomal DNA; relief of inhibition from the core histone amino-termini, facilitated binding of multiple factors to nucleosomes, and the stimulatory effect of accessory proteins which increase factor binding.

A. Relief of Inhibition from the Histone Amino-Terminal Tails

A striking finding regarding factor binding to nucleosomes is that much of the inhibition of binding is due to the core histone amino termini. These domains of the core histone proteins are highly conserved and undergo numerous modifications, yet they are not essential for nucleosome core structure and are relatively disordered (reviewed in Turner, 1991). Lee and colleagues (1993) found that tryptic removal of the amino termini brought about the binding of TFIIIA to nucleosome

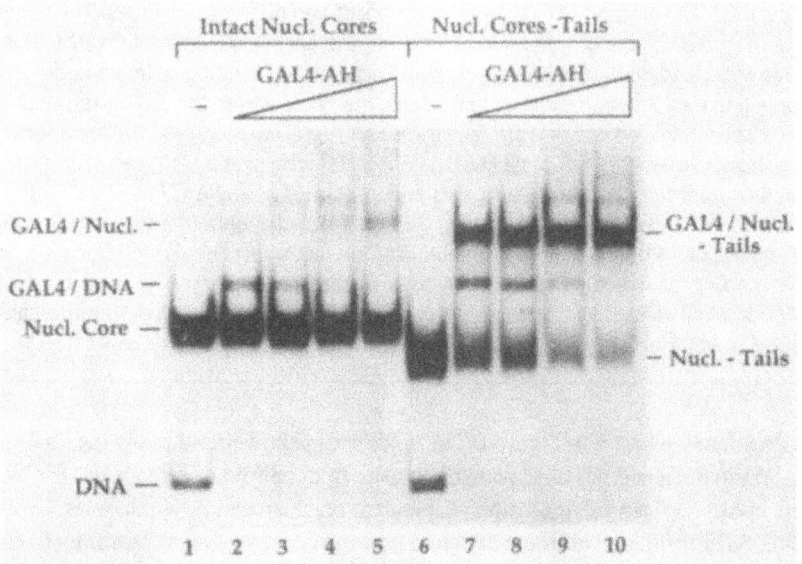

Figure 2. Stimulation of GAL4–AH binding to nucleosome cores by removal of the core histone amino-terminal tails. The nucleosome used contained a GAL4 site at the Intermediate position described in Figure 1. Shown is a mobility shift gel of GAL4–AH binding at increasing concentrations of factor to intact nucleosome cores or nucleosome cores lacking the amino terminal tails. The migrations of naked DNA, nucleosome cores, GAL4–AH/DNA complexes, and GAL4–AH/nucleosome ternary complexes are indicated. The concentrations of GAL4–AH used were 0, lanes 1 and 6; 5nM, lanes 2 and 7; 15nM, lanes 3 and 8; 50nM, lanes 4 and 9; and 150nM, lanes 5 and 10. Adapted from Figure 7 of Vettese–Dadey et al., 1994, Mol. Cell Biol. 14, 970–981.

cores reconstituted onto DNA fragments containing either the Xenopus borealis or Xenopus laevis 5S RNA genes (the location of the histone octamer differs by 28 bp on these two DNA fragments). Similarly, core histones which were hyperacetylated on lysine residues within the amino termini also allowed TFIIIA binding (Lee et al., 1993). These data suggest that the core histone tails restrict transcription factor access and that acetylation of lysine residues within the amino termini (which reduces their positive charge) relieves this inhibition.

Removal of the core histone tails also stimulates the binding of GAL4–AH to nucleosome cores (Vettese-Dadey et al., 1994). Figure 2 illustrates the binding of GAL4–AH to intact nucleosome cores and to nucleosome cores lacking the core histone tails. This probe contained a single GAL4 site, the center of which was located between 26 and 40 bp into the reconstituted nucleosome cores. GAL4–AH binding to these nucleosomes was stimulated up to 22 fold by removal of the amino

[GAL4-AH]

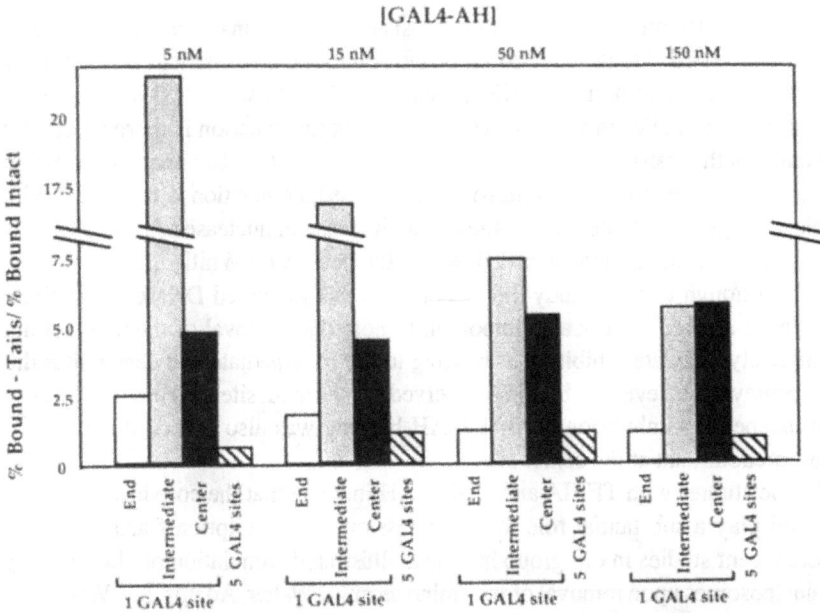

Figure 3. Stimulation of GAL4–AH binding to nucleosomes by removing the core histone tails is dependent on the position and number of binding sites. The fold stimulation resulting from removal of the core histone tails (%bound-tails/%bound intact) is graphed for single GAL4 sites at different positions in the nucleosome core and for five GAL4–site nucleosomes (as described in Figure 1) at different GAL–AH concentrations. Maximum stimulation was observed at the intermediate position which is reduced at higher GAL4–AH concentrations as a larger fraction of the nucleosomes become bound. A consistent five-fold stimulation was observed at the most central GAL4 site. By contrast, very little stimulation was observed at the end site, where nucleosome inhibition of GAL4–AH binding is the lowest, and with the five site nucleosomes since facilitated binding overcame inhibition of binding even when the core histone tails were present (Figure 1).

termini. The degree of stimulation observed by removal of the amino terminal tails, however, is dependent on the location of the binding site within the nucleosome core (Figure 3). Stimulation was greatest at an intermediate position (between 26 and 40 bp) while near the end of the core particle (between 7 and 20 bp into the nucleosome) very little stimulation was observed (1–2.5 fold). Additionally, at the center of the nucleosome core (between 60 and 74 bp into the nucleosome) a consistent 5–6 fold stimulation was observed.

These data indicate that the amino termini are responsible in part for nucleosome position effects. At the end of the core particle (end site), GAL4–AH binding is relatively uninhibited and not significantly stimulated by the removal of the amino terminal tails (Figures 1 and 3). This is consistent with thermal denaturation studies which suggested that the first 20 bp of DNA into the core particle is less tightly constrained (Weischet et al., 1978; Simpson, 1979; Morse et al., 1987) and with recent *in vivo* studies in yeast which indicate that this location is more accessible to dam methyltransferase (Kladde and Simpson, 1994). The large stimulation observed upon removal of the tails at the intermediate position is consistent with DNase I digestion studies which showed an increase in nuclease sensitivity in this region upon removal of the tails (Lilley and Tatchell, 1977; Whitlock and Simpson, 1977) although another study did not observe this increased DNase I sensitivity (Hayes et al., 1991). It is also important to note that removal of the tails did not completely alleviate inhibition as binding to the intermediate and center sites did not achieve the level of binding observed to the end site (Figure 1). Thus, a tail-independent inhibition of GAL4–AH binding was also observed which was most predominant at the center site.

These studies with TFIIIA and GAL4–AH indicate that the core histone amino termini play a substantial role in the repression of transcription f actor binding. More recent studies in our group have also illustrated stimulation of USF binding to nucleosomes upon removal of the amino termini (Walter, Adams, and Workman, unpublished). As with TFIIIA (Lee et al., 1993) GAL4–AH and USF binding was also stimulated by hyperacetylation of core histones, albeit never to an extent equal to that observed by removal of the tails (Vettese-Dadey, Walter, and Workman, unpublished). Thus, stimulation of factor binding by relief of inhibition (i.e., by acetylation) from the histone tails appears to be a general mechanism applicable to the binding of several factors.

B. Facilitated Binding of Multiple Factors to Nucleosomes

A hallmark of most enhancers and promoters is the occurrence of multiple binding sites for various factors. The often close grouping of such binding sites may play an important role in the mechanism by which factors access these sites in chromatin. Studies analyzing the binding of GAL4 derivatives to nucleosomes have illustrated a nucleosome-dependent cooperative effect on factor binding. Binding of GAL4 derivatives to nucleosomes bearing five binding sites occurs at approximately a tenfold lower concentration of factor than to a single GAL4 site (Taylor et al., 1991; Juan et al., 1994). This cooperative effect, termed facilitated binding, is apparent in mobility shift assays where after the initial binding of one GAL4–AH dimer the bound complexes jump to five bound dimers (Vettese-Dadey et al., 1994; Juan et al., 1993). Facilitated binding is illustrated in Figure 4 (lanes 1–5) and is not apparent upon GAL4–AH binding to naked DNA (lanes 6–8) where all the intermediate complexes, representing 1–5 bound dimers, are apparent. Footprinting experiments have illustrated that the first GAL4–AH dimer bound to five-site nucleosome cores occurs at one of the sites nearest the end of the core particle, as predicted from the position effects on GAL4 binding described previously. Follow-

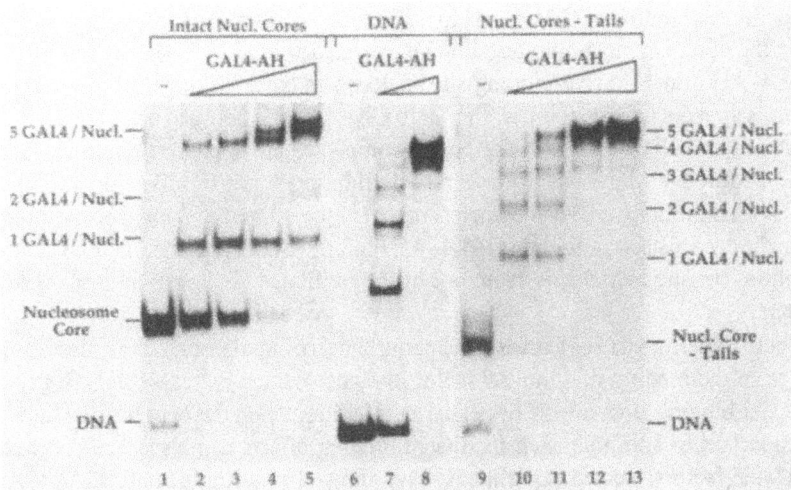

Figure 4. Facilitated binding of GAL4–AH to nucleosome cores occurred in response to inhibition from the core histone amino terminal tails. The binding of GAL4–AH to intact nucleosome cores (lanes 1–5), naked DNA (lanes 6–8), and nucleosome cores lacking the core histone tails (lanes 9–13) all bearing five GAL4 sites is shown in this mobility shift gel. Following the binding of one GAL4–AH dimer to the intact nucleosome cores the complex jumps to five bound dimers due to the cooperative nature of GAL4–AH binding. By contrast, upon binding naked DNA all of the intermediate complexes representing 1–5 bound GAL4–AH dimers are observed indicating that binding is noncooperative. When the core histone tails were removed from the nucleosome cores, binding of GAL4–AH to the nucleosome proceeded in a noncooperative manner similar to naked DNA. The concentrations of GAL4–AH included in each reaction were; 0, lanes 1, 6, and 9; 5 nM, lanes 2, 7, and 10; 15nM, lanes 3, 8, and 11; 50nM, lanes 4 and 12; 150nM, lanes 5 and 13. Adapted from Figures 2 and 3 of Vettese-Dadey et al., 1994, Mol. Cell Biol. 14, in press.

ing the binding of the initial dimers the remaining GAL4 sites fill in a cascade to bring about complete occupancy (Vettese-Dadey et al., 1994).

Facilitated binding overcomes nucleosome position effect inhibition. The five-site nucleosome cores utilized in Figure 4 contain five GAL4 sites covering the central 95 bp of the nucleosome core and, thus, includes sites in the center of the nucleosome. Therefore, for complete occupancy of GAL4 sites to occur, binding must also be achieved at the center of the nucleosome where binding is most difficult. However, as shown in Figure 1, binding of five dimers of GAL4–AH to nucleosome cores occurs with an affinity similar to that observed for a single site near the end of the nucleosome. Thus, facilitated binding increases the affinity of GAL4–AH to centrally located binding sites to a level similar to that of sites located near the end of the nucleosome alleviating nucleosome position effects.

Facilitated binding of GAL4–AH to nucleosome cores also overcomes inhibition from the core histone amino termini (Vettese-Dadey et al., 1994). The binding of GAL4–AH dimers to five-binding-site nucleosome cores lacking the amino terminal tails appears noncooperative similar to GAL4–AH binding to naked DNA (i.e., all the intermediate GAL4 dimer bound complexes are apparent; Figure 4, lanes 9–13). Moreover, removal of the amino termini does not significantly stimulate GAL4–AH binding to five-site nucleosomes (Figures 1 and 3). Thus, the inhibition from the core histone tails upon GAL4–AH binding to both intermediate and central positions (Figures 1 and 2) is overcome by the facilitated binding of multiple GAL4 dimers.

These experiments suggest that the perturbation of nucleosome structure resulting from the binding of an initial factor to a nucleosome increases the affinity of adjacent binding sites in part by relieving inhibition from the core histone tails. It is important to note that such facilitated binding effects can also occur between unrelated factors. Recent experiments have shown that a single bound GAL4–AH dimer increases the affinity of USF for an adjacent site by approximately two orders of magnitude (Adams and Workman, 1994). Thus, nucleosome binding by factors like USF (a basic/helix-looop-helix protein), which alone has a very low affinity for nucleosomal DNA, may be highly dependent on facilitated binding with neighboring factors.

C. Function of Accessory Proteins in Stimulating Factor Binding to Nucleosomes

Sequence-specific activators are usually able to recognize and bind their sequence motifs on naked DNA by themselves. However, the structural constraints placed upon their binding by nucleosomes and higher order chromatin structures raise the possibility that additional activities stimulate factor binding *in vivo*. Interest in this possibility has been fueled by genetic studies in yeast which have revealed several gene products that are involved in transcription activation and linked to chromatin function, yet are apparently not sequence-specific DNA binding proteins (reviewed in Winston and Carlson, 1992).

Since the histones in nucleosome cores inhibit transcription factor binding, proteins which interact with histones could play an important role in increasing factor access. Such a mechanism has been demonstrated *in vitro* in studies using the acidic histone-binding protein, nucleoplasmin (Chen et al., 1994). Nucleoplasmin is a 29 kDa protein which exists as a homopentamer and specifically interacts with histones H2A/H2B *in vivo* (Kleinschmidt et al., 1985; Dilworth et al., 1987; Kleinschmidt et al., 1990 and references therein). Nucleoplasmin has been immuno-localized to transcribed regions of amphibian lampbrush chromosomes consistent with a potential role in transcription (Moreau et al., 1986). Nucleoplasmin stimulates the binding of GAL4–AH, USF, and SP1 to nucleosomes cores approximately tenfold. By contrast, nucleoplasmin does not affect the binding of these factors to naked DNA. Stimulation of factor binding by nucleoplasmin appears to be linked to nucleosome disassembly (also see in next section). Nucleoplasmin depleted nucleosome cores bound by five dimers of GAL4–AH of histones H2A/H2B while not affecting the histone composition of unbound nucleosome cores (Chen et al., 1994). Thus, nucleoplasmin stimulated factor binding which induced H2A/H2B dimer loss onto nucleoplasmin.

Recently we have initiated studies of the yeast SWI/SNF complex with Craig Peterson's laboratory at the University of Massachusetts Medical Center (for a detailed review of the SWI/SNF complex see Chapter 8). These experiments have revealed an ATP-dependent stimulation of GAL4–AH binding to nucleosomes by the purified SWI/SNF complex (Côte et al., 1994). The SWI/SNF complex is a 2 × 10⁶ dalton complex which includes the SWI1, SWI2, SWI3, SNF5, and SNF6 gene products and is required for the function of numerous activators *in vivo* and is genetically linked to chromatin function (reviewed in Winston and Carlson, 1992; also see Peterson and Herskowitz, 1992). It has been suggested that the SWI/SNF complex may function by perturbation of H2A/H2B dimers since many swi/snf mutations can be suppressed in yeast by deletion of one of the H2A/H2B gene pairs (Hirschhorn et al., 1992). Further biochemical experiments will reveal the extent to which SWI/SNF complex activity in stimulating GAL4–AH binding resembles or differs from that of nucleoplasmin.

IV. CONSEQUENCES OF FACTOR BINDING TO NUCLEOSOMES

In the previous sections we have considered the inhibition of factor binding resulting from nucleosome cores and mechanisms by which factors might overcome this inhibition to gain access to their binding sites. However, an equally important question is the effect of factor binding on nucleosome structure and stability. The occurrence of nuclease hypersensitive sites over active enhancers and promoters indicates the absence of archetypal nucleosomes over these regions (reviewed in Gross and Garrard, 1988; Elgin, 1988). This raises the important question as to whether the histones have actually been displaced or are still present but structurally altered (i.e., disrupted) due to the interaction of sequence-specific binding factors (reviewed in Adams and Workman, 1993).

A. Formation of Transcription Factor/Nucleosome Ternary Complexes

A striking observation from *in vitro* studies is that in all instances where factor binding to nucleosome cores has been observed, the factors did not displace the underlying nucleosome. Instead, factor binding results in the formation of ternary complexes containing bound factor, histones, and DNA. This has been observed for the glucocorticoid receptor (Perlmann and Wrange, 1988; Pina et al., 1990; Archer et al., 1991), GAL4 derivatives (Taylor et al., 1991; Workman and Kingston, 1992; Vettese-Dadey et al., 1994; Juan et al., 1993), TFIIIA (Lee et al., 1993), USF

(Chen et al., 1994) and Sp1 (Li et al., 1994). An example of ternary complex formation is illustrated in Figure 2 where it is clear that the complex formed upon GAL4–AH binding to nucleosome cores is supershifted relative to GAL4–AH binding to naked DNA indicating that the histones remain in the GAL4–AH/nucleosome complex. In fact, analysis of proteins in purified complexes illustrated that the binding of up to five dimers of GAL4 derivatives does not result in the displacement of the underlying core histones (Workman and Kingston, 1992).

One important function of ternary complex formation was discussed above, namely the facilitated binding of additional factors to nucleosomes. As illustrated in the studies with GAL4 derivatives, formation of a ternary complex with the bound factors increases the affinity of additional factors. Thus, the affinity of subsequent activators for a preformed factor/nucleosome ternary complex is greater than the affinity of an activator for an unoccupied nucleosome. A similar mechanism has been proposed from *in vivo* protein/DNA:crosslinking studies of the mouse mammary tumor virus promoter. Bresnick and colleagues (1992) suggest that glucocorticoid receptor binding to the MMTV promoter induces NF1 binding to the promoter without actually displacing the underlying core histones.

B. Nucleosome Destabilization and Displacement by Transcription Factor Binding

Since the presence of the histone octamer reduces the affinity of transcription factors for nucleosomal DNA, it follows that the presence of bound factors should conversely reduce the affinity of the histones. This reduction in nucleosome stability can be measured by competition experiments with nonspecific plasmid DNA (Workman and Kingston, 1992). As shown in Part A of Figure 5, nucleosome cores alone are extremely stable to high concentrations of naked DNA competitors at physiological ionic strength. When five dimers of GAL4–AH are bound to a nucleosome core, it results in the formation of a ternary complex with the core histones (Figure 5, Part B). In contrast to unbound nucleosome cores, however, the ternary complex is not stable to the presence of naked DNA competitor. In the presence of the competitor the ternary complex broke down into one of two complexes: either the original nucleosome core when GAL4–AH dissociated from the ternary complex, or GAL4–AH/DNA complexes when the histone octamer dissociated. Since the majority of the probe DNA began as nucleosome cores, the latter instance illustrates the displacement of nucleosome cores as a result of factor binding and histone competition.

The appearance of both GAL4/DNA and nucleosome cores upon the addition of nonspecific DNA to GAL4–AH/nucleosome ternary complexes suggests that either histones or GAL4–AH were lost from the ternary complexes at similar rates. However as stated above, the binding of five GAL4–AH dimers to nucleosome cores in the presence of nucleoplasmin results in the specific loss of H2A/H2B dimers from the ternary complex onto nucleoplasmin. The subsequent addition of

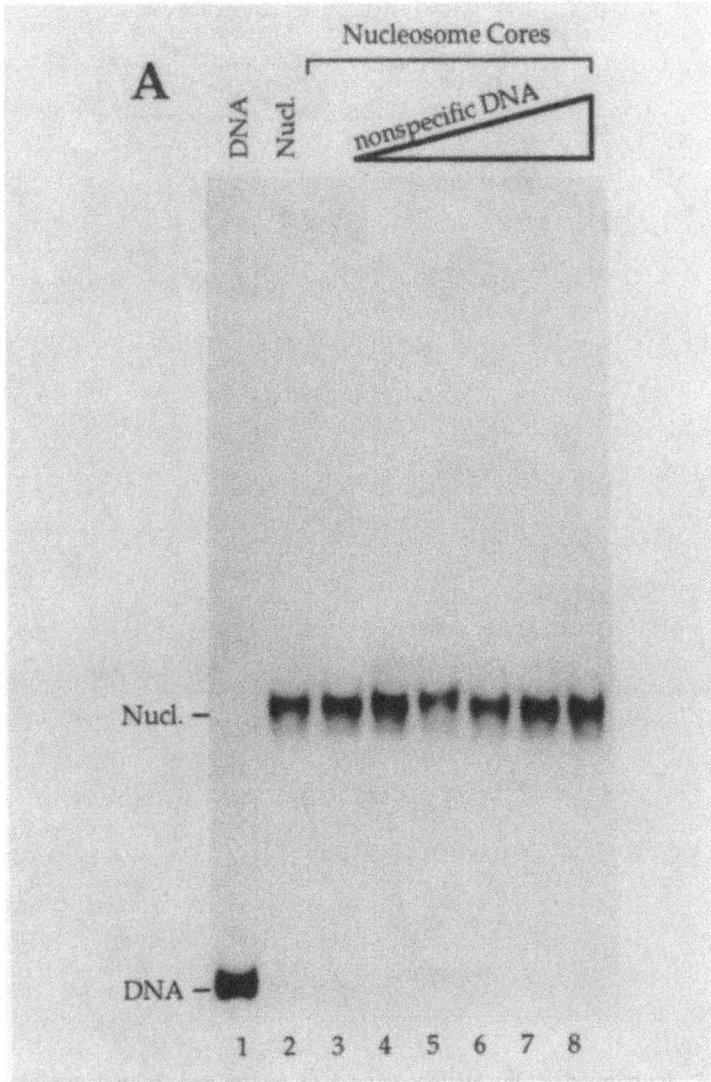

Figure 5. Dissociation of histones from GAL4–AH/nucleosome complexes in the presence of competitor DNA. Reconstituted nucleosome cores, Part A, or nucleosome cores bound by five dimers of GAL4–AH, Part B, were challenged with increasing concentrations of nonspecific competitor DNA and assayed on a mobility shift gel. Part A illustrates that the nucleosome cores in the absence of GAL4–AH were stable in the presence of increasing amounts of competitor DNA as indicated by the fact that the nucleosome core complex was not converted to naked DNA (the naked DNA migration is shown in lane 1). In Part B, lane 1 illustrates the migration of five dimers of GAL4–AH bound to the naked DNA probe, while lane 2 illustrates the migration

(continued)

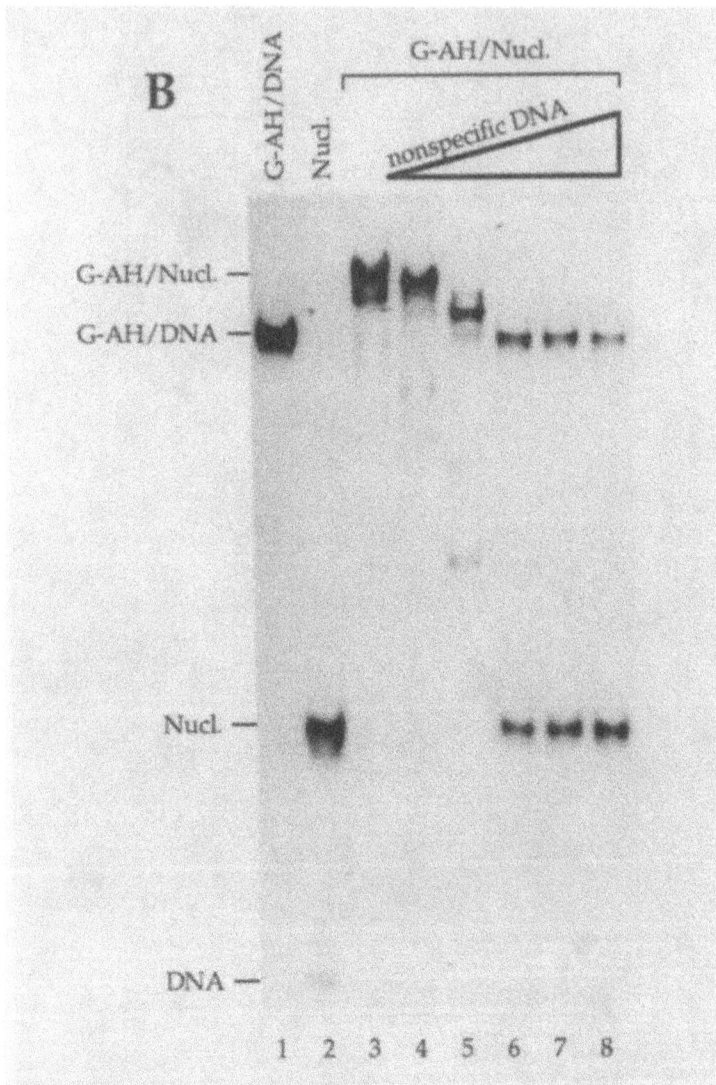

Figure 5. (continued) of the nucleosome core reconstituted probe. Upon binding of five GAL4–AH dimers, a ternary complex (containing GAL4–AH, core histones and DNA) was formed (lane 3) which was supershifted relative to the GAL4–AH bound DNA. However, following the addition of increasing amounts of competitor DNA this complex dissociated into either the original nucleosome core (i.e., the GAL4–AH dimers were lost) or GAL4–AH/DNA complexes (i.e., the core histones were lost). The latter instance illustrates nucleosome displacement as a result of factor binding and histone competition. The amounts of competitor DNA added to each reaction was in both panels; 0, lanes 1–3; 40ng, lane 4; 120ng, lane 5; 400ng, lane 6; 1.2μg, lane 7; and 4μg, lane 8. Reproduced from Figure 2 of Workman and Kingston, 1992, Science 258, 1780–1784.

nonspecific DNA to these reactions results in the sole production of GAL4–AH/DNA complexes (Chen et al., 1994). Therefore, in the presence of nucleoplasmin, GAL4–AH binding resulted in complete nucleosome disassembly by the displacement of H2A/H2B onto nucleoplasmin followed by the displacement of H3/H4 onto competitor DNA.

These initial studies of nucleosome displacement implicate a crucial role of sequence-specific binding factors in destabilizing nucleosome cores and a role of histone binding proteins in stimulating factor binding and removing histones from the ternary complexes. Thus, nucleosome displacement may occur as a result of factor binding through the transfer of histones from factor binding sites onto histone chaperones via a mechanism of nucleosome disassembly.

V. SUMMARY

In this paper we have summarized *in vitro* experiments analyzing the interactions of regulatory transcription factors with nucleosome cores. We have focused on general principles which have emerged and promise to play a role in the regulation of numerous class II promoters and enhancers. These studies indicate that the ability of a particular factor to bind its recognition site within a nucleosome is dependent on 1) its inherent nucleosome-binding ability (most likely a property of its DNA binding domain), 2) the location of its binding site within the core particle, 3) the presence of other factors binding the same nucleosomes (i.e., resulting in facilitated binding), and 4) the action of accessory proteins which stimulate its binding (i.e., by interacting with histones). In addition, these studies implicate factor binding in initiating nucleosome disassembly as a mechanism to replace nucleosomes at regulatory elements with sequence-specific binding proteins. Further studies of additional regulatory factors, and promoter and enhancer elements will reveal the extent to which these principles apply to the regulation of specific eukaryotic genes.

ACKNOWLEDGMENTS

We are grateful to Orjan Wrange, Peter Becker, Mike Kladde, and Robert Simpson for communicating results prior to publication. The work presented from our laboratory was supported by a grant from the Public Health Service to Jerry L. Workman (GM47867) and funding from the Leukemia Society of America. Workman is a Leukemia Society Scholar, Christopher C. Adams is an NIH postdoctoral fellow and Jacques Cote is a postdoctoral fellow of the Medical Research Council of Canada.

REFERENCES

Adams, C.C., & Workman, J.L. (1993). Nucleosome displacement in transcription. Cell 72, 305–308.
Adams, C.C., & Workman, J.L. (1995). The binding of disparate transcriptional activators to nucleosomal DNA is inherently cooperative. Mol. Cell. Biol. 15, 1407–1421.
Almouzni, G., & Wolffe, A.P. (1993). Replication-coupled chromatin assembly is required for the repression of basal transcription *in vitro*. Genes and Development 7, 2033–2047.

Archer, T.K., Cordingley, M.G., Wolford, R.G., & Hager, G.L. (1991). Transcription factor access is mediated by accurately positioned nucleosomes on the mouse mammary tumor virus promoter. Mol. Cell. Biol. 11, 688–698.

Bresnick, E.H., Bustin, M., Marsaud, V., Richard-Foy, H., & Hager, G.L. (1992). The transcriptionally-active MMTV promoter is depleted of histone H1. Nucl. Acids. Res. 20, 273–278.

Brown, D.D. (1984). The role of stable complexes that repress and activate eucaryotic genes. Cell 37, 359–365.

Chen, H., Li, B., & Workman, J.L. (1994). A histone-binding protein, nucleoplasmin, stimulates transcription factor binding to nucleosomes and factor-induced nucleosome disassembly. EMBO J. 13, 380–390.

Côte, J., Quinn, J., Workman, J.L., & Peterson, C.L. (1994). Stimulation of GAL4 derivatives binding to nucleosomal DNA by the yeast SWI/SOF complex. Science 265, 53–60.

Croston, G.E., Kerrigan, L.A., Lira, L.M., Marshak, D.R., & Kadonaga, J.T. (1991). Sequence-specific antirepression of histone H1-mediated inhibition of basal RNA polymerase II transcription. Science 251, 643–649.

Croston, G.E., Laybourn, P.J., Paranjape, S.M., & Kadonaga, J.T. (1992). Mechanism of transcriptional antirepression by GAL4–VP16. Genes and Development 6, 2270–2281.

Dilworth, S.M., Black, S.J., & Laskey, R.A. (1987). Two complexes that contain histones are required for nucleosome assembly in vitro: Role of nucleoplasmin and N1 in Xenopus egg extracts. Cell 51, 1009–1018.

Dusserre, Y., & Mermod, N. (1992). Purified cofactors and histone H1 mediate transcriptional regulation by CTF/NF1. Mol. Cell. Biol. 12, 5228–5237.

Elgin, S.C.R. (1988). The formation and function of DNase hypersensitive sites in the process of gene activation. J. Biol. Chem. 263, 19259–19262.

Eriksson, P., & Wrange, O. (1993). The glucocorticoid receptor acts as an antirepressor in receptor-dependent in vitro transcription. Eur. J. Bio. 215, 505–511.

Felsenfeld, G. (1992). Chromatin as an essential part of the transcriptional mechanism. Nature 355, 219–224.

Gross, D.S., & Garrard, W.T. (1988). Nuclease hypersensitive sites in chromatin. Annu. Rev. Biochem. 57, 159–197.

Hayes, J.J., Clark, D.J., & Wolffe, A.P. (1991). Histone contributions to the structure of DNA in the nucleosome. Proc. Natl. Acad. Sci. USA 88, 6829–6833.

Hirschhorn, J.N., Brown, S.A., Clark, C.D., & Winston, F. (1992). Evidence that SNF2/SWI2 and SNF5 activate transcription in yeast by altering chromatin structure. Genes and Development 6, 2288–2298.

Juan, L., Walter, P., Taylor, I.C.A., Kingston, R.E., & Workman, J. L. (1993). The role of nucleosome cores and histone H1 in the binding of GAL4 derivatives and the reactivation of transcription from nucleosome templates in vitro. Cold Spring Harbor Symposium on Quantitative Biology: DNA and Chromosomes 58, 213–224.

Kamakaka, R.T., Bulger, M., & Kadonaga, J.T. (1993). Potentiation of RNA polymerase II transcription by GAL4–VP16 during but not after DNA replication and chromatin assembly. Genes and Development 7, 1779–1795.

Kerrigan, L.A., Croston, G.E., Lira, L.M., & Kadonaga, J.T. (1991). Sequence-specific antirepression of the Drosophila Kruppel gene by the GAGA factor. J. Biol. Chem. 266, 574–582.

Kladde, M.P., & Simpson, R.T. (1994). Positioned nucleosomes inhibit dam methylation in vivo. Proc. Natl. Acad. Sci. USA 91, 1361–1365.

Kleinschmidt, J.A., Fortklamp, E., Krohne, G., Zentgraf, H., & Franke, W.W. (1985). Coexistence of two different types of soluble histone complexes in nuclei of Xenopus laevis oocytes. J. Biol. Chem. 260, 1166–1176.

Kleinschmidt, J.A., Seiter, A., & Zentgraf, H. (1990). Nucleosome assembly *in vitro:* Separate histone transfer and synergistic interaction of native histone complexes purified from nuclei of *Xenopus laevis* oocytes. EMBO J. 9, 1309–1318.

Knezetic, J.A., & Luse, D.S. (1986). The presence of nucleosomes on a DNA template prevents initiation by RNA polymerase II *in vitro.* Cell 45, 95–104.

Knezetic, J.A., Jacob, G.A., & Luse, D.S. (1988). Assembly of RNA polymerase II preinitiation complexes before assembly of nucleosomes allows efficient initiation of transcription on nucleosomal templates. Mol. Cell. Biol. 8, 3114–3121.

Laybourn, P.J., & Kadonaga, J.T. (1991). Role of nucleosome cores and histone H1 in regulation of transcription by RNA polymerase II. Science 254, 238–254.

Lee, D.Y., Hayes, J.J., Pruss, D., & Wolffe, A.P. (1993). A positive role for histone acetylation in transcription factor access to nucleosomal DNA. Cell 72, 73–84.

Li, B., Adams, C.C., & Workman, J.L. (1994). Nucleosome-binding by the constitutive transcription factor Sp1. J. Biol. Chem. 269, 7756–7763.

Li, Q., & Wrange, O. (1994). Translational positioning of a nucleosomal glucocorticoid response element modulates glucocorticoid receptor affinity. Genes and Development 7, 2471–2482.

Lilley, D.M., & Tatchell, K. (1977). Chromatin core particle unfolding induced by tryptic cleavage of histones. Nucl. Acids Res. 4, 2039–2055.

Lorch, Y., LaPointe, J.W., & Kornberg, R.D. (1992). Initiation on chromatin templates in a yeast RNA polymerase II transcription system. Genes and Development 6, 2282–2287.

Lorch, Y., LaPointe, J.W., & Kornberg, R.D. (1987). Nucleosomes inhibit the initiation of transcription but allow chain elongation with the displacement of histones. Cell 49, 203–210.

Matsui, T. (1987). Transcription of adenovirus 2 major late and peptide IX genes under conditions of *in vitro* nucleosome assembly. Mol. Cell. Biol. 7, 1401–1408.

Mitchell, P.J., & Tjian, R. (1989). Transcriptional regulation in mammalian cells by sequence-specific DNA binding proteins. Science 245, 371–378.

Moreau, N., Angelier, N., Bonnanfant-Jais, M.-L., & Kubisz, P. (1986). Association of nucleoplasmin with transcription products as revealed by immunolocalization in the amphibian oocyte. J. Cell Biol. 103, 683–690.

Morse, R.H., Pederson, D.S., Dean, A., & Simpson, R.T. (1987). Yeast nucleosomes allow thermal untwisting of DNA. Nucl. Acids Res. 15, 10311–10330.

Pederson, D.S., Thomas, F., & Simpson, R.T. (1986). Core particle, fiber, and transcriptionally active chromatin structure. Annu. Rev. Cell Biol. 117–147.

Perlmann, T. (1992). Glucocorticoid receptor DNA-binding specificity is increased by the organization of DNA in nucleosomes. Proc. Natl. Acad. Sci. USA 89, 3884–3888.

Perlmann, T., & Wrange, O. (1988). Specific glucocorticoid receptor binding to DNA reconstituted in a nucleosome. EMBO J. 7, 3073–3079.

Peterson, C.L., & Herskowitz, I. (1992). Characterization of the Yeast SWI1, SWI2, and SWI3 Genes, Which Encode a Global Activator of Transcription. Cell 68, 573–583.

Pina, B., Bruggemeier, U., & Beato, M. (1990). Nucleosome positioning modulates accessibility of regulatory proteins to the mouse mammary tumor virus promoter. Cell 60, 719–731.

Rhodes, D., & Laskey, R.A. (1989). Assembly of nucleosomes and chromatin *in vitro.* Methods in Enzymology 170, 575–585.

Saltzman, A.G., & Weinmann, R. (1989). Promoter specificity and modulation of RNA polymerase II transcriptions. FASEB J. 3, 1723–1733.

Sandoltzopoulos, R., Blank, T., & Becker, P.B. (1994). Transcriptional repression by nucleosomes but not H1 in reconstituted preblastodern Drosophila chromatin. EMBO J. 13, 373–379.

Sawadogo, M., & Sentenac, A. (1990). RNA polymerase B(II) and general transcription factors. Annu. Rev. Biochem. 59, 711–754.

Simpson, R.T. (1979). Mechanism of a reversible, thermally induced conformational change in chromatin core particles. J. Biol. Chem. 254, 10123–10127.

Simpson, R.T. (1990). Nucleosome positioning can affect the function of a cis-acting DNA element *in vivo*. Nature 343, 387–389.

Simpson, R.T. (1991). Nucleosome positioning: occurrence, mechanisms and functional consequences. Prog. in Nucl. Acids Res. Mol. Biol. 40, 143–184.

Taylor, C.A., Workman, J.L., Schuetz, T.J., & Kingston, R.E. (1991). Facilitated binding of GAL4 and heat shock factor to nucleosomal templates: Differential function of DNA-binding domains. Genes Dev. 5, 1285–1298.

Turner, B.M. (1991). Histone acetylation and control of gene expression. J. Cell Science 99, 13–20.

Vettese-Dadey, M., Walter, P., Chen, H., Juan, L.-J., & Workman, J.L. (1994). Role of the histone amino termini in facilitated binding of a transcription factor, GAL4–AH, to nucleosome cores. Mol. Cell Biol. 14, 970–981.

Weischet, W.O., Tatchell, K., van Holde, K.E., & Klump, H. (1978). Thermal denaturation of nucleosomal core particles. Nucl. Acids Res. 5, 139–160.

Whitlock, J.P., & Simpson, R.T. (1977). Localization of the sites along nucleosomal DNA which interact with NH2-terminal histone regions. J. Biol. Chem. 252, 6516–6520.

Winston, F., & Carlson, M. (1992). Yeast SNF/SWI transcriptional activators and the SPT/SIN chromatin connection. Trends Genet. 8, 387–391.

Wolffe, A.P. (1992). New insights into chromatin function in transcriptional control. FASEB J. 6, 3354–3361.

Wolffe, A.P., & Brown, D.D. (1988). Developmental regulation of two 5S ribosomal RNA genes. Science 241, 1626–1632.

Wolffe, A.P., & Drew, H.R. (1989). Initiation of transcription on nucleosomal templates. Proc. Natl. Acad. Sci. USA 86, 9817–9822.

Wolffe, A.P., & Hayes, J.J. (1993). Transcription factor interactions with model nucleosome templates. Methods in Molecular Genetics 2, 314–330.

Workman, J.L., & Kingston, R.E. (1992). Nucleosome core displacement *in vitro* via a metastable transcription factor:nucleosome complex. Science 258, 1780–1784.

Workman, J.L., & Roeder, R.G. (1987). Binding of transcription factor TFIID to the major late promoter during *in vitro* nucleosome assembly potentiates subsequent initiation by RNA polymerase II. Cell 51, 613–622.

Workman, J.L., Abmayr, S.M., Cromlish, W.A., & Roeder, R.G. (1988). Transcriptional regulation by the immediate early protein of pseudorabies virus during *in vitro* nucleosome assembly. Cell 55, 211–219.

Workman, J.L., Roeder, R.G., & Kingston, R.E. (1990). An upstream transcription factor, USF (MLTF), facilitates the formation of preinitiation complexes during *in vitro* chromatin assembly. EMBO J. 9, 1299–1308.

Workman, J.L., Taylor, I.C.A., & Kingston, R.E. (1991). Activation domains of stably bound GAL4 derivatives alleviate repression of promoters by nucleosomes. Cell 64, 533–544.

Yaniv, M., & Cereghini, S. (1986). Structure of transcriptionally active chromatin. CRC Crit. Rev. Biochem. 21, 1–26.

Zawel, L., & Reinberg, D. (1993). Initiation of transcription by RNA polymerase II: A multi-step process. Prog. Nucleic Acids Res. Mol. Biol. 44, 67–108.

Chapter 5

Nucleosome Positioning by a Yeast Repressor Complex

SHARON Y. ROTH

The Nucleus
Volume 1, pages 101–119.
Copyright © 1995 by JAI Press Inc.
All rights of reproduction in any form reserved.
ISBN: 1-55938-940-0

ABSTRACT

Repression of yeast **a** cell type specific genes in α cells is tightly linked to the positioning of nucleosomes adjacent to the α2/MCM1 operator. In the presence of α2, nucleosomes are stably and precisely positioned 15 bp from the edge of the operator, incorporating promoter regions of the **a** cell specific genes. Positioning is propagated downstream for several nucleosomes, reflecting the packaging of these genes into repressive chromatin subdomains. Moreover, mutations in histone H4 which alter nucleosome stability partially derepress **a** cell specific gene expression, affirming the importance of these nucleosomes to repression. Recent data indicate that several global repressors, including SSN6 and TUP1, are also involved in the organization of chromatin. Positioning of nucleosomes adjacent to the α2/MCM1 operator provides a model system for understanding how multiple proteins cooperate in the formation or maintenance of regulatory chromatin structures.

I. INTRODUCTION

Because packaging of promoter regions into chromatin represses transcription initiation both *in vitro* and *in vivo* (see Wolffe, 1992, for review), much attention has been focused on the interplay between transcriptional activators and chromatin. Indeed, current models of transcriptional activation often invoke nucleosome destabilization via formation of ternary complexes between activator proteins, histone octamers, and DNA (Felsenfeld, 1992; Adams and Workman, 1993). Conversely, constitutive repression could be achieved through the organization of regulatory elements into inaccessible conformations. The establishment of small repressive chromatin subdomains (relative to higher order, heterochromatic regions) by sequence-specific repressors would provide an efficient mechanism for repression of individual genes flanked by genes which are active.

Such a mechanism appears operative in the repression of a set of yeast cell type specific genes. In haploid α cells and diploid **a**/α cells, expression of **a** cell-specific genes is repressed by an α cell-specific repressor, α2, in cooperation with a second, ubiquitous protein, MCM1 (see Table 1) (see Herskowitz, 1989, for review). The α2/MCM1 complex binds to a 32 bp sequence located upstream of the **a** cell specific genes. While this binding is necessary, it is not sufficient for repression. At least

Table 1. Cell Type Distribution of Factors which Regulate Yeast **a** Cell Specific Gene Expression

	α *cells*	**a** *cells*	**a**/α *cells*
a cell specific genes	off	on	off
α2	+	−	+
MCM1, SSN6, TUP1	+	+	+

two other proteins, SSN6 and TUP1, are also required (Lemontt et al., 1980; Schultz and Carlson, 1987; Keleher et al., 1992). Neither of these proteins appear to bind DNA directly, and it has been suggested that they are "recruited" to specific promoters through interactions with sequence specific binding proteins such as α2/MCM1 (Keleher et al., 1992). Recent data indicate that SSN6 and TUP1 are also important in the organization of chromatin. This paper will review the structural characteristics of chromatin adjacent to the α2/MCM1 operator in α cells and **a** cells, and will discuss the role of "global repressors" such as SSN6 and TUP1 in the establishment of these repressive regions.

II. NUCLEOSOME POSITIONING ADJACENT TO THE α2/MCM1 OPERATOR

The α2/MCM1 operator sequence is located ~200 bp upstream of the mRNA start sites of the five **a** cell-specific genes (Wilson and Herskowitz, 1986). Repression occurs when the operator is inserted either upstream or downstream of heterologous upstream activating sequences (UAS) and is observed over distances of several hundred base pairs (Keleher et al., 1988, 1992). These features are consistent with a mechanism of repression involving chromatin, and early studies sought to compare the packaging of sequences adjacent to the operator in the presence (in α cells) or absence (in **a** cells) of repression.

A. Nucleosome Positioning in TRP1/ARS1 Derivatives

Initial experiments investigating the interactions of α2/MCM1 with chromatin made use of derivatives of a yeast episome, TRP1/ARS1 (Roth et al., 1990). This plasmid provides a useful model system for determining factors which affect nucleosome position *in vivo*. The chromatin structure of the original TRP1/ARS1 plasmid was studied in detail by Thoma and colleagues (1984), who discovered two nucleosomal domains in the plasmid chromatin (Figure 1). One domain consists of four nucleosomes which incorporate the bulk of the TRP1 gene. These nucleosomes appear to be relatively unstable as they exhibit a high sensitivity to nuclease digestion. A second nucleosomal domain is present on the opposite side of the plasmid, consisting of three very stably positioned nuclesosomes (I, II, and III, Figure 1). Nuclease hypersensitive regions are also observed in TRP1/ARS1 chromatin, including one region (HSR B) upstream of the TRP1 gene, and a second region (HSR A) which contains the 3′ end of the TRP1 gene, as well as elements required for autonomous replication (ARS1 elements). These hypersensitive sites may provide "boundaries" to nucleosome formation, thereby contributing to the positioning of nucleosomes over adjacent sequences (Thoma and Simpson, 1985). Indeed, insertion of a 95 bp sequence into HSR B causes a randomization of nucleosome positions (Roth et al., 1990), consistent with a disruption of a positioning signal in this hypersensitive domain.

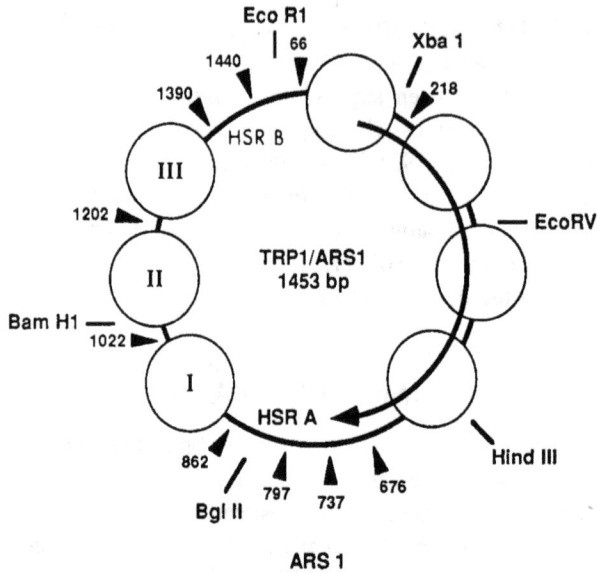

Figure 1. Chromatin structure of TRP1/ARS1. Four unstable nucleosomes span the TRP1 gene (curved arrow), while three very stable nucleosomes (labeled I, II, and III) incorporate sequences on the opposite side of the plasmid (Thoma et al., 1984). Two nuclease hypersensitive regions separate these nucleosomal domains, HSR A and HSR B. The positions of micrococcal nuclease cut sites in chromatin (in map units) are indicated by arrowheads. The Eco R1 site marks map unit 1, and map units continue in a clockwise direction around the plasmid. The positions of other unique restriction sites are also indicated.

Insertion of the α2/MCM1 operator into the disrupted HSR B has no effect on nucleosome location in **a** cells, where α2 is not expressed. Indirect end-labeling experiments indicate that nucleosomes are located randomly on this construct (termed TALS) in **a** cells. In α cells, however, a striking organization of nucleosomes is observed (Roth et al., 1990). In this case, the entire plasmid, excluding only HSR A (which contains replication elements) is packaged into a stable array of precisely positioned nucleosomes (Figure 2). The α2/MCM1 operator appears as a discrete nuclease hypersensitive site, and binding of the repressor complex to this site apparently causes organization of nucleosomes over adjacent sequences which is then propagated around the plasmid. A similar organization of nucleosomes is observed in α cells, but not **a** cells, when the operator is inserted at a different position in TRP1/ARS1. Higher resolution (primer extension) maps indicate that nucleosomes adjacent to the operator (IV and V) are positioned both translationally and rotationally (Shimizu et al., 1991).

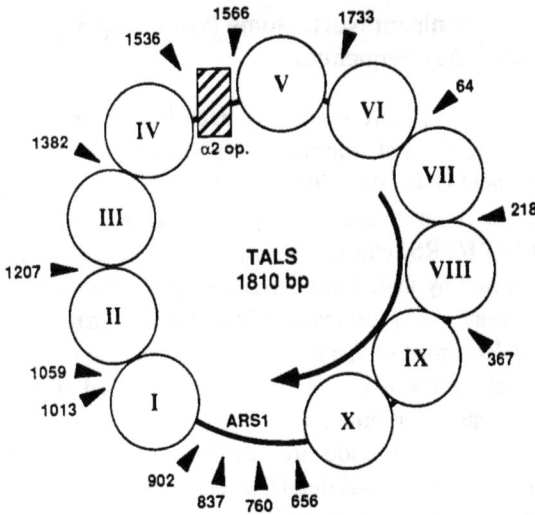

Figure 2. Chromatin structure of TALS in α cells. The plasmid TALS is a derivative of TRP1/ARS1 in which sequences containing the α2/MCM1 operator have been inserted at the unique Eco R1 site. Insertion of these sequences causes a randomization of nucleosome positions in **a** cells, where α2 is not expressed. In α cells and **a**/α cells, binding of α2/MCM1 to the operator causes nucleosome positioning adjacent to the operator which is then propagated around the plasmid. Only sequences in the region of HSR A which contains ARS1 replication elements remain nuclease sensitive in the presence of α2. Positions of micrococcal nuclease cut sites are indicated as in Figure 1.

The organization of nucleosomes in the presence of α2 is transdominant (Roth et al., 1990). When **a** cells carrying the TALS plasmid are mated to nontransformed α cells, the structure of TALS in the resulting **a**/α diploid cells is identical to that observed in haploid α cells. Nucleosomes which are randomly positioned in **a** cells become stably and precisely positioned in the presence of α2.

Since each nucleosome imparts one negative supercoil to DNA, the topoisomer density of plasmid chromatin can reflect the density of nucleosomes on the plasmid. The topoisomer density of the TALS plasmid is altered in the presence of α2 (Roth et al., 1990). Chloroquine gel analysis indicates that the plasmid contains one extra negative supercoil in α cells. This shift in topoisomer density is not observed in plasmids which do not contain the α2/MCM1 binding site. The presence of the extra negative supercoil in α cells is consistent with the presence of an extra nucleosome in the presence of α2, indicating the repressor may not only direct nucleosome position, but may also facilitate nucleosome formation.

B. α2 Provides a Dominant Nucleosome Positioning Signal Relative to Endogenous TRP1/ARS1 Sequences

The ARS1 elements in HSR A are organized into three domains (Campbell, 1988). The 11 bp core element required for replication (domain A) is flanked by a region containing bent DNA (domain B) which extends into the TRP1 gene, and a third region on the opposite side (domain C) which overlaps the position of nucleosome I in TRP1/ARS1 chromatin (Figure 3). The position of this nucleosome appears to be directed by underlying DNA elements which are not well defined (Thoma and Simpson, 1985). By making deletions in domain C, Simpson shifted the position of nucleosome I so that it incorporated the ARS1 A element (Simpson, 1990). When this element was located near the pseudodyad of the nucleosome, its replication ability was compromised, resulting in a ~40X drop in plasmid copy number. These experiments provide direct evidence that nucleosomes can limit the accessibility and function of cis-acting DNA elements *in vivo*.

To further test this hypothesis, the α2 operator was inserted 60 bp upstream of the edge of the nucleosome incorporating the ARS1 A element (Simpson, 1990).

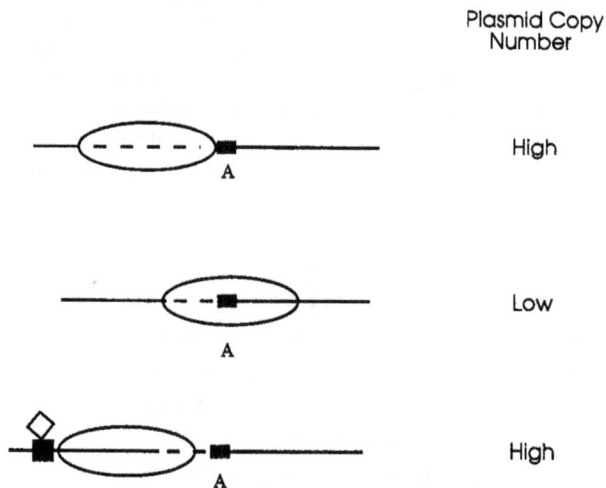

Figure 3. The α2/MCM1 operator provides a dominant nucleosome positioning signal in TRP1/ARS1. Nucleosome I (ellipse) is stably positioned in TRP1/ARS1 chromatin and incorporates domain C (dashed line) of the ARS1 replication elements. The essential ARS1 A box (black box) required for replication is located at the edge of nucleosome I. Deletion of sequences within domain C shifts the position of nucleosome I toward the A element. When the A element is within 40 nucleotides of the pseudodydad of the nucleosome, replication is inhibited, resulting in a drop in plasmid copy number (Simpson, 1990). Insertion of the α2/MCM1 operator (grey box) downstream re-positions the nucleosome in the presence of α2/MCM1 (diamond), freeing the ARS1 A element and rescuing replication.

When this construct was introduced into α cells, replication was rescued, resulting in a restoration of plasmid copy number. Examination of the chromatin structure of this plasmid revealed a shift in nucleosome position such that the nucleosome was now located adjacent to the α2 operator, and the ARS1 A element was again exposed in a nuclease sensitive region. Importantly, this shift in nucleosome position and rescue of replication only occurred in α cells. In **a** cells, nucleosome location did not change and plasmid copy number remained low.

These experiments indicate that α2 can provide a dominant nucleosome positioning signal relative to endogenous signals present in TRP1/ARS1. This dominance, together with the precision and stability of nucleosome positioning adjacent to the α2/MCM1 operator, suggest that these nucleosomes are directly positioned by α2/MCM1 and/or associated proteins.

C. Nucleosome Positioning in the Promoters of **a** Cell-Specific Genes

If the organization of nucleosomes observed in the minichromosomes described above is important to the function of α2 as a transcriptional repressor, the same organization should also be present in the promoter regions of the **a** cell-specific

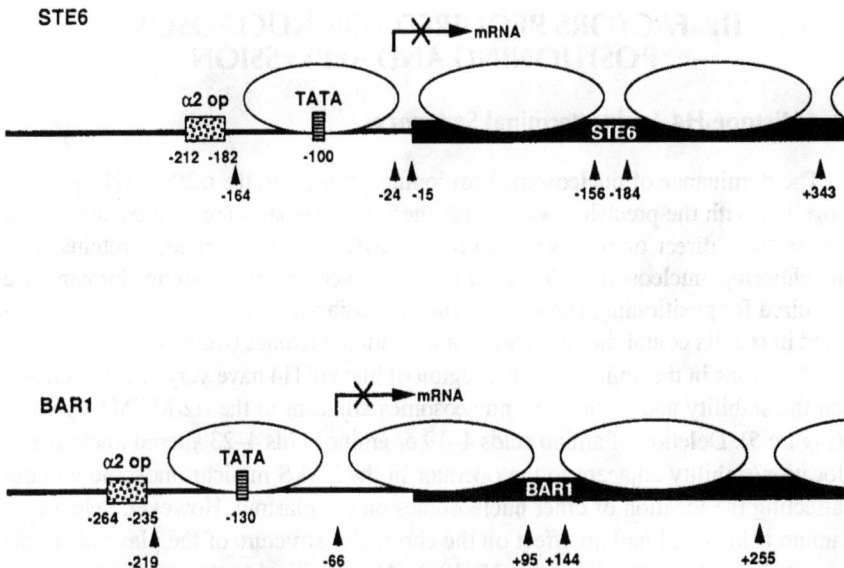

Figure 4. Chromatin structure of STE6 and BAR1. Nucleosome positions are inferred from primer extension analysis of micrococcal nuclease sensitive (arrowheads) and protected regions (ellipses; nucleosomes) (Shimizu et al., 1991). The locations of TATA elements, the α2/MCM1 operator, and transcription start sites (bent arrows) are as indicated. Positioned nucleosomes are only observed in the presence of α2/MCM1, when these genes are repressed.

genes regulated by α2. The chromatin structures of two of these genes, STE6 and BAR1, have been examined in detail (Figure 4). In both cases, nucleosomes are precisely located downstream of the operator in α cells, but not in **a** cells (Shimizu et al., 1991). Nucleosome positioning begins 15 bp. downstream of the operator and is propogated over neighboring sequences into the coding regions of STE6 and BAR1. Further experiments indicate that positioning continues for more than 20 nucleosomes through the STE6 gene (Simpson). Repression extends beyond the first nucleosome. Movement of the TATA element into a short (10 bp) linker region between the first two nucleosomes fails to relieve repression of a STE6/LACZ fusion gene (Patterton and Simpson, 1994).

Interestingly, nucleosomes are only positioned downstream of the α2/MCM1 operator in STE6 and BAR1 (Shimizu et al., 1991), in contrast to the apparent bidirectional positioning observed in the TALS minichromosome (Roth et al., 1990). The unidirectional organization at STE6 and BAR1 might reflect a pattern of nucleosome formation established during replication. The absence of nucleosomes upstream of the operator in these genes could also reflect the presence of sequences in these regions which do not favor nucleosome formation or the occupation of upstream sequences by other trans-acting factors.

II. FACTORS REQUIRED FOR NUCLEOSOME POSITIONING AND REPRESSION

A. Histone H4 Amino-terminal Sequences

The dominance of nucleosome positioning adjacent to the α2/MCM1 operator, together with the precision with which these nucleosomes are located suggest an interaction, direct or indirect, between α2/MCM1 (or associated proteins) and neighboring nucleosomes. To determine whether specific histone domains are required for positioning, chromatin structures adjacent to the operator were examined in α cells containing mutations in individual histones (Roth et al., 1992).

Mutations in the amino terminal region of histone H4 have very specific effects on the stability and position of nucleosomes adjacent to the α2/MCM1 operator (Figure 5). Deletion of amino acids 4–19 or amino acids 4–23 altered nucleosome location/stability adjacent to the operator in the TALS minichromosome without affecting the location of other nucleosomes on the plasmid. However, deletion of amino acids 4–14 had no effect on the chromatin structure of the plasmid. These results indicate that amino acids 15–19 in H4 are critical to the stable positioning of nucleosomes adjacent to the α2 operator. The loss of one negative supercoil from the TALS plasmid in α cells containing the longer H4 amino terminal deletions further indicates the importance of this region to nucleosome stability (Roth et al., 1992).

Changes in the chromatin structure of the chromosomal STE6 gene were also observed in the presence of the H4 amino terminal mutations, including deletion

A.

B.

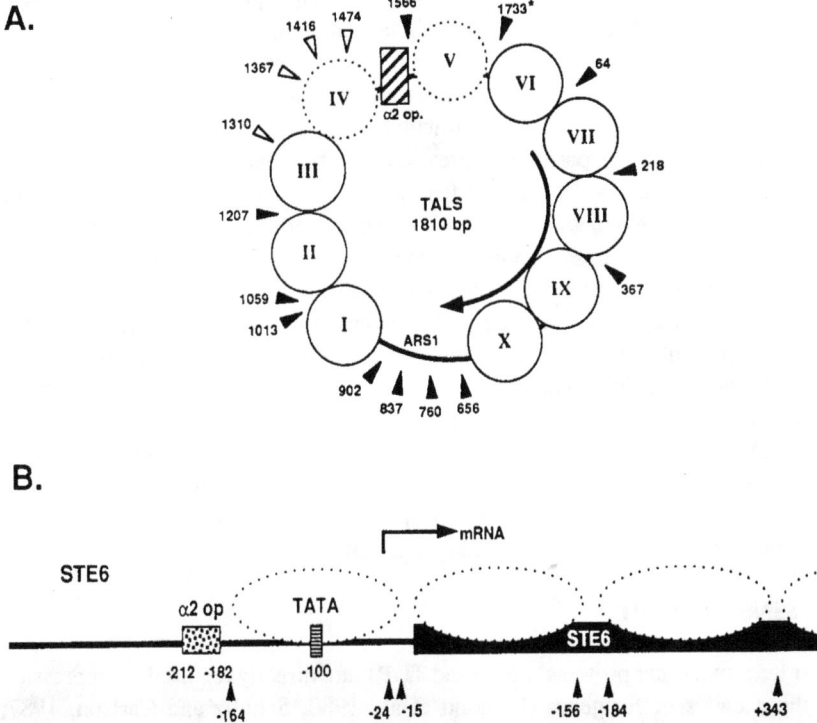

Figure 5. Nucleosome positioning adjacent to the α2/MCM1 operator is destabilized in the presence of H4 amino-terminal mutations. A) The chromatin structure of the TALS plasmid in the presence of deletion or point mutations of amino acids 4–19 in histone H4, as inferred from indirect end-labeling experiments (Roth et al., 1992). Nucleosomes IV and V (dashed ellipses) adjacent to the α2/MCM1 operator are destabilized while other nucleosomes on the plasmid are unaffected. White arrowheads indicate micrococcal nuclease cut sites which are exposed in the presence of the H4 mutations. These sites are normally completely protected in chromatin in α cells. Asterisks indicate cut sites (1566 and 1733) which are weakened in the presence of the H4 mutations. B). The chromatin structure of STE6 in the presence of the H4 mutations. Positions of several nucleosomes downstream of the operator are destabilized, as indicated by the dashed lines.

of amino acids 4–19, 4–23 or point mutation of residues 16 (K to G), 17 (R to G) or 18 (H to G) (see Figure 5). These mutations yielded a STE6 chromatin structure in α cells with combined features of that present in wild type α and wild type **a** cells (Roth et al., 1992). Nuclease hypersensitive sites reflecting linker regions between nucleosomes persisted, but cleavage was markedly increased in regions normally fully protected by nucleosomes in α cells. Moreover, the pattern of cleavage in these regions matched that present in wild type **a** cell chromatin, consistent with nucleosome destabilization. These disruptions in nuclease protection were observed not only for the nucleosome immediately adjacent to the α2/MCM1 operator, but for downstream nucleosomes as well, indicating a requirement for the H4 amino terminal region throughout the repressed domain.

The partial change in chromatin structure in the presence of the H4 mutations was accompanied by a partial derepression of a reporter gene under α2 control (Roth et al., 1992). The MFa2/lacZ fusion gene was expressed at a level ~5–15 percent of that present in wild type a cells in α cells harboring these mutations. This level of expression represents a modest but significant derepression since expression of this gene is undetectable in wild type α cells. Indeed, the partial derepression observed correlates well with the partial changes in chromatin structure described above. Importantly, *in vivo* footprinting studies demonstrate that the α2/MCM1 operator remains fully occupied in the presence of the H4 mutations, indicating that binding of α2/MCM1 is not sufficient for repression in the presence of this altered chromatin structure. The incomplete loss of nucleosome positioning and partial derepression may reflect the involvement of other regions of H4, other histones, or other regulatory factors in the establishment of the fully organized, repressive chromatin structure observed in wild type α cells.

B. SSN6 and TUP1

At least two other proteins, SSN6 and TUP1, are directly required for repression of the a cell specific genes (Lemontt et al., 1980; Schultz and Carlson, 1987; Keleher et al., 1992). These proteins are also required for repression of haploid specific genes (such as RME1) and glucose repressible genes (such as SUC2), leading to the designation of these proteins as "global" repressors (see Table 3). Both SSN6 and TUP1 have been cloned and sequenced, and both are homologous to proteins in other systems. SSN6 contains 10 copies of TPR (tetratricopeptide repeat) motifs present in proteins associated with diverse cellular processes such as mitosis, transcription, RNA splicing, protein import, and neurogenesis (Schultz et al., 1990; Sikorski et al., 1990; Goebl and Yanagida, 1991). These repeats are thought to be important in protein-protein interactions, and at least 3–4 of the repeats present in SSN6 are required for function (Schultz et al., 1990). TUP1 is homologous to β-transducins (Fong et al., 1986). The significance of this homology is not clear, but this domain is essential for TUP1 function (Williams and Trumbly, 1990).

SSN6 and TUP1 can be coimmunoprecipitated from yeast cell extracts as part of a large protein complex (Williams et al., 1991). An *in vivo* association of SSN6 and TUP1 is consistent with their corequirement in the regulation of several genes as described above. Interestingly, TPR-containing proteins have been functionally linked to β-transducin like proteins in other systems as well (Goebl and Yanagida, 1991).

Neither SSN6 or TUP1 appear to bind DNA directly (Schultz and Carlson, 1987; Williams and Trumbly, 1990). However, fusion of SSN6 to the DNA binding domain of the bacterial lex A protein facilitates repression of reporter genes bearing the lex A operator sequence (Keleher et al., 1992). Moreover, repression by the SSN6-lex A fusion protein requires TUP1. These data suggest that SSN6/TUP1 complexes repress target promoters via association with specific DNA binding proteins such as α2/MCM1 (Keleher et al., 1992).

SSN6/TUP1 complexes might affect repression through interactions with general transcription factors, inhibiting their function and/or association with other transcription proteins. Such interactions would explain the ability of SSN6/TUP1 to regulate many different genes. Alternatively, SSN6/TUP1 might function in the organization of chromatin such as that accompanying repression of the **a** cell specific genes.

To directly assess the involvement of SSN6 and TUP1 in nucleosome positioning, the chromatin structure of STE6 was examined in α cells bearing disruptions of the gene for either protein (Cooper et al., 1994). In both cases, alterations in nucleosome position/stability downstream of the α2 operator were observed. In the ssn6⁻ cells, the chromatin structure of STE6 resembled that present in the presence of the H4 amino-terminal mutations. Nuclease hypersensitive sites corresponding to linker regions between nucleosomes were still evident, but regions normally protected from digestion by positioned nucleosomes were exposed. In the tup1⁻ cells, STE6 exhibited a chromatin structure virtually identical to that present in wild type **a** cells, with a complete loss of nucleosome positioning.

Similar results were obtained upon examination of the chromatin structure of the TALS minichromosome in the absence of SSN6 or TUP1(Cooper et al., 1994). The loss of either protein resulted in alterations of nucleosome stability/position adjacent to the α2 operator, and more severe effects were observed in the tup1⁻ cells than in the ssn6⁻ cells.

The severity of alterations in chromatin structure in the absence of either SSN6 or TUP1 reflected the degree of derepression of the **a** cell specific genes in these cells (Table 2) (Cooper et al.,1994). The tup1⁻ cells exhibited a higher expression (~8X) of a STE6/LAC Z fusion reporter gene (scored by β Gal assays) than did the ssn6⁻ cells. Expression of secreted **a** pheromones (MFa1 and MFa2) was increased

Table 2. Nucleosome Positioning is Inversely Correlated with Gene Expression

Strain	Positioned Nucleosomes	Expression
WTα	+	–
WTa	–	+++
ssn6⁻,α	+/–	+
tup1⁻,a	–	+

Note: Nucleosomes are stably positioned in wild type α cells, where **a** cell specific gene expression is fully repressed. In **a** cells, nucleosome positioning is not observed, and **a** cell specific genes are maximally expressed. In α cells lacking either SSN6 or TUP1, nucleosome positioning is partially disrupted, and the disruptions are accompanied by a partial derepression of the **a** cell specific genes. The severity of the alterations in nucleosome positioning correlates with the degree of derepression observed.

in the tup1⁻ cells as well (scored by halo assay). Thus, changes in chromatin structure adjacent to the α2/MCM1 operator correlated to changes in gene expression, and the extent of change correlated to the extent of derepression.

C. Alterations in Nucleosome Positioning Are Not Dependent upon Transcription

Mutations in the TATA element of the STE6 promoter inhibit expression of a STE6/LAC Z reporter gene by at least 95% (Cooper et al., 1994). Nucleosomes are positioned on this inactive promoter in wild type α cells, but not in a cells. Since this gene is not transcribed in either cell type, the absence of positioning in a cells indicates that a lack of transcription is not sufficient to induce nucleosome positioning. Moreover, nucleosomes are not positioned on this promoter in tup1⁻ α cells, demonstrating that the disruption in positioning observed in the absence of TUP1 is not dependent on transcription. Taken together, these experiments provide strong evidence that α2/MCM1:SSN6/TUP1 complexes function to organize repressive regions of chromatin.

D. Other Global Regulators

Four other factors, SIN3 (RPD1,UME4) (Vidal et al., 1991), SIN4 (TSF3) (Jiang and Stillman, 1992; Chen et al., 1993), and RPD3 (Vidal and Gaber, 1991) are required for repression of the a cell specific genes. These genes were all identified in genetic screens for factors regulating other cellular processes, and mutations in each have pleiotropic effects (Table 3). Like SSN6 and TUP1, SIN3, SIN4, and RPD3 have been termed "global" regulators because they are involved in the regulation of diverse genes.

The SIN3, SIN4, and RPD3 genes have all been cloned and sequenced, but few clues as to their functions can be deduced from this information. Only SIN3 resembles transcriptional activators; it contains four helix-loop-helix motifs as well as acidic, glutamine rich, and proline rich domains (Vidal et al., 1991). SIN4 and RPD3 bear no homologies to other proteins currently in the data base (Jiang and Stillman, 1992; Vidal and Gaber, 1991; Chen et al., 1993), and TOP3 is a topoisomerase with homology to bacterial topoisomerase I (Wallis et al., 1989).

Given their involvement in the regulation of several genes, including the haploid specific genes, glucose repressible genes, an acid phosphatase gene (PHO5), as well as the a cell specific genes (Table 3), it is attractive to hypothesize that at least some of the above proteins function through modulations of chromatin structure. Indeed, mutations in the amino terminus of histone H4 affect both the activation and repression of genes regulated by these factors (Durrin et al., 1991), indicating that chromatin structure is important to this regulation.

To date, only SIN4 has been linked to changes in chromatin structure (Jiang and Stillman, 1992). Mutations in SIN4 alter the topoisomer density of plasmids in yeast. Also, an increased sensitivity to nuclease digestion has been observed in

Table 3. Global Regulators of Yeast Transcription

Factor	**a** *cell-specific*	*haploid-specific*	*glucose repressible*	*PHO5*
SSN6	+	+	+	?
TUP1	+	+	+	?
SIN3	+	+	?	+
SIN4	+	+	+	+
RPD3	+	+	?	+
TOP3	+	?	+	?

Note: A (+) symbol indicates factors which are required for proper regulation of a particular gene or set of genes.
A (?) indicates that the consequences of the absence or mutation of a given factor in the regulation of these
genes has not yet been reported.

chromatin isolated from strains bearing a disruption of the SIN4 gene (Roth, 1994). The molecular basis for these changes is currently unknown, but an increased accessibility of chromatin in sin4⁻ cells, reflected by increased nuclease sensitivity, might explain the activation of UAS-less promoters in these cells.

III.   A WORKING MODEL FOR REPRESSION OF a CELL-SPECIFIC GENES

The above experiments suggest a mechanism for repression for the **a** cell specific genes (Figure 6). Binding of the α2/MCM1 complex to its operator recruits SSN6/TUP1 to the promoter region. This complex causes a nucleosome to be positioned downstream, incorporating cis-acting promoter elements, through a process requiring H4 amino-terminal residues. Positioning is propagated beyond this first nucleosome, creating a region of repressive chromatin.

The propagation of positioned nucleosomes may reflect folding of these nucleosomes into a higher order structure. This folding is likely to involve the H4 amino terminus since nucleosome stability is compromised throughout the STE6 locus in the presence of the H4 amino terminal mutations.

Formation of a higher order structure might be facilitated by association with a nuclear "anchor." One intriguing possibility, for example, is that SSN6/TUP1 (or associated proteins) are localized to a specific nuclear structure, providing a focal point for the organization of chromatin. Other TPR proteins appear to be associated with the nuclear matrix (Goebl and Yanagida, 1991). However, SSN6/TUP1 containing complexes are soluble in crude cellular extracts, and SSN6 does not cofractionate with components of the nuclear matrix (Williams et al., 1991). Therefore, association of SSN6/TUP1 with the matrix, or any other nuclear structure, is completely speculative at present. Additional studies are needed to discern the subnuclear localization, if any, of SSN6/TUP1 and whether such localization is important to SSN6/TUP1 function.

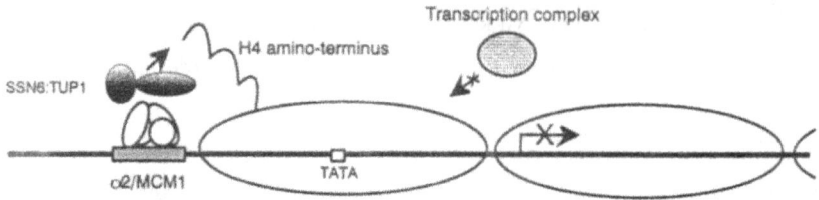

Figure 6. A working model for repression of **a** cell specific gene expression. Binding of α2/MCM1 to its operator upstream of the **a** cell specific genes "recruits" SSN6/TUP1 complexes to these promoters. This macromolecular complex then causes nucleosomes to be positioned 15 bp downstream of the operator through a process facilitated by H4 amino-terminal sequences. Nucleosome positioning is propagated for several nucleosomes downstream, suggesting that the **a** cell specific genes may be folded into higher order chromatin structures.

A. Repression of Transcription by RNA Polymerases I, II, and III

The above model should be considered in context with what is known about the ability of α2/MCM1 to confer repression on various promoters, and the organization of chromatin at these promoters.

In addition to the five **a** cell specific genes regulated by α2/MCM1, heterologous pol II promoters are uniformly repressed by α2/MCM1 in the presence of the α2/MCM1 operator, regardless of whether the operator is located upstream or downstream of UAS elements. The chromatin structure of only a few of these genes has been examined, but in each of these cases, nucleosome positioning accompanies repression (Figure 7).

The α2/MCM1 complex is also able to repress transcription by RNA polymerase I (Herschbach and Johnson, 1993). A single gene is transcribed by this promoter in yeast, that encoding the 35S rRNA gene. Recently, Herschbach and Johnson (1993) demonstrated that a plasmid borne 35S rRNA minigene is subject to α2/MCM1 repression and that this repression also requires SSN6 and TUP1. As suggested by Herschbach and Johnson, these findings may indicate that a shared component of pol I and pol II basal transcriptional apparati is targeted by the α2/MCM1 repression complex. However, these results are also consistent with the formation of repressive chromatin. Although the chromatin structure of the repressed rRNA minigene was not determined, inactive rRNA genes in other organisms are packaged into ordered chromatin structures which are not associated with the active forms of these genes (Prior et al., 1983; Conconi et al., 1989).

In contrast to genes transcribed by pol I or pol II, genes transcribed by RNA polymerase III can escape repression by chromatin, usually by prevention or disruption of nucleosome formation (see Wolffe, 1992 for review). For example, Morse and colleagues (1992) directly demonstrated that an actively transcribed tRNA gene interferes with nucleosome positioning *in vivo*, including those nor-

Pol II

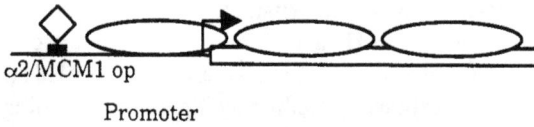

Nucleosomes Positioned

Repressed

α2/MCM1 op

Promoter

Pol III

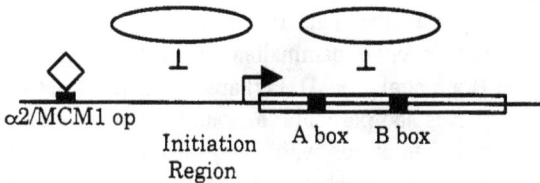

Nucleosomes Not Positioned

Not repressed

α2/MCM1 op

Initiation   A box   B box
Region

Figure 7. Nucleosome positioning and repression by α2/MCM1 are ineffective at pol III promoters. When the α2/MCM1 operator is located upstream of a pol II promoter, repression is accompanied by nucleosome positioning for all genes examined to date. In contrast, nucleosomes are not positioned when the operator is located upstream of a pol III transcribed gene (Morse et al., 1992), and pol III promoters are not repressed by α2/MCM1 (Morse et al., 1992; Herschbach and Johnson, 1993).

mally positioned adjacent to the α2/MCM1 operator (Figure 7). Not surprisingly, this tRNA gene is refractory to α2/MCM1 repression (Morse et al., 1992), as are two other pol III transcribed genes (another tRNA gene and the yeast U6 RNA gene) (Herschbach and Johnson, 1993). How these genes escape nucleosomal repression is unclear. Perhaps the abundance of pol III transcription factors, such as TFIIIB and TFIIIC, facilitate a rapid assembly of transcription complexes which "out competes" nucleosome formation and repression.

Genes transcribed by pol III do not always escape repression. A tRNA gene can be silenced by integration into the yeast silent mating loci (Schnell and Rine, 1986). This result may reflect an overall difference in the organization of chromatin at *silenced* loci versus loci which are constitutively *repressed*. Silenced genes are, by definition, never expressed under any conditions or in any cell type. Repressed genes, however, can be re-programmed for expression in response to developmental or environmental signals. A higher degree of chromatin organization almost certainly is associated with silenced genes to ubiquitously prohibit transcription.

IV. OUTSTANDING QUESTIONS

Obviously several questions regarding the mechanism of nucleosome positioning adjacent to the α2 operator remain unanswered. What, for example, is the role of the H4 amino-terminus in this process? This H4 domain could directly interact with the repressor complex to establish nucleosome location. Alternatively, the H4 "tail" may simply stabilize nucleosome locations by facilitating higher order folding of the repressed regions. In either case, acetylation/deacetylation of the H4 amino terminus could easily influence both nucleosome positioning and repression. For example, the H4 amino terminus is important to silencing at the yeast silent mating loci and telomeres, and a hypoacetylation of H4 is observed at the silent loci (Braunstein et al., 1993). Moreover, H4 acetylated at LYS16 is nonrandomly distributed in Drosophila (Turner et al., 1992) and mammalian chromatin (Jeppesen et al., 1992), being enriched in some transcriptionally active loci (such as the Drosophila male X chromosomes, see Turner et al., 1992) and absent in transcriptionally inactive regions (such as the inactive mammalian female X chromosome, see Jeppesen and Turner, 1993; Bone et al., 1994). Perhaps interactions between regulatory macromolecular complexes and particular histone H4 domains set the stage for the formation of higher order structures, which may then be further folded into even more complex domains by silencing or heterochromatin-specific proteins.

Another question relevant to the establishment and maintenance of repressive regions of chromatin is the importance (or unimportance) of nucleosome formation and placement during DNA replication. The yeast PHO5 gene, for example, is organized into four positioned nucleosomes under conditions of repression (Almer and Horz, 1986). How these nucleosomes are positioned is unknown, but they are selectively removed upon binding of activator proteins, and are re-established when these proteins are no longer present (Almer et al., 1986). The loss and "retrieval" of these nucleosomes occurs independently of replication (Schmid et al., 1992). Since PHO5 is an inducible gene, subject to environmental signals (i.e., phosphate concentrations) rapid changes in chromatin structure are consistent with rapid changes in gene expression. However, constitutively repressed genes, such as the **a** cell-specific genes, might be expected to require a more extensive "reprogramming" involving replication to allow subsequent transcription.

V. RELEVANCE TO OTHER LOCI AND OTHER SYSTEMS

The organization of chromatin by SSN6/TUP1 may well be important to the regulation of other genes repressed by these factors, including the glucose repressible genes and the haploid specific genes. Examination of the chromatin structure and regulation of SUC2, a glucose repressible gene, supports this idea (Perez-Ortin et al., 1987). Moreover, two specific proteins, SNF2 (SWI2) and SNF5, are required to activate SUC2, and mutations in SNF2 are suppressed by mutations in histone H3 (Peterson and Herskowitz, 1992). These data are consistent with a model wherein SSN6/TUP1 are recruited to the SUC2 promoter (probably by the sequence specific repressor MIG1) to organize nucleosome locations important to repression. This repressive chromatin structure must then be disrupted by SNF2/SNF5 before SUC2 can be activated.

SSN6, TUP1, α2, and MCM1 are all homologous to proteins in other organisms. The proteins α2 and MCM1 are related to DNA binding proteins containing a homeodomain (α2, see Porter and Smith, 1986) or a MADS box motif (MCM1, see Passmore et al., 1989). Like SSN6 and TUP1, TPR-containing proteins and β-transducin-like proteins also appear to be functionally linked in other organisms (Goebl and Yanagida, 1991). The recruitment of SSN6 and TUP1 by specific DNA binding proteins, and the manipulation of chromatin, truly a "global repressor," by these factors in yeast may provide important pardigms for the regulation of diverse processes in other systems.

ACKNOWLEDGMENTS

The author would like to thank Robert Simpson, in whose lab the work described in this review was performed. I would also like to thank R. Simpson, J. Cooper, and R. Morse for reading of the manuscript, and Alisha Tizenor for graphics presented in figures 1, 2, 4, and 5.

REFERENCES

Adams, C.C., & Workman, J.L. (1993). Nucleosome displacement in transcription. Cell 72, 305–308.

Almer, A., & Horz, W. (1986). Nuclease hypersensitive regions with adjacent positioned nucleosomes mark the gene boundaries of the PHO5/PHO3 locus in yeast. EMBO J. 5, 2681–2687.

Almer, A., Rudolph, H., Hinnen, A. & Horz, W. (1986). Removal of positioned nucleosomes from the yeast PHO5 promoter upon PHO5 induction releases additional upstream activating DNA elements. EMBO J. 5, 2689–2696.

Bone, J.R., Lavender, J., Richman, R., Palmer, M.J., Tyrner, B.M., & Kuroda, M.I. (1994). Acetylated histone H4 on the male X chromosome is associated with dosage compensation in Drosophila. Genes and Dev. 8, 96–104.

Braunstein, M., Rose, A.B., Holmes, S.G., Allis, C.D., & Broach, J.R. (1993). Transcriptional silencing in yeast is associated with reduced nucleosome acetylation. Genes and Dev. 7, 592–604.

Campbell, J.L. (1988). Eukaryotic DNA replication: yeast bares its ARSs. Trends in Biochem. Sci. 13, 212–217.

Chen, S., West, R.W.J., Johnson, S.L., Gans, H., Kruger, B., & Ma, J. (1993). TSF3, a global regulatory protein that silences transcription of yeast GAL genes, also mediates repression by α2 repressor and is identical to SIN4. Mol. Cell Biol. 13, 831–840.

Conconi, A., Widmer, R.M., Koller, T., & Sogo, J.M. (1989). Two different chromatin structures coexist in ribosomal RNA genes throughout the cell cycle. Cell 57, 753–761.

Cooper, J.P., Roth, S.Y., & Simpson, R.T. (1994). The global transcriptional regulators, Ssn6p and Tup1p, play distinct roles in the establishment of a repressive chromatin structure. Genes and Dev. 8, 1400–1410.

Durrin, L.K., Mann, R.K., Kayne, P.S., & Grunstein, M. (1991). Yeast histone H4 N-terminal sequence is required for promoter activation *in vivo*. Cell 65, 1023–1031.

Felsenfeld, G. (1992). Chromatin as an essential part of the transcriptional mechanism. Nature 355, 219–224.

Fong, H.K.W., Hurley, J.B., Hopokins, R.S., Miake-Lye, R., Johnson, M.S., Doolittle, R.F., & Simon, M.I. (1986). Repetitive segmental structure of the transducin beta subunit: Homology with the CDC4 gene and identification of related mRNAs. Proc. Natl. Acad. Sci. USA 83, 2162–2166.

Goebl, M., & Yanagida, M. (1991). The TPR snap helix: a novel protein repeat from mitosis to transcription. Trends Biochem. Sci. 16, 173–177.

Herschbach, B.M., & Johnson, A.D. (1993). The yeast α2 protein can repress transcription by RNA polymerases I and II but not III. Mol. Cell Biol. 13, 4029–4038.

Herskowitz, I. (1989). A regulatory hierarchy for cell specialization in yeast. Nature 342, 749–757.

Jeppesen, P., & Turner, B.M. (1993). The inactive X chromosome in female mammals is distinguished by a lack of histone H4 acetylation, a cytogenetic marker for gene expression. Cell 74, 281–289.

Jiang, Y.W., & Stillman, D.J. (1992). Involvement of the SIN4 global transcriptional regulator in the chromatin structure of Saccharomyces cerevisiae. Mol. Cell Biol. 12, 4503–4514.

Keleher, C.A., Goutte, C., & Johnson, A.D. (1988). The yeast cell type specific repressor α2 acts cooperatively with a non cell type specific protein. Cell 53, 927–936.

Keleher, C.A., Redd, M.J., Schultz, J., Carlson, M., & Johnson, A.D. (1992). SSN6-TUP1 is a general repressor of transcription in yeast. Cell 68, 709–719.

Lemontt, J.F., Fugit, D.R., & Mackay, V.L. (1980). Pleiotropic mutations at the TUP1 locus that affect the expression of mating type dependent functions in Saccharomyces cerevisiae. Genetics 94, 899–920.

Morse, R.H., Roth, S.Y., & Simpson, R.T (1992). A transcriptionally active tRNA gene interferes with nucleosome positioning *in vivo*. Mol. Cell Biol. 12, 4015–4025.

Passmore, S., Randolph, E., & Tye, B.-K. (1989). A protein involved in minichromosome maintenance in yeast binds a transcriptional enhancer conserved in eukaryotes. Genes and Dev. 3, 921–935.

Patterton, H.G., & Simpson, R.T. (1994). Nucleosomal location of the STE6 TATA box and mat Matαzp-mediated repression. Mol. Cell Biol. 14, 4002–4010.

Perez-Ortin, J.E., Estruch, F., Matanllana, E., & Franco, L. (1987). Structure of the SUC2 gene3 changes upon derepression. Comparison between chromosomal and plasmid-inserted genes. Nucl. Acids Res. 15, 6937–6956.

Peterson, C.L., & Herskowitz, I. (1992). Characterization of the yeast SWI1, SWI2 and SWI3 genes, which encode a global activator of transcription. Cell 68, 573–583.

Porter, S.D., & Smith, M. (1986). Homeo-domain homology in yeast Mata2 repressor is essential for repressor activity. Nature 320, 766–768.

Prior, C.P., Cantor, C.R., Johnson, E.M., Littau, V.C., & Allfrey, V.G. (1983). Reversible changes in nucleosome structure and histone H3 accessibility in transcriptionally active and inactive states of rDNA chromatin. Cell 34, 1033–1042.

Roth, S.Y., Dean, A., & Simpson, R.T. (1990). Yeast α2 repressor positions nucleosomes in TRP1/ARS1 chromatin. Mol. Cell Biol. 10, 2247–2260.

Roth, S.Y., Shimizu, M., Johnson, L., Grunstein, M., & Simpson, R.T. (1992). Stable nucleosome positioning and complete repression by the yeast α2 repressor are disrupted by amino-terminal mutations in histone H4. Genes and Dev. 6, 411–425.

Roth, S.Y. (1994). Unpublished observations.

Schmid, A., Fascher, K., & Horz, W. (1992). Nucleosome disruption at the yeast PHO5 promoter upon PHO5 induction occurs in the absence of DNA replication. Cell 71, 853–864.

Schnell, R., & Rine, J. (1986). A position effect on the expression of a tRNA gene mediated by the SIR genes in Saccharomyces cerevisiae. Mol. Cell. Biol. 6, 494–501.

Schultz, J., & Carlson, M. (1987). Molecular analysis of SSN6, a gene functionally related to SNF1 kinase of Saccharomyces cerevisiae. Mol. Cell Biol. 7, 3637–3645.

Schultz, J., Marshall-Carlson, L., & Carlson, M. (1990). The N terminal TPR region is the functional domain of SSN6, a nuclear phosphoprotein of Saccharomyces cerevisiae. Mol. Cell Biol. 10, 4744–4756.

Shimizu, M., Roth, S.Y., Szent-Gyorgyi, C., & Simpson, R.T. (1991). Nucleosomes are positioned with base pair precision adjacent to the α2 operator in Saccharomyces cerevisiae. EMBO J. 10, 3033–3041.

Sikorski, R.S., Boguski, M.S., Goebl, M., & Hieter, P. (1990). A repeating amino acid motif in CDC23 defines a family of proteins and a new relationship among genes required for mitosis and RNA synthesis. Cell 60, 307–317.

Simpson, R.T. (1990). Nucleosome positioning can effect the function of a cis-acting DNA element *in vivo*. Nature 343, 387–389.

Simpson, R.T. (1994). Personal communication.

Thoma, F., & Simpson, R.T. (1985). Local protein-DNA interactions may determine nucleosome positions on yeast plasmids. Nature (London) 315, 250–252.

Thoma, F., Bergman, W., & Simpson R.T. (1984). Nuclease digestion of circular TRP1/ARS1 chromatin reveals positioned nucleosomes separated by nuclease sensitive regions. J. Mol. Biol. 177, 715–733.

Turner, B.M., Birley, A.J., & Lavender, J. (1992). Histone H4 isoforms acetylated at specific lysine residues define individual chromosomes and chromatin domains in Drosphila polytene nuclei. Cell 69, 375–384.

Vidal, M., & Gaber, R.F. (1991). RPD3 encodes a second factor required to achieve maximum positive and negative transcription states in *Saccharomyces cerevisiae*. Mol. Cell Biol. 11, 6317–6327.

Vidal, M., Strich, R., Esposito, R.E., & Gaber, R.F. (1991). RPD1 (SIN3/UME4) is required for maximal activation and repression of diverse yeast genes. Mol. Cell Biol. 11, 6306–6316.

Wallis, J.W., Chrebet, G., Brodsky, G., Rolfe, M., & Rothstein, R. (1989). Hypercombination mutation in *S. cerevisiae* identifies a novel eukaryotic topoisomerase. Cell 58, 409–419.

Williams, F.E., & Trumbly, R.J. (1990). Characterization of TUP1, a mediator of glucose repression in *Saccharomyces cerevisiae*. Mol. Cell Biol. 10, 6500–6511.

Williams, F.E., Varanasi, U., & Trumbly, R.J. (1991). The CYC8 and TUP1 proteins involved in glucose repression in *Saccharomyces cerevisiae* are associated in a protein complex. Mol. Cell Biol. 11, 3307–3316.

Wilson, K.L., & Herskowitz, I. (1986). Sequences upstream of the STE6 gene required for its expression and regulation by the mating type locus of *Saccharomyces cerevisiae*. Proc. Natl. Acad. Sci. USA 83, 2536–2540.

Winston, F., & Carlson, M. (1992). Yeast SNF/SWI transcriptional activators and the SPT/SIN chromatin connection. Trends in Genetics 11, 387–391.

Wolffe, A.P. (1992). Chromatin: structure and function. Academic Press, London.

Part III

Transcriptional Activation from Nucleosomal
Templates

Chapter 6

Modulation of Transcription Factor Access and Activity at the MMTV Promoter *In Vivo*

TREVOR K. ARCHER and JOSEPH S. MYMRYK

The Nucleus
Volume 1, pages 123–150.
Copyright © 1995 by JAI Press Inc.
All rights of reproduction in any form reserved.
ISBN: 1-55938-940-0

ABSTRACT

The Mouse Mammary Tumor Virus (MMTV) system offers an excellent opportunity to address the complex interplay between transcription factors and chromatin in that it adopts a highly reproducible chromatin structure *in vivo* and *in vitro* (Richard-Foy and Hager, 1987; Archer et al., 1991). Within the MMTV long terminal repeat (LTR) there is a coincidence between an extended hormone-dependent hypersensitivity region (HSR), encompassing the steroid receptor, nuclear factor 1 (NF1) and octamer transcription factor (OTF) binding sites, and a phased nucleosome (Nuc B) (Zaret and Yamamoto, 1984; Richard-Foy and Hager, 1987). This hormone-dependent remodeling of chromatin is accompanied by the loading of ubiquitous transcription factors and the establishment of an initiation complex (Cordingley et al., 1987b). These results were combined into a hypothesis that predicts that the prior assembly of promoter sequences into chromatin acts to restrict transcription factor binding (Cordingley et al., 1987a; Hager, 1988; Archer et al., 1989). To assess the veracity of this proposal, we have undertaken a rigorous examination of nucleosomal and non-nucleosomal MMTV templates *in vivo* (Archer et al., 1992; Lee and Archer, 1994). Using transient transfection and *in vivo* footprinting techniques, we demonstrate that the access of transcription factors to the LTR *in vivo* is determined by the nucleosomal organization of the DNA. Further, the organization of the promoter as chromatin serves to restrict its hormone responsive phenotype (Smith et al., 1993; Archer et al., 1994a, 1994b). Finally, our results suggest that the transcription factor complement, including steroid receptors, of a cell has a profound influence on the *in vivo* organization of the LTR as chromatin (Mymryk et al., 1995).

I. INTRODUCTION

The organization of information in eukaryotic cells is complex. In the nucleus there is a juxtaposition of a large number of transacting proteins and DNA packaged as chromatin (van Holde, 1988; Wolffe, 1992b). This chromatin arrangement not only provides a mechanism whereby DNA sequences may be efficiently and economically compacted, but also an opportunity to regulate the access of transacting proteins (Elgin, 1988; Hager, 1989; Wolffe, 1991; Felsenfeld, 1992; Workman and Buchman, 1993).

Much of our knowledge of how transacting proteins act in the context of chromatin has come from genetic analysis of DNA in lower eukaryotes such as *Sacchromyces cerevisiae* (Pederson and Simpson, 1988; Grunstein, 1990a). Detailed characterization of chromatin has been derived from biochemical fractionation of histone and nonhistone proteins and their analysis using small animal viruses, such as SV40 and polyoma virus, in cell culture to study replication and transcription (Oudet et al., 1975; Hager, 1988; van Holde, 1988; DePamphilis, 1993). Further advances have come from studies utilizing *in vitro* assembly systems derived from *Xenopus laevis* oocyte and eggs, *Drosophila melanogaster* embryos and mammalian cells (Shimamura et al., 1988; Smith and Stillman, 1989; Kadonaga, 1990; Becker et al., 1991; Almouzni and Wolffe, 1993a).

A particularly fruitful approach has been the creation of high copy number cell lines utilizing animal viruses, such as the bovine papilloma virus (BPV) and the mouse mammary tumor virus (MMTV), by Hager and colleagues (Ostrowski et al., 1983; Cordingley et al., 1987b; Richard-Foy and Hager, 1987; Archer et al., 1989). The multiplicity of copy number improves the "signal to noise" ratio and allows one to conduct analysis that would not be possible on single copy genes (the advent of polymerase chain reaction [PCR] methods has significantly reduced this problem). Our studies on the transcriptional regulation of the MMTV LTR have reinforced the concept that the organization of regulatory sequences as chromatin modulates the access of ubiquitous and inducible transcription factors to the promoter *in vivo* (Archer et al., 1992; Lee & Archer, 1994).

We have limited the scope of this review to recent developments in our understanding of the mechanisms by which chromatin influences transcription factor interactions *in vivo*, focusing largely on our recent work during the last two years, using the MMTV model system (Archer et al., 1992, 1994a, 1994b; Archer, 1993; Smith et al., 1993; Lee and Archer, 1994; Mymryk et al., 1995). The reader is directed to recent reviews for a more detailed exposition of previous work in this system as well as a discussion of the roles of histone modifications, histone H1 and mechanisms of chromatin remodeling in the process of activation (Archer et al., 1989; Hager and Archer, 1991; Hager et al., 1993).

A. Organization of DNA in the Nucleus of Eukaryotes

In eukaryotic cells, DNA is packaged in a chromatin hierarchy that begins with the initial wrapping around core histones and ends with the fully condensed structure of chromosomes (Kornberg, 1974; van Holde, 1988; Arents et al., 1991; Wolffe, 1992b). The fundamental unit of this packaging system is the nucleosome, which contains 146 base pairs (bp) of DNA wrapped on the octamer core (McGhee and Felsenfeld, 1980; Richmond et al., 1984; Arents and Moudrianakis, 1993). As DNA is polymerized, parental and daughter strands are quickly organized into linear arrays of nucleosomes, the well-known "beads on a string" model of chromatin (Olins and Olins, 1974). This structure provides eukaryotes with a level of genetic regulation unavailable to prokaryotes, and a direct role for chromatin in regulating gene expression has been proposed (Wolffe, 1990; Felsenfeld, 1992). In support of this hypothesis, deletion of the genes for one or more of the core histones prevents normal chromatin formation and deregulates a variety of inducible genes (Grunstein, 1990b).

Many lines of evidence suggest that highly compact regions of DNA are transcriptionally silent, with transcription originating near areas of disorder in the chromatin (Stalder et al., 1980; McGhee et al., 1981; Groudine and Weintraub, 1982; Elgin, 1988). These regions of active gene expression are characterized by an increased accessibility to various nucleases, a property indicative of a more open conformation of the chromatin. In certain cases, regions of promoters are coincident with a hypersensitivity to DNase I, and are often correlated with a loss of nucleosomes and binding by transcription factors (Wu, 1980; McGhee et al., 1981; Elgin, 1984, 1988; Zaret and Yamamoto, 1984; Almer and Hörz, 1986; Richard-Foy and Hager, 1987; Gross and Garrard, 1988). One such system that exhibits many of these characteristics, and which is amenable to detailed biochemical analysis, is the MMTV promoter (Hager, 1988).

B. The MMTV LTR as a Model for Regulated Transcription

MMTV is a B-type retrovirus with a single-stranded RNA genome that integrates into the host genetic material as a double-stranded DNA provirus following reverse transcription (Panganiban, 1985). Glucocorticoids stimulate the production of virus particles and viral mRNA accumulation via an increase in transcription initiation (Ringold et al., 1975; Scolnick et al., 1976; Young et al., 1977). A series of studies using partially purified glucocorticoid receptor (GR) and purified MMTV DNA have demonstrated that the receptor binds specifically to sites between 100 and 200 bp upstream from the cap site (Chandler et al., 1983; von der Ahe et al., 1985; Otten et al., 1988; Beato, 1989) (see Figure 1, Part I). Transfection experiments and mutational analysis have revealed that this relatively small region of DNA is sufficient to confer hormonal regulation of transcription to a fused heterologous gene (Huang et al., 1981; Chandler et al., 1983; Majors and Varmus, 1983). However, it is important to note that this sequence functions as a promiscuous hormone response element (HRE), such that progestins, mineralocorticoids, and androgens will also activate transcription (von der Ahe et al., 1985; Cato et al., 1986, 1987; Darbre et al., 1986; Arriza et al., 1987; Ham et al., 1988; Otten et al.,1988). Indeed, the fact that expression of MMTV in the mammary gland is elevated during pregnancy suggests that progesterone may be more biologically relevant than glucocorticoids for regulation of MMTV gene expression (Munoz-Bradham and Bolander, 1989).

To regulate transcription, steroid receptors must communicate with other components of the transcription machinery to establish the initiation complex (Green and Chambon, 1988). A detailed analysis of the MMTV LTR has revealed that, along with HREs, binding sites for several transcription factors are located in the proximal portion of the promoter (Huang et al., 1981; Chandler et al., 1983; von der Ahe et al., 1985; Beato, 1989; Hager and Archer, 1991) (Figure 1, Part I). The most extensively characterized member of this group of regulatory proteins is NF1; a protein originally examined for its role in adenovirus replication and activation of the chicken lysozyme gene (Rosenfeld and Kelly, 1986; Jones et al., 1987; Cordingley and Hager, 1988; Mermod et al., 1989). In addition, binding sites for the ubiquitous OTF and the TFIID complex are present (Kühnel et al., 1986; Beato,1989). *In vitro* transcription and gene transfer experiments have delineated important roles for each of these transacting proteins (Schule et al., 1988; Gowland and Buetti, 1989; Tsai et al., 1990; Brüggemeier et al., 1991). Importantly, coop-

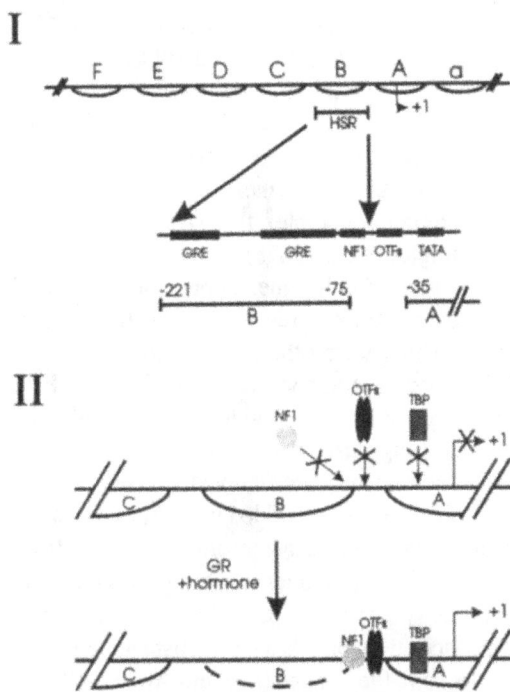

Figure 1. I. Schematic representation of the nucleosomal organization of the MMTV LTR. The MMTV LTR is reproducibly assembled into a phased array of six nucleosomes (A–F) *in vivo* (Richard-Foy and Hager, 1987). The hormone inducible hypersensitive region (HSR) is positioned over nucleosome B (position -221 to -75)(Zaret and Yamamoto, 1984; Richard-Foy and Hager, 1987). The proximal promoter region consists of binding sites for the steroid receptors (GRE), NF1, the octamer transcription factors (OTFs), and the TATA binding protein (TBP).

II. Model of hormone activated transcription from the MMTV LTR. Prior to hormone addition the nucleosomal structure of the LTR excludes binding by the transcription factors NF1, OTF, and TBP. Hormone activation of the GR leads to a change in the organization of nucleosome B (indicated by the dotted outline) that permits loading of NF1, the OTFs, and the assembly of the transcription initiation complex (Cordingley et al., 1987a; Hager, 1988; Archer et al., 1989).

erative interactions between NF1 and the GR are required for maximal glucocorti-coid stimulation (Buetti and Kühnel, 1986; Buetti et al., 1989) while the OTF sites are necessary for basal but not induced transcription (Buetti, 1994).

The choice of the MMTV promoter for analysis of transcription factor binding to chromatin *in vivo* flows from several features of this system. Glucocorticoid stimulation of transcription initiation commences with the generation of a HSR which encompasses the HREs in the proximal promoter (Zaret and Yamamoto,

1984; Richard-Foy and Hager, 1987). Analysis of MMTV BPV chimeras in mouse cells demonstrated that the MMTV LTR adopted a phased-array of six nucleosomes in the absence of hormonal stimulation, and that the second or B nucleosome of this phased assay was positioned over the HSR (Figure 1, Part I) (Richard-Foy and Hager, 1987). Further experiments demonstrated that the hormonal induction of MMTV was accomplished by the hormone-dependent loading of transcription factor NF1 and a preinitiation complex that was coincident with the region of hypersensitivity (Cordingley et al., 1987b).

These studies led Hager and colleagues to propose that the organization of the MMTV promoter as chromatin was responsible for the exclusion of transacting factors, such as NF1, which were otherwise abundant and competent for DNA binding in the nuclei prior to hormone treatment (Figure 1, Part II). Our demonstration that the ability of chromatin to modulate transcription factor access could be reproduced *in vitro* using purified components clearly supports this concept (Archer et al., 1991). Subsequently, we demonstrated that transient introduction of DNA into cells, such that it is not assembled into a specific chromatin, permits the promiscuous binding of transcription factors otherwise excluded from their cognate sites by the organization of the promoter into chromatin (Archer et al., 1992).

In addition, using this system it was possible to show that the cessation, as well as the initiation of transcription follows a mechanism not observed with transiently transfected non-nucleosomal templates (Lee and Archer, 1994). Further, using *in vivo* analysis, we demonstrate that the specific chromatin arrangement of the LTR is apparently dependent upon a given cell type and transcription factor complement (Archer, 1994; Archer et al., 1994a, 1994b; Mymryk et al., 1995). Before turning to these investigations in detail, we will briefly recap observations made using *in vitro* reconstitution experiments (Archer et al., 1991) as well as the initial demonstrations of the differences between chromatin and nonchromatin templates *in vivo* (Archer et al., 1992).

II. EXPERIMENTAL OBSERVATIONS

A. Transcription Factor I Nucleosome Interactions on the MMTV Promoter *In Vitro*

The ability of glucocorticoids to stimulate transcription from the MMTV promoter *in vivo* predicts that the GR must be able to interact with its site assembled as chromatin (Ringold et al., 1975; Richard-Foy and Hager, 1987). This prediction was borne out in experiments demonstrating that the GR was able to interact with its cognate binding site on mononucleosome templates as well as naked DNA templates (Perlmann and Wrange, 1988). Subsequently, we and others demonstrated that the GR DNA binding domain and the progesterone receptor (PR) could interact with the HREs when assembled into a dinucleosome (Archer et al., 1991) or a mononucleosome (Piña et al., 1990). This latter set of experiments also

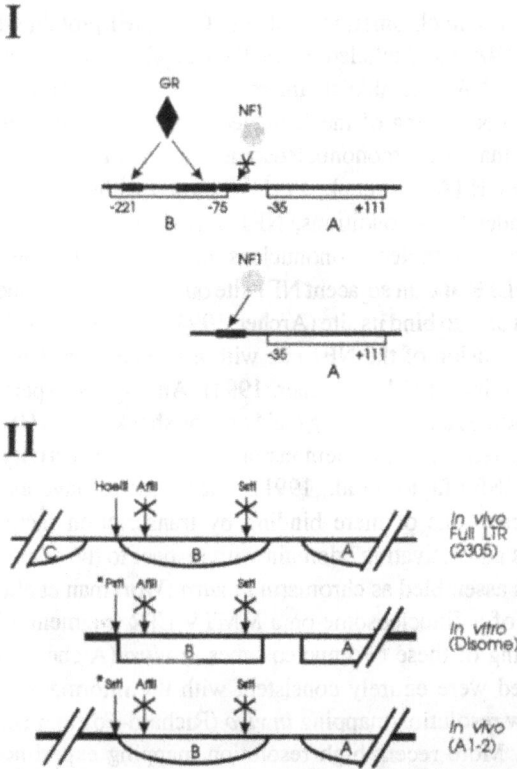

Figure 2. I. Analysis of factor binding to an *in vitro* assembled dinucleosome. *In vitro* assembly of a fragment of the MMTV LTR encompassing nucleosomes A and B into a dinucleosome excludes binding of NF1 but not the GR. *In vitro* assembly of a fragment of the MMTV LTR encompassing nucleosome A and the NF1 site, but not nucleosome B, into a mononucleosome fails to exclude NF1 binding (Archer et al., 1991; Archer, 1994). This suggests that the assembly of the NF1 site into chromatin is responsible for the exclusion of NF1. II. Assembly of the MMTV LTR into chromatin restricts the access of restriction enzymes to nucleosome B. The ability of restriction enzymes to access cleavage sites on the LTR was accessed *in vivo* on the full LTR in mouse 2305 cells, *in vitro* on a reconstituted disome, and *in vivo* on a disome in human A1–2 cells (Archer et al., 1991, 1994b; Mymryk et al., 1995). Restriction enzymes with sites in nucleosome B (Afl II and SstI) cleaved inefficiently. Restriction enzymes with sites adjacent to nucleosome B in the LTR (HaeIII) or on non-LTR DNA (PstI and *SstI) cleave efficiently.

demonstrated that transcription factor NF1 was unable to interact with its site under the same conditions (Figure 2, Part I). This result was significant for three reasons. First, it reproduced the *in vivo* observation that NF1 was excluded from the site in chromatin. Second, it demonstrated that transcription factors could have differential abilities to interact with chromatin templates. Thus, NF1 protein which has a high affinity for free DNA was excluded, while the GR which has a substantially lower affinity for free DNA was able to interact with its site. Third, in these latter experiments, the positioning of the B nucleosome differed by approximately 30 nucleotides from that of the mononucleosome, such that the NF1 site was partially within nucleosome B (Archer et al., 1991) or buried within the nucleosome (Piña et al., 1990). Under both conditions, NF1 was excluded from its site. In later experiments using assembled mononucleosomes containing the A nucleosome from the MMTV LTR and an adjacent NF1 site outside this positioned nucleosome, purified NF1 was able to bind its site (Archer, 1994) (Figure 2, Part I). This suggests that the partial inclusion of the NF1 site within nucleosome B is responsible for preventing its binding to DNA (Archer, 1994). Analogous experiments utilizing synthetic nucleosomes encompassing Gal4 or heat-shock factor (HSF) binding sites within a mononucleosome have demonstrated a similar selectivity between Gal4 derivatives and HSF (Taylor et al., 1991). These studies have also been used to distinguish contributions of mere binding by transcription factors and binding which introduces an "activating" domain with respect to the alleviation of repression of templates assembled as chromatin *in vitro* (Workman et al., 1991).

The assembly of a dinucleosome on a MMTV LTR fragment allowed for high resolution mapping of these two nucleosomes *in vitro* (Archer et al., 1991). The positions obtained were entirely consistent with the information generated previously using low resolution mapping *in vivo* (Richard-Foy and Hager, 1987) (see Figure 1 Part I). More recent high resolution mapping experiments using PCR techniques (Shimizu et al., 1991; Archer et al., 1992; Bresnick et al., 1992) confirmed the *in vitro* assignments of nucleosome positions (Archer et al.,1991). These experiments are consistent with the concept that information for the positioning of nucleosomes resides within the MMTV LTR. This premise was further supported by our recent characterization of a human cell line (T47D-A1-2) which contains a stably integrated MMTV LTR fragment that is similar to the truncated MMTV LTR used in the disome reconstitution experiments (Archer et al., 1994b). Restriction enzyme accessibility studies indicate that nucleosomes A and B present in this stably integrated LTR have adopted positions consistent with those observed using the dinucleosome *in vitro* (Archer et al., 1991) (see Figure 2, Part II).

B. *In Vivo* Transcription Factor Interactions on Chromatin and Nonchromatin Templates

To define more fully the contributions that chromatin makes to MMTV regulation, we have pursued a series of experiments utilizing the stable BPV MMTV cell lines in conjunction with transient transfection of an identical LTR driving a different reporter gene (Archer et al., 1992) (see Figure 3). The cells can then be selectively assayed for the transcriptional activation of each reporter gene, in most cases CAT (Gorman et al., 1982; Ostrowski et al., 1983) and luciferase (de Wet et

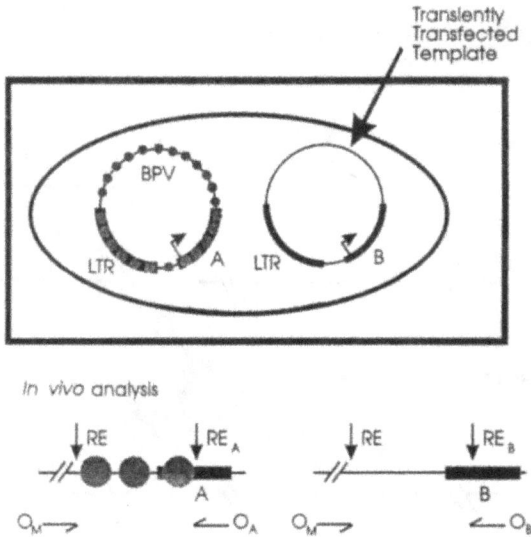

Figure 3. Experimental approach to simultaneously examine transcription factor loading on stable and transient templates. Cell lines stably maintaining the MMTV LTR driving expression of reporter gene A on a multicopy BPV episome are transfected with an identical LTR driving expression of a second reporter gene B. *In vivo* analysis of each template is possible by using restriction sites unique to each template (RE_A and RE_B) and an oligonucleotide O_M common to both LTRs or by oligonucleotides specific for each reporter (O_A and O_B) and a restriction enzyme (RE) common to both LTRs (Archer et al., 1992; Archer, 1993).

al., 1987; Lefebvre et al., 1991), and independently monitored for restriction enzyme hypersensitivities and transcription factor interactions in the presence and absence of the stimulating steroid, dexamethasone. The design of this experiment allows for the analysis of the transient and endogenous templates within the same nuclei providing a direct comparison between the stable and transient templates. The otherwise identical LTRs are distinguished in two PCR based *in vivo* Exo III footprinting protocols (Wu, 1985; Shimizu et al., 1991; Archer et al., 1992). In the first case we used oligonucleotides specific for either of the reporter genes (O_A or O_B), hence we were able to monitor transcription factor binding using an entry site common to the LTR for both reporter systems. In the second case, we used selective enzyme entry sites specific for either the transient or stable template (RE_A or RE_B) and an oligonucleotide which is complementary to both LTRs (O_M). We have previously demonstrated the absolute requirement for the restriction enzyme cleaved entry site in order to detect bound proteins by Exo III (Wu, 1985; Cordingley et al., 1987b; Archer et al., 1992). These experiments make use of a

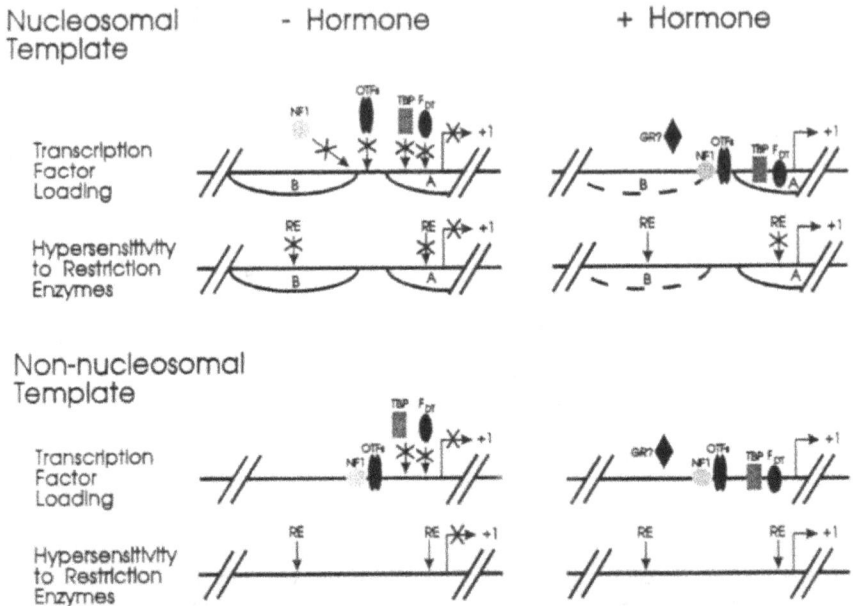

Figure 4. A comparison of nucleosomal and non-nucleosomal templates. Nucleosomal templates *in vivo*: Transcription factors NF1, OTFs, TBP, and F_{DT} bind to the promoter in a hormone dependent fashion. Hormone addition results in a remodeling of nucleosome B, such that it forms a HSR, allowing efficient cleavage by restriction enzymes cutting in this region. No hormone dependent changes in cleavage are observed for nucleosome A. Non-nucleosomal templates: Prior to hormone addition NF1 and OTFs bind to the template. Hormone treatment results in the binding of additional factors including TBP and F_{DT}. The template is hypersensitive to restriction enzyme cleavage in the absence or presence of hormone (Archer et al., 1991, 1992).

linear PCR protocol and thereby differ from traditional *in vivo* footprinting or ligation mediated PCR (Mueller and Wold, 1989). This approach is thus devoid of complications that might arise through geometric amplification of signals when comparing control versus induced states.

Analysis with micrococcal nuclease, DNAse I, or restriction enzymes failed to demonstrate an ordered array of nucleosomes on transiently introduced copies of the MMTV LTR (Archer et al., 1992; Archer, 1994). Indeed, the transiently introduced template appears to be constitutively hypersensitive to cleavage by restriction enzymes (Figure 4). Similar results have been obtained with calcium phosphate, electroporation, and DEAE dextran transfection protocols (Archer, 1994). One prediction from these experiments would be that this hypersensitive template would be available for binding by ubiquitous transcription factors such as NF1. This prediction was indeed borne out as *in vivo* footprinting experiments demonstrated that NF1 bound the transient template in the presence or absence of hormone (Archer et al., 1992). The observation that NF1 was present on the transient template under all conditions, while confirming the prediction from the hypersensitivity experiments, presented a paradox. Namely, the transiently transfected template was still hormone inducible although the transcription factor most firmly identified with hormone induction of transcription from this promoter was constitutively loaded! This was resolved by additional experiments showing that the formation of the preinitiation complex required steroid hormone, as the binding of the transcription factor TATA binding protein (TBP) was only detected in the presence of hormone (Archer et al., 1992). Thus, these experiments provide firm support for the original model that the exclusion of transcription factors from the LTR results from the presence of nucleosome B.

Subsequent analysis on the rat tyrosine amino transferase (TAT) promoter in hepatoma cells has produced a picture which closely parallels that of the MMTV promoter. Glucocorticoids induce a HSR that overlaps two positioned nucleosomes located 2–3 kb upstream from the transcription start site (Carr and Richard-Foy, 1990). Again in similarity to MMTV, *in vivo* footprinting experiments revealed the hormone dependent binding of hepatic transcription factor HNF5 at the hypersensitive sites (Rigaud et al., 1991). In the next series of studies we have built on our previous investigations to begin to dissect the mechanism by which transcription complexes are assembled and removed from the promoter *in vivo*.

C. Kinetics of the MMTV Response to Glucocorticoids Using Chromatin and Nonchromatin Templates

The induction of transcription by glucocorticoids from the MMTV promoter is extremely rapid such that an increase in mRNA initiation is observed within 5–10 minutes of hormone induction (Ucker and Yamamoto, 1984; Archer et al., 1994a). However, while the induction of transcription is extremely rapid, it is also transient such that initiation peaks by approximately one hour of hormone treatment and then subsequently declines to near basal levels within 8–10 hours. The formation of the HSR over nucleosome B follows similar kinetics (Archer et al., 1994a). What is surprising about this rapid deinduction is that it occurs in the presence of active hormone and the promoter becomes refractory to glucocorticoids for the next 48 hours. This suggests that some feature of chromatin remodeling requires an activity or protein that becomes unavailable for the reactivation of this promoter by glucocorticoids (Lee and Archer, 1994).

The rapid cessation of transcription seems to preclude a hormone related down regulation of receptor synthesis (Okret et al., 1986) as a mechanism for the loss of the activity at the promoter. Previous studies have suggested that oncogenes such as *ras* and *mos* may act to inhibit transcription from glucocorticoid responsive promoters such as that of the tyrosine aminotransferase (TAT) gene and the MMTV LTR (Jaggi et al., 1989; Qi et al., 1989). This is thought to occur via the down regulation of the activated receptor resulting in a similar transient activation. However, we observe the same kinetics in cell lines that do not express these oncogenes, indicating that this is an intrinsic feature of this promoter as chromatin.

We considered two alternative mechanisms for the generation of this phenomena. This "refractory phase" could be due either to a change of state involving the well-characterized chromatin transition (template-related), or to a phenomenon involving receptor cycling or modification (receptor-related). In an attempt to distinguish between these mechanisms, we investigated the kinetics of induction in response to a second hormone, progesterone, which was previously shown to be active at the same HRE via the PR (Gowland and Buetti, 1989).

D. Glucocorticoids but not Progestins Activate the Endogenous MMTV Promoter in Mouse Cells

We cotransfected mouse cell lines normally expressing the GR with a chicken PR expression vector, and asked whether concurrent or subsequent addition of both hormones were able to modulate the kinetics observed with glucocorticoid alone. The results of these experiments were quite surprising (Archer et al., 1994a)(see Figure 5). Under the conditions examined, the PR was unable to influence transcription from the MMTV promoter assembled as chromatin in our mouse cells. Consistent with this result, the PR was unable to induce hypersensitivity to restriction enzymes at nucleosome B in the endogenous MMTV template. However, it is well established that the MMTV promoter responds to progesterone when transiently transfected into mouse cells. Given the differences between transient and stable MMTV templates, we repeated these experiments by cotransfecting a MMTV luciferase reporter along with the PR expression vector. The results of these experiments confirm both previous observations. Namely, the MMTV LTR on a transient template was fully capable of responding to the PR while in the same nuclei the endogenous chromatin template was relatively inert to this agent (see Figure 5) (Archer et al., 1994a).

Because the transient transfection experiments would introduce PR into only a subpopulation of cells (5–20%), there exists the possibility that our failure to detect a response in the endogenous template was due simply to the inefficiency of introducing the receptor. To overcome this problem we utilized fluorescence-acti-vated cell sorting (FACS) to enrich cells that acquired and expressed the receptor cDNA (Smith et al., 1993). In agreement with the preceding experiments these cells do not respond to progesterone. Thus, the failure to respond does not result from a flaw of the experimental design, but rather suggests differences between the PR and GR.

40 ┬

Fold Induction Over Control

30 —

20 —

10 —

0 —

Dex   R5020   Dex   R5020

Transient (Luc)   Stable (CAT)

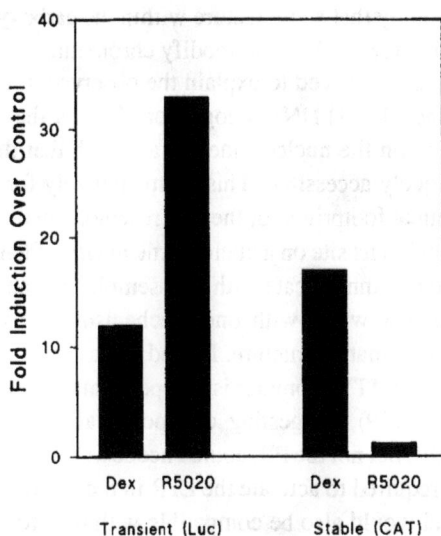

Figure 5. Progesterone and glucocorticoid induction of expression from the MMTV LTR from nucleosomal and non-nucleosomal templates. Mouse mammary cells which contain endogenous GR and a stably maintained MMTV LTR driving CAT expression, were transiently transfected with a MMTV luciferase (luc) reporter plasmid and chicken progesterone receptor. Cells were treated with dexamethasone (10^{-7} M) or the synthetic progestin R5020 (10^{-8} M) for 24 hours prior to harvesting. Whole cell extracts were used for CAT and luciferase assays (Archer, 1994).

These results revealed that the assembly of LTR sequences into stable nucleoprotein converts a promiscuous LTR into a promoter with restricted hormone responsiveness. This suggests that the GR may possess an activity or domain which by itself, or in concert with other factors, is necessary for the alteration in chromatin structure associated with activation of this promoter. Taken together, they support a template related model for the appearance of this "refractory" phase.

As an initial approach to understanding the potential differences between the glucocorticoid and progesterone receptor activities, we have examined the ability of a glucocorticoid receptor mutant (GR_{556}) to activate a chromatin template (Godowski et al.,1987). GR_{556} has previously been shown to be a constitutive activator in transient transfection experiments by virtue of the elimination of its hormone binding domain. The results of our assays confirm that GR_{556} is able to transactivate the transiently transfected MMTV template (Godowski et al., 1987). However, as observed with the PR, it is unable to stimulate transcription from the endogenous MMTV LTR or induce hypersensitivity (Archer et al., 1994a). This experiment argues strongly that some feature within the carboxy-terminal domain of the GR is responsible for its ability to modify chromatin.

Several models can be advanced to explain the observed differential activation potential of the GR and PR: (1) DNA recognition sites for the receptors could be differentially organized on the nucleosome surface such that the binding site for one receptor was uniquely accessible. This seems unlikely for the MMTV LTR, given the almost identical footprints for the two receptors on this template and the fact that the PR can bind to its site on a nucleosome *in vitro* (Pham et al., 1992). (2) The two receptors could communicate with or assemble the transcription initiation complex via different pathways, with one mechanism involving a component uniquely sensitive to chromatin structure. Indeed, it has been argued that progesterone activation of the MMTV promoter is independent of NF1 (Cato et al., 1986; Gowland and Buetti, 1989), suggesting one potential difference in activation mechanism. (3) The GR, but not the PR, could encode a previously uncharacterized activity or "domain" required to activate the LTR in the stable chromatin configuration. This last model would also be compatible with the aforementioned observations concerning the non-nucleosomal structure of the transient template and constitutive binding of NF1 (Archer et al., 1992). The lack of this putative "domain" may render the receptor unable to alter chromatin but still allow it to activate the transient template. If this is correct we can then formally separate the processes of chromatin remodeling from transcriptional activation.

E. Transcription Preinitiation Complex Assembly and Disassembly *In Vivo*

The failure of progesterone to stimulate the MMTV promoter in chromatin but activate a transient template suggests that a comparison of the kinetics of activation of the chromatin and transient templates may provide information on the mechanism by which transcription was initiated and then rapidly shut off. To elucidate the mechanism of this transient activation, we evaluated two models: (1) the transient nature of the response results from alterations of protein levels or activities of factors such as NF1, TBP, or the GR; and (2) the transient character of the preinitiation complex was a feature of the chromatin structure, that is, reformation of the nucleosome eliminates the complex, and thus would only be apparent on a stable chromatin template.

For these experiments we used the transfection protocol, outlined previously (Figure 3), that allows the two templates to be examined by a PCR/Exo III *in vivo* footprinting assay that distinguishes transient from stable templates. Plasmids were introduced by transient transfection at various times following hormone treatment to assess differences with respect to transcription factor levels and/or modifications that occur in response to hormone. As an example, cells containing endogenous copies of the LTR/CAT construct were treated for 24 hours with hormone, at which time the chromatin template is refractory to stimulation, and then transiently transfected with the LTR/luciferase template. The ability of the preinitiation complex to form and activate transcription from these two templates was then assessed.

As a prerequisite for these experiments, we have determined whether the activation process was influenced by the calcium phosphate transfection protocol. To do this we transiently transfected cells for periods as short as one hour in the presence of dexamethasone and asked: (1) Did the transfection process influence the hormone inducibility of the resident stable template? and (2) Would transcription factors load onto a newly introduced template? One hour after transfection, in the presence of calcium phosphate precipitate, cells were able to exhibit a full hormone response as indicated by changes in chromatin structure and the assembly of a preinitiation complex (PIC) on stable templates. Significantly, even within this short time frame, the transiently transfected template was fully occupied by NF1, indicating that binding of this transcription factor was rapid and stable with or without hormone (Lee and Archer, 1994).

An examination of the PIC *in vivo*, revealed that glucocorticoid stimulation led to the rapid appearance of NF1 on the stable promoter by one hour but it was absent at 24 hours. The results with the transient template were dramatically different as NF1 was bound under all conditions. This result was observed whether the transient template was maintained in cells treated with hormone for 24 hours or introduced into cells previously treated for 24 hours. Examination of other members of the PIC revealed a more complicated picture (Figure 6). In agreement with the analysis of NF1, we detected the OTFs and TBP on the stable template only at one hour of dexamethasome treatment. On the transient template OTFs behaved as did NF1 in that they were present under all conditions. The binding of TBP and F_{DT} (a novel transcription factor designated Factor Downstream of TBP) was only observed on the transient template in the presence of dexamethasone and occurred at both one hour and 24 hours.

We observed dramatically different results in the *in vivo* Exo III footprinting experiments if we introduced Exo III upstream or downstream of the cap site. Initiation of Exo III digestion with an entry site 5′ to the NF1 site detects only NF1. However, if we initiate Exo III digestion downstream of the cap site, we observed several additional members of the preinitiation complex, namely F_{DT}, TBP, and OTF besides NF1. The surprising ability to detect multiple factors in this assay could result from a variety of transcription factor template distributions, two of which are illustrated (Figure 7). First, it may derive from heterogeneity of the protein DNA complex due to multiple templates present in these cells. Alternatively, it may arise from the ability of Exo III nuclease to displace a fraction of the bound F_{DT}, TBP, and OTFs during digestion, but not NF1. We favor the second scenario for two reasons. First, if templates were present as a heterogenous population, we would expect to observe the OTFs and the TBPs in experiments where the Exo III entry site is 5′ to the NF1 site. In all cases we detect only NF1. Second, we have never observed Exo III digestion through the NF1 stop under the conditions used in our current (Archer et al., 1992; Lee and Archer, 1994) or previous experiments

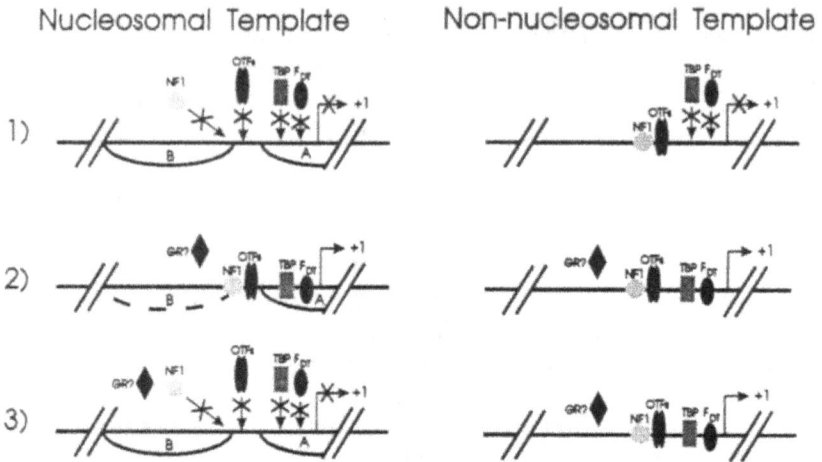

Figure 6. Kinetics of transcription factor loading on nucleosomal and non-nucleosomal templates. Nucleosomal templates: (1) In the absence of hormone, transcription factors NF1, OTFs, TBP, and F_{DT} are excluded from the LTR. (2) Within minutes of hormone addition, the B nucleosome is remodeled such that it becomes hypersensitive to restriction enzyme cleavage, NF1, OTFs, TBP, and F_{DT} binding occurs and transcription is activated. (3) By 24 hours after hormone addition, the "closed" organization of the B nucleosome has been re-established. The template is no longer hypersensitive, NF1, OTFs, TBP, and F_{DT} are no longer detected on the template, and transcription from the promoter has returned to basal levels. Non-nucleosomal templates: (1) Prior to hormone addition NF1 and OTFs bind to the template, but TBP and F_{DT} are not detected. Within one hour of hormone treatment TBP and F_{DT} binding is detectable and the transcription is activated. Unlike the stable chromatin template, 24 hours after hormone treatment NF1, OTFs, TBP, and F_{DT} binding is still detected on the transient template and transcription is still activated (Lee and Archer, 1994).

(Cordingley et al., 1987b; Archer et al., 1991). We suggest that the high affinity of NF1 binding to its site may make it a significantly better block to Exo III digestion than the other transcription factors on the promoter.

Previous *in vitro* experiments (Archer et al., 1991) have allowed us to subdivide transacting factors into those to which the DNA is "transparent" when in a nucleosome (i.e., GR) (Wolffe, 1992a), and those where the sites are "opaque" and thereby prevented from binding (i.e., NF1). Our *in vivo* footprinting experiments allow us to further subdivide the second category into two groups. The first group (i.e., NF1 and OTF) is able to interact with their sites provided they are not assembled as chromatin *in vivo*. The second group (i.e., TBP and F_{DT}) requires the action of ancillary factors such as the GR to occupy their sites for both nucleosomal and non-nucleosomal templates. For these latter factors, the absence of a specific chromatin structure alone is not sufficient to promote tight binding. However, in contrast to what is observed on the chromatin template, TBP and F_{DT} remain bound

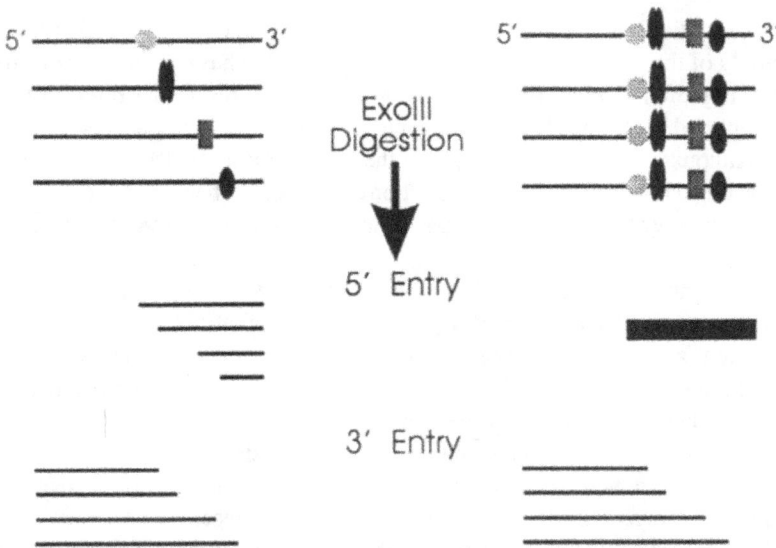

Figure 7. Schematic of alternative ways of generating multiple ExoIII stops. Multiple Exo III stops corresponding to bound protein factors could result from either of two mechanisms; either the templates are heterogeneous, or Exo III is capable of displacing the less strongly bound factors. Experimental evidence favors the latter model as the introduction of Exo III 5′ to the NF1 site detects only NF1, a result unlikely to be observed with a heterogeneous population of templates.

to the transient template even after 24 hours of hormone treatment. Thus, the establishment of a specific chromatin structure both prevents the promiscuous binding of transcription factors as well as removes them to re-establish the basal state.

Our Exo III footprinting assay does not reproducibly detect the GR bound to DNA suggesting that either the binding of the receptor to its target is transitory (i.e., a hit-and-run mechanism) or that it may be labile to this nuclease (Cordingley et al., 1987b; Archer et al., 1992). However, others have suggested that the GR once bound to a hormone remains in the nucleus under a similar time frame that we use for our experiments (0.5–24 hours) (Picard and Yamamoto, 1987). While we do not address the issue directly, our data on the hormone dependent loading of TBP and F_{DT} on the transient template, at times when the stable template is refractory, are consistent with the presence of active receptor in the nucleus. More direct evidence comes from the fact that transcription from the MMTV reporter gene transfected into "refractory" cells (cells treated for hormone for 24 hours and exhibiting no activity at the endogenous chromatin template) can be activated by hormone to a similar degree as in naive cells (Lee and Archer, 1994). Thus, our analysis of transiently transfected templates demonstrated that the activities of NF1, OTF, TBP, GR, and F_{DT} were neither down-regulated or diminished.

These studies link the loss of hypersensitivity and transcriptional activity to the assembly of the promoter as chromatin in a manner that has been done previously for the initiation of transcription. In this respect, these results are reminiscent of ones reported using the *Xenopus* 5S gene *in vitro* (Almouzni et al., 1990) and *in vivo* (Almouzni and Wolffe, 1993b). In their experiments Wolffe and colleagues demonstrated that pre-existing transcription complexes at the 5S gene are disrupted by the assembly of physiologically spaced nucleosomes on the promoter (Almouzni et al., 1991).

Our experiments have allowed us to eliminate the down-regulation, or loss of a specific transcription factor necessary for stimulation of mRNA synthesis from this promoter as a mechanism for the refractory period. They suggest that there may be a requirement for some "activity" or associated factor needed for the GR mediated chromatin disruption or nucleosome displacement. This putative "activity" or factor would not be required for transcription from the transient template, but could be rate limiting or down-regulated after the initial chromatin disruption event. This would then prevent the continued activation or the reactivation of the stable template as seen at 24 hours of hormone treatment. This could potentially include proteins such as the recently described coactivators that interact with NF1 and/or the basal transcription factors, alternately they may be mammalian homologous of the yeast *SWI* proteins that influence chromatin structure and may interact with the steroid receptors (Dynlacht et al., 1991; Meisterernst and Roeder, 1991; Dusserre and Mermod, 1992; Peterson and Herskowitz, 1992; Yoshinaga et al., 1992).

Another possibility may be some type of replication mediated phenomenon as has been suggested for the *Xenopus* 5S gene (Wolffe and Brown, 1986). In this case, a competitive process occurs between the binding of TFIIIA and histones to establish an active or inactive gene. Our transient transfection experiments suggest that transcription factors, for example, NF1 and OTF, bind first and "prevent" the assembly of the specific chromatin structure observed on stably introduced MMTV LTR sequences. This result is complimentary to our previous experiments where the binding of NF1 and the presence of Nuc B are never observed on the same template, that is, they are mutually exclusive. Further support for this argument is provided by the observations that purified NF1 and its cofactors antagonize histone mediated repression of transcription *in vitro* (Dusserre and Mermod, 1992). Thus, one possibility may be that, as our transfected plasmids do not replicate, they are never able to displace NF1 and OTF to allow the histone octamer access to the transient templates. With respect to replication, our previous experiments suggest that replication *per se* is not necessary for this activation process and indeed transcription occurs quite efficiently even if replication is blocked by the drug aphidicolon (Archer et al., 1989). We are currently addressing the question of whether inhibiting replication will have an effect on the kinetics of the response. A similar replication independent activation of transcription has also been recently described for the PHO5 (Schmid et al., 1992) and TAT genes (Reik et al., 1991). In the case of TAT, hormone removal resulted in a rapid loss of activity and chromatin remodeling at the promoter. It will be interesting to determine the kinetics of the response in the continued presence of hormone for TAT as well as the effects of withdrawal on the MMTV promoter.

F. Progesterone Regulation of MMTV Transcription in Human Breast Cancer Cells

In order to extend our analysis of progesterone regulation, we have constructed and isolated a series of human T47D breast cancer cell lines that stably maintain the MMTV LTR mobilized on the BPV episome. The plasmid (pJ83d) used to transfect these cells is a BPV based construct carrying the MMTV LTR attached to the bacterial CAT gene, and the large T Antigen of SV40 attached to the rat POMC proximal promoter (Charron et al., 1989). The cells contain high levels of PR, eliminating the need to transfect the receptor, and represent a model system which allows progesterone regulation to be analyzed in the absence of a functional GR. Characterization of one of these cell lines, designated 2963.1, indicates that these cells also differ from many of the previous mouse BPV cell lines analyzed, in that they have a relatively low copy number (5–10 copies per cell versus 100–200). In conjunction with the murine cell line 2305 (Charron et al., 1989) which contains an identical BPV MMTV chimera also present in low copy number (15 copies per cell), we have compared progesterone and glucocorticoid regulation of MMTV transcription *in vivo*.

We began our analysis by investigating hypersensitivity to restriction endonucleases (Archer et al., 1991) in both cell lines. Dexamethasone treatment of 2305 cells resulted in substantially elevated cleavage by restriction enzymes Afl II, Mbo I, Dde I, and Sst I, which cleave within Nuc B, but not by Hae III, which cleaves adjacent to the Nuc B boundary. In contrast, the T47D derived 2963.1 cells were hypersensitive even in the absence of progesterone (Mymryk et al., 1995). This constitutive accessibility to restriction enzymes is reminiscent of our observations with the transiently transfected template (Archer et al., 1992). However, unlike the transient template, this region of hypersensitivity is restricted to the region encompassing nucleosome B. Thus, with this human cell line we arrived at a novel organization of the MMTV promoter. In this case the promoter now contains a constitutive hypersensitive site around nucleosome B amid an apparently normal chromatin arrangement on either side at the flanking nucleosomes A and C (Mymryk et al., 1995). In control experiments, the growth of the cells in the presence of hormone antagonist or media supplemented with charcoal stripped serum for 24 hours did not affect this hypersensitivity, ruling out the trivial explanation of residual hormone from the serum (Mymryk et al., 1995). This increased access of restriction enzymes within the region occupied by Nuc B in 2963.1 cells suggests that in the absence of hormone, the MMTV LTR in these cells has already adopted an "open" conformation. *In vivo* footprinting experiments are also consistent with this region being hypersensitive or open, as we observe a substantial level of NF1

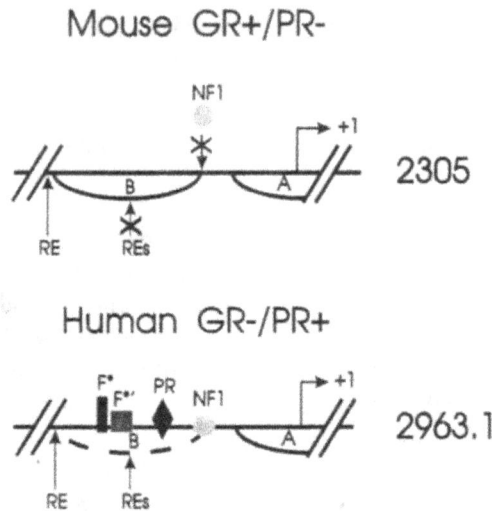

Figure 8. Transcription factor loading and restriction enzyme hypersensitivity of the MMTV LTR in cells containing different receptor complements. In mouse 2305 cells containing the GR but not the PR, the B nucleosome excludes NF1 binding and restriction enzymes from their sites prior to hormone treatment. In human T47D–2963.1 cells which contain the PR but not a functional GR, the organization of the B nucleosome differs dramatically (as indicated by the dotted outline). Transcription factors including NF1, the PR, and two other unidentified proteins bind to the promoter even in the absence of hormone treatment and this region appears constitutively hypersensitive to restriction enzyme cleavage. Thus the nucleosomal organization of the LTR appears to depend on the receptor complement (Mymryk et al., 1995).

binding on the template in the absence or presence of progestins (Figure 8). In addition, we detected a band that likely corresponds to the PR at the proximal progesterone response elements (PRE) at position -126 of the LTR. This Exo III stop is independent of hormone addition and unaffected by removal of endogenous steroids in the serum by charcoal treatment, suggesting that if it is the receptor, it is bound constitutively at the site in the absence of hormone (Mymryk et al., 1995). In addition, we also reproducibly detected two additional stops immediately downstream from the proximal HRE corresponding to -164 and -158. These proteins likely represent proteins previously identified in mammary cell extracts by *in vitro* footprinting experiments (Langer and Ostrowski, 1988).

The discovery of an open chromatin region bound by transcription factors that appears to be stably inherited with each cell division raises a question of which event precedes the other? (That is, *has the MMTV promoter in these cell lines adopted an open chromatin structure prior to transcription factor binding? or is it the binding of these transcription factors which maintains the open chromatin structure?*) The differences in chromatin organization between PR^+/GR^- and $PR^-/$ GR^+ cells may be related to the occupancy of the MMTV LTR by progesterone receptor and NF1 in the absence of ligand, as suggested by Exo III footprinting (Figure 8). This occupancy could compete with chromatin assembly following DNA replication, resulting in the altered nucleosomal structure we observe, and is consistent with the observation that unliganded PR is predominantly localized in the nucleus (Guiochon-Mantel et al., 1989), while unliganded GR is primarily found in the cytosol (Picard and Yamamoto, 1987). Alternatively, as shown for the mouse proliferin gene (Diamond et al., 1990; Pearce and Yamamoto, 1993) and human collagenase gene (Jonat et al., 1990; Yang-Yen et al., 1990), we recognize that nonreceptor differences between our cell lines may contribute to the observations reported here.. Indeed, it has been recently reported that transcriptional activation by the GR is linked to the presence of a human homologue of the *Sacchromyces cerevisiae* SNF2/SWI2 (Peterson and Herskowitz, 1992; Yoshinaga et al., 1992) and *Drosophila brama* (Muchardt and Yaniv, 1993) gene products, which may serve roles in alleviating nucleosome mediated repression of gene expression. In addition, the recently described effects of the protein calreticulin in modifying or suppressing steroid hormone action reveal a potential additional mode by which the organization of the promoter could be altered (Dedhar et al., 1994; Burns et al., 1994).

III. SUMMARY

Taken together, these data suggest that the MMTV LTR chromatin structure is plastic and the precise arrangement of nuclear proteins is dependent upon the cell environment. How then is transcription regulated at the MMTV promoter? Our results strongly suggest that the exclusion of NF1 and OTF from high affinity sites can be attributed directly to the organization of these sites as chromatin. The loading of TBP and F_{DT} onto the promoter requires activated GR to facilitate this binding, but at present we do not know whether this recruitment of TBP and F_{DT} is direct or indirect. Although our *in vivo* footprinting studies demonstrate that TBP is loaded onto the promoter at a site within nucleosome A, the HSR does not extend to this nucleosome suggesting that binding of TBP and F_{DT} would have to occur on this unmodified nucleosome. Thus, unlike our observations with the GR, or results by others with Gal4 (Morse, 1993), but as reported for the heat shock transcription factor (Pederson and Fidrych, 1994), transcription factor binding need not necessarily be preceded by the disruption of a bound nucleosome. This is somewhat reminiscent of the result of Zaret and colleagues on the albumin gene in hepatoma cells where they observe a precisely positioned array of nucleosomes that are assembled only in the presence of liver specific transcription factors (McPherson et al., 1993).

Despite the fact that the GR is absolutely required to initiate this cascade of events, our footprinting experiments have failed to reproducibly detect it. We have proposed that this may be due to the GR:chromatin association being labile to Exo III or that its mechanism of action involves a transient interaction with the template (Archer et al., 1989), a similar proposal has been made for TAT (Rigaud et al., 1991). Support for the second idea comes from our experiments with human 2963.1 cells. In this case the promoter is always hypersensitive and a group of transcription factors, including the PR, are always present on the template. It is reasonable to expect that the affinity or lability of the GR and the PR, for the HRE and to Exo III respectively, would be almost identical given the homology of their DNA binding domains. Thus our ability to detect the PR suggests that the GR is not labile, but rather that it is not stably present on the template even in the presence of hormone. If this were correct, it would argue that the kinetics of GR induced transcription and chromatin remodeling result from this feature of GR:chromatin binding. Thus, the loss of the PIC that is observed in mouse cells may result from the dissociation of the GR and the subsequent re-establishment of the repressive chromatin structure. In the human cells the stable binding of the PR would inhibit the "rechromatinization" of the B nucleosome and NF1 would remain bound. While we know little of the mechanism by which the PIC is lost, it clearly results from "rechromatinization" rather than down-regulation of transcription factors. Thus, once NF1, OTF, and TBP are bound to transient "nonchromatin" template they remain bound even as they are lost from the chromatin template. This leads to the prediction that in human 2963.1 cells, the stable binding of the PR and constitutively "open" organization of Nuc B would eliminate the rapid deinduction of expression observed for the chromatin template in GR+ mouse cells.

Thus the organization of the promoter as chromatin not only restricts the access of transcription factors (i.e., NF1, OTFs, etc.) but also serves to regulate the occupancy of the PIC on the template. Indeed this chromatin architecture provides a mechanism for the PIC disassembly that generates the unique kinetics of the hormone response from this promoter. These findings are most easily accommodated within the concept that an accurate understanding of eucaryotic transcriptional regulation requires its analysis in the context of chromatin.

ACKNOWLEDGMENTS

We wish to thank Dr. Gordon Hager for introducing us to the MMTV system, stimulating discussions, and for making available many of the reagents we have used independently and in collaboration with his group. We would also like to thank colleagues in the Archer lab, especially Huay-Leng Lee and Eva Zaniewski for allowing us to quote from unpublished work. Finally, we acknowledge the expert secretarial skills of Denise Hynes in the preparation of this manuscript. The work reported here was supported in part by grants to TKA from the National Cancer Institute of Canada (NCIC), Victoria Hospital Corporation, and London

Regional Cancer Centre. Trevor K. Archer is a NCIC Career Scientist. Joseph S. Mymryk was supported in part by a Wyeth Fellowship.

REFERENCES

Almer, A., & Hörz, W. (1986). Nuclease hypersensitive regions with adjacent positioned nucleosomes mark the gene boundaries of the PHO5/PHO3 locus in yeast. EMBO J. 5, 2681–2687.

Almouzni, G., Clark, D.J., Mechali, M., & Wolffe, A.P. (1990). Chromatin assembly on replicating DNA *in vitro*. Nucl. Acids Res. 18, 5767–5774.

Almouzni, G., Mechali, M., & Wolffe, A.P. (1991). Transcription complex disruption caused by a transition in chromatin structure. Mol. Cell. Biol. 11, 655–665.

Almouzni, G., & Wolffe, A.P. (1993a). Nuclear assembly, structure, and function: The use of *Xenopus in vitro* systems. Exp. Cell Res. 205, 1–15.

Almouzni, G., & Wolffe, A.P. (1993b). Replication-coupled chromatin assembly is required for the repression of basal transcription *in vivo*. Genes Dev. 7, 2033–2047.

Archer, T.K. (1993). Nucleosomes modulate access of transcription factor to the MMTV promoter *in vivo* and *in vitro*. Ann. NY Acad. Sci. 684, 196–199.

Archer, T.K. (1994). Unpublished data.

Archer, T.K., Cordingley, M.G., Marsaud, V., Richard-Foy, H., & Hager, G.L. (1989). Steroid transactivation at a promoter organized in a specifically-positioned array of nucleosomes. In Proceedings: Second International CBT Symposium on the Steroid/Thyroid Hormone Receptor Family and Gene Regulation (Gustafsson, J.A., Eriksson, H., & Carlstedt-Duke, J., eds.), pp. 221–238. Birkhauser Verlag AG, Berlin.

Archer, T.K., Cordingley, M.G., Wolford, R.G., & Hager, G.L. (1991). Transcription factor access is mediated by accurately positioned nucleosomes on the mouse mammary tumor virus promoter. Mol. Cell. Biol. 11, 688–698.

Archer, T.K., Lefebvre, P., Wolford, R.G., & Hager, G.L. (1992). Transcription factor loading on the MMTV promoter: A bimodal mechanism for promoter activation. Science 255, 1573–1576.

Archer, T.K., Lee, H.-L., Cordingley, M.G., Mymryk, J.S., Fragoso, G., Berard, D.S., & Hager, G.L. (1994a). Differential steroid hormone induction of transcription from the mouse mammary tumor virus promoter. Mol. Endo. 8, 568–576.

Archer, T.K., Zaniewski, E., Moyer, M., & Nordeen, S.K. (1994b). The differential capacity of glucocorticoids and progestins to alter chromatin structure and induce gene expression in human breast cancer cells. Mol. Endo. 8, 1154–1162.

Arents, G., Burlingame, R.W., Wang, B.-C., Love, W.E., & Moudrianakis, E.N. (1991). The nucleosomal core histone octamer at 3.1 Å resolution: A tripartite protein assembly and a left-handed superhelix. Proc. Natl. Acad. Sci. USA 88, 10148–10152.

Arents, G., & Moudrianakis, E.N. (1993). Topography of the histone octamer surface: Repeating structural motifs utilized in the docking of nucleosomal DNA. Proc. Natl. Acad. Sci. USA 90, 10489–10493.

Arriza, J.L., Weinberger, C., Cerelli, G., Glaser, T.M., Handelin, B.L., Housman, D.E., & Evans, R.M. (1987). Cloning of human mineralocorticoid receptor complementary DNA: Structural and functional kinship with the glucocorticoid receptor. Science 237, 268–275.

Beato, M. (1989). Gene regulation by steroid hormones. Cell 56, 335–344.

Becker, P.B., Rabindran, S.K., & Wu, C. (1991). Heat shock-regulated transcription *in vitro* from a reconstituted chromatin template. Proc. Natl. Acad. Sci. USA 88, 4109–4113.

Bresnick, E.H., Rories, C., & Hager, G.L. (1992). Evidence that nucleosomes on the mouse mammary tumor virus promoter adopt specific translational positions. Nucleic Acids Res. 20, 865–870.

Brüggemeier, U., Kalff, M., Franke, S., Scheidereit, C., & Beato, M. (1991). Ubiquitous transcription factor OTF-1 mediates induction of the MMTV promoter through synergistic interaction with hormone receptors. Cell 64, 565–572.

Buetti, E. (1994). Stably integrated mouse mammary tumor virus long terminal repeat DNA requires the octamer motifs for basal promoter activity. Mol. Cell. Biol. 14, 1191–1203.

Buetti, E., Kühnel, B., & Diggelmann, H. (1989). Dual function of a nuclear factor I binding site in MMTV transcription regulation. Nucl. Acids Res. 17, 3065–3078.

Buetti, E. & Kühnel, B. (1986). Distinct sequence elements involved in the glucocorticoid regulation of the mouse mammary tumor virus promoter identified by linker scanning mutagenesis. J. Mol. Biol. 190, 379–389.

Burns, K., Duggan, B., Atkinson, E.A., Famulski, K.S., Nemer, M., Bleackley, R.C., & Michalak, M. (1994). Modulation of gene expression by calreticulin binding to the glucocorticoid receptor. Nature 367, 476–480.

Carr, K.D., & Richard-Foy, H. (1990). Glucocorticoids locally disrupt an array of positioned nucleosomes on the rat tyrosine aminotransferase promoter in hepatoma cells. Proc. Natl. Acad. Sci. USA 87, 9300–9304.

Cato, A.C., Henderson, D., & Ponta, H. (1987). The hormone response element of the mouse mammary tumour virus DNA mediates the progestin and androgen induction of transcription in the proviral long terminal repeat region. EMBO J. 6, 363–368.

Cato, A.C.B., Miksicek, R., Schütz, G., Arnemann, J., & Beato, M. (1986). The hormone regulatory element of mouse mammary tumour virus mediates progesterone induction. EMBO J. 5, 2237–2240.

Chandler, V.L., Maler, B.A., & Yamamoto, K.R. (1983). DNA sequences bound specifically by glucocorticoid receptor in vitro render a heterologous promoter hormone responsive in vivo. Cell 33, 489–499.

Charron, J., Richard-Foy, H., Berard, D.S., Hager, G.L., & Drouin, J. (1989). Independent glucocorticoid induction and repression of two contiguous responsive genes. Mol. Cell. Biol. 9, 3127–3131.

Cordingley, M.G., & Hager, G.L. (1988). Binding of multiple factors to the MMTV promoter in crude and fractionated nuclear extracts. Nucl. Acids Res. 16, 609–628.

Cordingley, M.G., Richard-Foy, H., Lichtler, L., Ostrowski, M.C., & Hager, G.L. (1987a). The hormone response element of the MMTV LTR: A complex regulatory region. In DNA: Protein Interactions and Gene Regulation (Thompson, E.B., & Papaconstantinou, J., eds.), pp. 233–243. Univ. of Texas Press, Galveston, TX.

Cordingley, M.G., Riegel, A.T., & Hager, G.L. (1987b). Steroid-dependent interaction of transcription factors with the inducible promoter of mouse mammary tumor virus in vivo. Cell 48, 261–270.

Darbre, P., Page, M., & King, R.J. (1986). Androgen regulation by the long terminal repeat of mouse mammary tumor virus. Mol. Cell. Biol. 6, 2847–2854.

de Wet, J.R., Wood, K.V., DeLuca, M., Helinski, D.R., & Subramani, S. (1987). Firefly luciferase gene: Structure and expression in mammalian cells. Mol. Cell. Biol. 7, 725–737.

Dedhar, S., Rennie, P.S., Shago, M., Leung Hagesteijn, C.-Y., Yang, H., Filmus, J., Hawley, R.G., Bruchovsky, N., Cheng, H., Matusik, R.J., & Giguère, V. (1994). Inhibition of nuclear hormone receptor activity by calreticulin. Nature 367, 480–483.

DePamphilis, M.L. (1993). Eukaryotic DNA replication: Anatomy of an origin. Annu. Rev. Biochem. 62, 29–63.

Diamond, M.I., Miner, J.N., Yoshinaga, S.K., & Yamamoto, K.R. (1990). Transcription factor interactions: Selectors of positive or negative regulation from a single DNA element. Science 249, 1266–1272.

Dusserre, Y., & Mermod, N. (1992). Purified cofactors and histone H1 mediate transcriptional regulation by CTF/NF-1. Mol. Cell. Biol. 12, 5228–5237.

Dynlacht, B.D., Hoey, T., & Tjian, R. (1991). Isolation of coactivators associated with the TATA-binding protein that mediate transcriptional activation. Cell 66, 563–576.

Elgin, S.C. (1984). Anatomy of hypersensitive sites [news]. Nature 309(5965), 213–214.

Elgin, S.C. (1988). The formation and function of DNase I hypersensitive sites in the process of gene activation. J. Biol. Chem. 263, 19259–19262.

Felsenfeld, G. (1992). Chromatin as an essential part of the transcriptional mechanism. Nature 355, 219–224.

Godowski, P.J., Rusconi, S., Miesfeld, R., & Yamamoto, K.R. (1987). Glucocorticoid receptor mutants that are constitutive activators of transcriptional enhancement [published erratum appears in Nature 1987 March 5–11;326(6108):105]. Nature 325, 365–368.

Gorman, C.M., Moffat, L.F., & Howard, B.H. (1982). Recombinant genomes which express chloramphenicol acetyltransferase in mammalian cells. Mol. Cell. Biol. 2, 1044–1051.

Gowland, P.L., & Buetti, E. (1989). Mutations in the hormone regulatory element of mouse mammary tumor virus differentially affect the response to progestins, androgens, and glucocorticoids. Mol. Cell. Biol. 9, 3999–4008.

Green, S., & Chambon, P. (1988). Nuclear receptors enhance our understanding of transcription regulation. Trends. Genet. 4, 309–314.

Gross, D.S., & Garrard, W.T. (1988). Nuclease hypersensitive sites in chromatin. Annu. Rev. Biochem. 57, 159–197.

Groudine, M., & Weintraub, H. (1982). Propagation of globin DNAase I-hypersensitive sites in absence of factors required for induction: A possible mechanism for determination. Cell 30, 131–139.

Grunstein, M. (1990a). Histones function in transcription. Annu. Rev. Cell Biol. 6, 643–678.

Grunstein, M. (1990b). Nucleosomes: Regulators of transcription. TIG 6, 395–400.

Guiochon-Mantel, A., Loosfelt, H., Lescop, P., Sar, S., Atger, M., Perrot-Applanat, M., & Milgrom, E. (1989). Mechanisms of nuclear localization of the progesterone receptor: Evidence for interaction between monomers. Cell 57, 1147–1154.

Hager, G.L. (1988). MMTV as a model for gene expression in mammary tissue. In Breast Cancer: Cellular and Molecular Biology (Lippman, M.E., & Dickson, R., eds.), pp. 267–281. Kluwer Academic Publishers, Boston.

Hager, G.L. (1989). Chromatin template remodeling and steroid receptor transactivation of MMTV. In Molecular Mechanisms of Hormone Action: 40th Mosbach Colloquium (Gehring, U., Helmreich, E., & Schultz, G., eds.), pp. 29–34. Springer-Verlag, Heidelberg.

Hager, G.L., & Archer, T.K. (1991). The interaction of steroid receptors with chromatin. In Nuclear Hormone Receptors (Parker, M.G., ed.), pp. 217–234. Academic Press, London.

Hager, G.L. et al. (1993). The influence of chromatin structure on the binding of transcription factors to DNA in Cold Spring Harbor Symp. Quant. Biol. LVIII: DNA and Chromosomes 58, 63–71.

Ham, J., Thomson, A., Needham, M., Webb, P., & Parker, M. (1988). Characterization of response elements for androgens, glucocorticoids and progestins in mouse mammary tumour virus. Nucleic Acids Res. 16, 5263.

Huang, A.L., Ostrowski, M.C., Berard, D., & Hager, G.L. (1981). Glucocorticoid regulation of the Ha-MuSV p21 gene conferred by sequences from mouse mammary tumor virus. Cell 27(2 Pt 1), 245–255.

Jaggi, R., Höck, W., Ziemiecki, A., Klemenz, R., Friis, R., & Groner, B. (1989). Oncogene mediated repression of glucocorticoid hormone response elements and glucocorticoid receptor levels. Cancer Res. (Suppl) 49, 2266s–2274s.

Jonat, C., Rahmsdorf, H.J., Park, K.-K., Cato, A.C.B., Gebel, S., Ponta, H., & Herrlich, P. (1990). Antitumor promotion and antiinflammation: Down-modulation of AP-1 (Fos/Jun) activity by glucocorticoid hormone. Cell 62, 1189–1204.

Jones, K.A., Kadonaga, J.T., Rosenfeld, P.J., Kelly, T.J., & Tjian, R. (1987). A cellular DNA-binding protein that activates eukaryotic transcription and DNA replication. Cell 48, 79–89.

Kadonaga, J.T. (1990). Assembly and disassembly of the *Drosophila* RNA polymerase II complex during transcription. J. Biol. Chem. 265, 2624–2631.

Kornberg, R.D. (1974). Chromatin structure: A repeating unit of histones and DNA. Science 184, 868–871.

Kühnel, B., Buetti, E., & Diggelmann, H. (1986). Functional analysis of the glucocorticoid regulatory elements present in the mouse mammary tumor virus long terminal repeat: A synthetic distal binding site can replace the proximal binding domain. J. Mol. Biol. 190, 367–378.

Langer, S.J., & Ostrowski, M.C. (1988). Negative regulation of transcription *in vitro* by a glucocorticoid response element is mediated by a *trans*-acting factor. Mol. Cell. Biol. 8, 3872–3881.

Lee, H.-L., & Archer, T.K. (1994). Nucleosome mediated disruption of transcription factor:chromatin initiation complexes at the Mouse Mammary Tumor Virus Long Terminal Repeat *in vivo*. Mol. Cell. Biol. 14, 32–41.

Lefebvre, P., Berard, D.S., Cordingley, M.G., & Hager, G.L. (1991). Two regions of the mouse mammary tumor virus LTR regulate the activity of its promoter in mammary cell lines. Mol. Cell. Biol. 11, 2529–2537.

Majors, J., & Varmus, H.E. (1983). A small region of the mouse mammary tumor virus long terminal repeat confers glucocorticoid hormone regulation on a linked heterologous gene. Proc. Natl. Acad. Sci. USA 80, 5866–5870.

McGhee, J.D., & Felsenfeld, G. (1980). Nucleosome structure. Annu. Rev. Biochem. 49, 1115–1156.

McGhee, J.D., Wood, W.I., Dolan, M., Engel, J.D., & Felsenfeld, G. (1981). A 200 base pair region at the 5' end of the chicken adult beta-globin gene is accessible to nuclease digestion. Cell 27, 45–55.

McPherson, C.E., Shim, E.-Y., Friedman, D.S., & Zaret, K.S. (1993). An active tissue-specific enhancer and bound transcription factors existing in a precisely positioned nucleosomal array. Cell 75, 387–398.

Meisterernst, M., & Roeder, R.G. (1991). Family of proteins that interact with TFIID and regulate promoter activity. Cell 67, 557–567.

Mermod, N., O'Neill, E.A., Kelly, T.J., & Tjian, R. (1989). The proline-rich transcriptional activator of CTF/NF–1 is distinct from the replication and DNA binding domain. Cell 58, 741–753.

Morse, R.H. (1993). Nucleosome disruption by transcription factor binding in yeast. Science 262, 1563–1566.

Muchardt, C., & Yaniv, M. (1993). A human homologue of *Saccharomyces cerevisiae* SNF2/SW12 and *Drosophila brm* genes potentiates transcriptional activation by the glucocorticoid receptor. EMBO J. 12, 4279–4290.

Mueller, P.R., & Wold, B. (1989). *In vivo* footprinting of a muscle specific enhancer by ligation mediated PCR. Science 246, 780–785.

Munoz-Bradham, B., & Bolander, F.F. (1989). The role of sex steroids in the expression of MMTV in the normal mouse mammary gland. Biochem. Biophys. Res. Commun. 159, 1020.

Mymryk, J.S., Berard, D., Hager, G.L., & Archer, T.K. (1995). Mouse mammary tumor virus chromatin in human breast cancer cells is constitutively hypersensitive and exhibits steroid hormone-independent loading of transcription factors in vivo. Mol. Cell. Biol. 15, 26–34.

Okret, S., Poellinger, L., Dong, Y., & Gustafsson, J.A. (1986). Down-regulation of glucocorticoid receptor mRNA by glucocorticoid hormones and recognition by the receptor of a specific binding sequence within a receptor cDNA clone. Proc. Natl. Acad. Sci. USA 83, 5899–5903.

Olins, A.L., & Olins, D.E. (1974). Spheroid chromatin units (ν–bodies). Science 183, 330–332.

Ostrowski, M.C., Richard-Foy, H., Wolford, R.G., Berard, D.S., & Hager, G.L. (1983). Glucocorticoid regulation of transcription at an amplified, episomal promoter. Mol. Cell. Biol. 3, 2045–2057.

Otten, A.D., Sanders, M.M., & McKnight, G.S. (1988). The MMTV LTR promoter is induced by progesterone and dihydrotesto sterone but not by estrogen. Mol. Endocrinol. 2, 143–147.

Oudet, P., Gross-Bellard, M., & Chambon, P. (1975). Electron microscopic and biochemical evidence that chromatin structure is a repeating unit. Cell 4, 281–300.

Panganiban, A.T. (1985). Retroviral DNA integration. Cell 42, 5–6.

Pearce, D., & Yamamoto, K.R. (1993). Mineralocorticoid and glucocorticoid receptor activities distinguished by nonreceptor factors at a composite response element. Science 259, 1161–1165.

Pederson, D.S., & Fidrych, T. (1994). Heat shock factor can activate transcription while bound to nucleosomal DNA in *saccharomyces cerevisiae*. Mol. Cell. Biol. 14, 189–199.

Pederson, D.S., & Simpson, R.T. (1988). Structural and regulatory hierarchies in transcriptionally active chromatin. I. S. I. Atlas of Science 1, 155–160.

Perlmann, T., & Wrange, O. (1988). Specific glucocorticoid receptor binding to DNA reconstituted in a nucleosome. EMBO J. 7, 3073–3079.

Peterson, C.L., & Herskowitz, I. (1992). Characterization of the yeast SWI1, SWI2, and SWI3 genes, which encode a global activator of transcription. Cell 68, 573–583.

Pham, T.A., McDonnell, D.P., Tsai, M.-J., & O'Malley, B.W. (1992). Modulation of progesterone receptor binding to progesterone response elements by positioned nucleosomes. Biochemistry 31, 1570–1578.

Picard, D., & Yamamoto, K.R. (1987). Two signals mediate hormone-dependent nuclear localization of the glucocorticoid receptor. EMBO J. 6, 3333–3340.

Piña, B., Brüggemeier, U., & Beato, M. (1990). Nucleosome positioning modulates accessibility of regulatory proteins to the mouse mammary tumor virus promoter. Cell 60, 719–731.

Qi, M., Hamilton, B.J., & DeFranco, D. (1989). *v-mos* oncoproteins affect the nuclear retention and reutilization of glucocorticoid receptors. Mol. Endo. 3, 1279–1288.

Reik, A., Schutz, G., & Stewart, A.F. (1991). Glucocorticoids are required for establishment and maintenance of an alteration in chromatin structure: Induction leads to a reversible disruption of nucleosomes over an enhancer. EMBO J. 10, 2569–2576.

Richard-Foy, H., & Hager, G.L. (1987). Sequence-specific positioning of nucleosomes over the steroid-inducible MMTV promoter. EMBO J. 6, 2321–2328.

Richmond, T.J., Finch, J.T., Rushton, B., Rhodes, D., & Klug, A. (1984). Structure of the nucleosome core particle at 7 A resolution. Nature 311, 532–537.

Rigaud, G., Roux, J., Pictet, R., & Grange, T. (1991). *In vivo* footprinting of rat TAT gene: Dynamic interplay between the glucocorticoid receptor and a liver-specific factor. Cell 67, 977–986.

Ringold, G.M., Yamamoto, K.R., Tomkins, G.M., Bishop, J.M., & Varmus, H.E. (1975). Dexamethasone-mediated induction of mouse mammary tumor virus RNA: A system for studying glucocorticoid action. Cell 6, 299–305.

Rosenfeld, P.J., & Kelly, T.J. (1986). Purification of nuclear factor I by DNA recognition site affinity chromatography. J. Biol. Chem. 261, 1398–1408.

Schmid, A., Fascher, K.-D., & Hörz, W. (1992). Nucleosome disruption at the yeast PH05 promoter upon PH05 induction occurs in the absence of DNA replication. Cell 71, 853–864.

Schule, R., Muller, M., Kaltschmidt, C., & Renkawitz, R. (1988). Synergism of several transcription factors with the progesterone or the glucocorticoid receptor. Science 242, 1418–1420.

Scolnick, E.M., Young, H.A., & Parks, W.P. (1976). Biochemical and physiological mechanisms in glucocorticoid hormone induction of mouse mammary tumor virus. Virology 69, 148–156.

Shimamura, A., Tremethick, D., & Worcel, A. (1988). Characterization of the repressed 5S DNA minichromosomes assembled *in vitro* with a high-speed supernatant of Xenopus laevis oocytes. Mol. Cell. Biol. 8, 4257–4269.

Shimizu, M., Roth, S.Y., Szent-Gyorgyi, C., & Simpson, R.T. (1991). Nucleosomes are positioned with base pair precision adjacent to the $\alpha2$ operator in *Saccaromyces cerevisiae*. EMBO J. 10, 3033–3041.

Smith, C.L., Archer, T.K., Hamlin-Green, G., & Hager, G.L. (1993). Newly-expressed progesterone receptor cannot activate stable, replicated MMTV templates but acquires transactivation potential upon continuous expression. Proc. Natl. Acad. Sci. 90, 11202–11206.

Smith, S., & Stillman, B. (1989). Purification and characterization of CAF-I, a human cell factor required for chromatin assembly during DNA replication *in vitro*. Cell 58, 15–25.

Stalder, J., Larsen, A., Engel, J.D., Dolan, M., Groudine, M., & Weintraub, H. (1980). Tissue-specific DNA cleavages in the globin chromatin domain introduced by DNAase I. Cell 20, 451–460.

Taylor, I.C., Workman, J.L., Schuetz, T.J., & Kingston, R.E. (1991). Facilitated binding of GAL4 and heat shock factor to nucleosomal templates: Differential function of DNA-binding domains. Genes Dev. 5, 1285–1298.

Tsai, S.Y., Srinivasan, G., Allan, G.F., Thompson, E.B., O'Malley, B.W., & Tsai, M.-J. (1990). Recombinant human glucocorticoid receptor induces transcription of hormone response genes *in vitro*. J. Biol. Chem. 265, 17055–17061.

Ucker, D.S., & Yamamoto, K.R. (1984). Early events in the stimulation of mammary tumor virus RNA synthesis by glucocorticoids. Novel assays of transcription rates. J. Biol. Chem. 259, 7416–7420.

van Holde, K.E. (1988). Chromatin, pp. 1–497. Springer-Verlag Heidelberg.

von der Ahe, D., Janich, S., Scheidereit, C., Renkawitz, R., Schutz, G., & Beato, M. (1985). Glucocorticoid and progesterone receptors bind to the same sites in two hormonally regulated promoters. Nature 313, 706–709.

Wolffe, A.P. (1990). New approaches to chromatin function. New Biol. 2, 211–218.

Wolffe, A.P. (1991). Activating chromatin. Current Biology 1, 366–368.

Wolffe, A.P. (1992a). New insights into chromatin function in transcriptional control. FASEB J. 6, 3354–3361.

Wolffe, A.P. (1992b). Chromatin Structure & Function. pp. 1–213. Academic Press, New York.

Wolffe, A.P., & Brown, D.D. (1986). DNA replication *in vitro* erases a *Xenopus* 5S RNA gene transcription complex. Cell 47, 217–227.

Workman, J.L., & Buchman, A.R. (1993). Multiple functions of nucleosomes and regulatory factors in transcription. TIBS 18, 90–95.

Workman, J.L., Taylor, I.C.A., & Kingston, R.E. (1991). Activation domains of stably bound GAL4 derivatives alleviate repression of promoters by nucleosomes. Cell 64, 533–544.

Wu, C. (1980). The 5' ends of *Drosophila* heat shock genes in chromatin are hypersensitive to DNase I. Nature 286, 854–860.

Wu, C. (1985). An exonuclease protection assay reveals heat-shock element and TATA box DNA-binding proteins in crude nuclear extracts. Nature 317, 84–87.

Yang-Yen, H.-F., Chambard, J.-C., Sun, Y.-L., Smeal, T., Schmidt, T.J., Drouin, J., & Karin, M. (1990). Transcriptional interference between c–Jun and the glucocorticoid receptor: Mutual inhibition of DNA binding due to direct protein-protein interaction. Cell 62, 1205–1215.

Yoshinaga, S.K., Peterson, C.L., Herskowitz, I., & Yamamoto, K.R. (1992). Roles of SWI1, SWI2, and SWI3 proteins for transcriptional enhancement by steroid receptors. Science 258, 1598–1604.

Young, H.A., Shih, T.Y., Scolnick, E.M., & Parks, W.P. (1977). Steroid induction of mouse mammary tumor virus: effect upon synthesis and degradation of viral RNA. J. Virol. 21, 139–146.

Zaret, K.S., & Yamamoto, K.R. (1984). Reversible and persistent changes in chromatin structure accompany activation of a glucocorticoid-dependent enhancer element. Cell 38, 29–38.

Chapter 7

Heat Shock Factor Potentiates the Promoter Chromatin Structure of the Yeast *HSP90* Genes

DAVID S. GROSS

The Nucleus
Volume 1, pages 151–182.
Copyright © 1995 by JAI Press Inc.
All rights of reproduction in any form reserved.
ISBN: 1-55938-940-0

ABSTRACT

Heat shock genes are poised for instantaneous transcriptional activation in response to environmental stress. Facilitating this hair-trigger response is the presence of DNase I hypersensitive, nucleosome-free regions over the promoters of such genes. To identify biochemical determinants potentiating the *HSP82* and *HSC82* heat shock promoters in *Saccharomyces cerevisiae*, we have introduced *in situ* mutations within sequences shown by chromatin footprinting to be engaged in constitutive protein/DNA interactions *in vivo*. Genomic footprinting of the *HSP82* promoter has revealed that the most proximal of four heat shock elements, termed HSE1, is uniquely occupied by protein ± heat shock; the other three are not stably bound by protein and are dispensable for both basal and induced function. Introduction of a double point mutation into HSE1 preferentially abolishes basal expression, whereas its deletion or substitution reduces both basal and induced transcription at least two orders of magnitude. Moreover, whereas a double point mutation within HSE1 has virtually no effect on the chromatin structure of the *hsp82* promoter, deletion or substitution of this sequence leads to a dramatic transition: the DNase I hypersensitive region is replaced by two stable, sequence-positioned nucleosomes. One of these is centered over the mutated upstream activation sequence (UAS), while the other is precisely positioned in a rotational sense over the TATA-initiation site. Overexpression of yeast heat shock factor (yHSF) suppresses the null phenotype of the induced *hsp82–ΔHSE1* gene and re-establishes DNase I hypersensitivity over its promoter. Such suppression is mediated through sequence disposed immediately upstream of HSE1 and containing two low-affinity heat shock elements, which map near the dyad of the UAS nucleosome. HSF-mediated suppression of the *hsp82–ΔHSE1* promoter can occur in the absence of DNA replication—implying that yHSF directly binds to the UAS nucleosome *in vivo*—but it is blocked by an extragenic mutation in the *SWI1* gene. Parallel experiments with the constitutively expressed *HSC82* gene demonstrate that its high-affinity HSF binding site, also termed HSE1, plays an analogous functional and structural role. In *HSC82*, however, point mutations within HSE1 are sufficient to abolish promoter function and lead to a loss of the nucleosome-free, DNase I hypersensitive region. Notably, a 6 bp substitution of the TATAAA motif, while reducing *HSC82* transcription an order of magnitude, has no effect on promoter chromatin structure. Taken together, our results imply a critical role for HSF in displacing stably positioned nucleosomes in *Saccharomyces cerevisiae* and suggest that HSF transcriptionally activates *HSP82* and *HSC82* at least partly through its ability to alleviate nucleosome repression of their core promoter regions. We propose the term *POWER* (Promoter Open Window Entry Regulator) to describe DNA-binding proteins such as yHSF that possess dual structural and functional roles in regulating gene expression.

I. INTRODUCTION

That nucleosomes can repress transcription initiation both *in vitro* and *in vivo* is well established (Grunstein, 1990a, 1990b; Kornberg and Lorch, 1991; Felsenfeld, 1992; Workman and Buchman, 1993). Such repression is generally the result of steric occlusion of the core promoter, whereby the presence of a stably positioned nucleosome severely restricts the accessibility of the TATA-initiation site to the TATA-binding protein (TBP), RNA polymerase II, and other components of the general transcriptional machinery. In addition, at some promoters, upstream promoter elements and enhancers which bind regulatory proteins that modulate the rate of transcription initiation must likewise lie within accessible regions (Wolffe, 1994). Such open regions serve as windows within the chromatin fiber through which activators, repressors, and components of the preinitiation complex gain access to the underlying DNA (Gross and Garrard, 1988; Elgin, 1988).

Our work over the past several years has addressed the biochemical determinants that potentiate heat shock promoters for rapid transcriptional activation. We have focused on the two members of the *HSP90* gene family of *Saccharomyces cerevisiae*, which encode a functionally identical protein of M_r 81–82 kD. *HSP82* is expressed at a low basal level and is strongly induced upon mild heat shock, while *HSC82* is expressed at a 10–fold higher constitutive level but is only slightly inducible (Borkovich et al., 1989). That these genes are principally regulated at the level of transcription initiation is suggested by measurements of their steady-state RNAs (Figure 1), which parallel those of the protein under both control (–) and heat shocked (+) conditions (Borkovich et al., 1989) as well as measurements of β-galactosidase levels in cells harboring *hsp82/lacZ* and *hsc82/lacZ* fusion genes (Adams and Gross, 1991). In this paper, I summarize selected results of our recent studies which have provided insight into the protein/DNA interactions that regulate transcription and potentiate the chromatin structure of each promoter.

II. STUDIES ON THE *HSP82* GENE

The chromatin structure of *HSP82* shows a remarkable degree of organization with respect to the underlying DNA sequence and is not detectably altered following heat shock (Gross et al., 1986; Szent–Gyorgyi et al., 1987). In particular, it contains a constitutive DNase I hypersensitive region within its promoter, a characteristic feature of all known eukaryotic heat shock genes (Wu, 1980; Gross and Garrard, 1988; Elgin, 1988). Additional features of the chromatin structure of the *HSP82* locus are schematically depicted in Figure 2.

A. Genomic Footprinting of the *HSP82⁺* Allele

To identify sequences upstream of *HSP82* that are specifically bound by proteins *in vivo*, we mapped sites of protein/DNA interaction and altered helical conformations within the proximal and core promoter regions using the genomic footprinting technique. Three complementary strategies were employed: (1) dimethyl sulfate (DMS) methylation of intact cells to probe tight protein/guanine interactions within the major groove; (2) DNase I digestion of spheroplast lysates to determine the accessibility and conformation of the minor groove; and (3) hydroxyl radical digestion of spheroplast

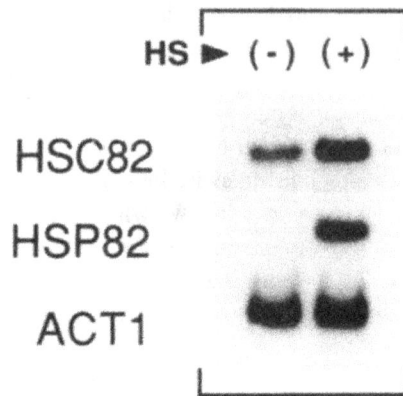

Figure 1. Steady-state *HSP90* mRNA levels in yeast cells before (–) and after (+) heat shock. Cells were cultivated at 30° C and shifted to 39° C for 15 minutes, RNA was isolated, electrophoretically separated, and blotted to a nylon membrane that was sequentially hybridized with probes specific to the *HSC82*, *HSP82*, and *ACT1* transcripts (from Erkine et al., 1995a).

lysates both to identify protein/DNA interactions at the sugar–phosphate backbone and to detect the periodicity of helical twist (Gross et al., 1990).

These experiments revealed that among four upstream sequences bearing homology to the heat shock consensus (consisting of tandem inverted repeats of *n*GAA*n*) (Xiao and Lis, 1988), the promoter-proximal heat shock element, termed HSE1, is preferentially and constitutively bound by protein (see Figure 3, wild-type [WT] panel), which we show below to be heat shock transcription factor (HSF). The HSE1-associated factor binds tightly within the major groove of DNA, as discerned by protection of five guanines from DMS methylation of intact cells (Gross et al., 1990). The upstream sequence of *HSP82* and sites protected from chemical and enzymatic cleavage *in vivo* are presented in Figure 4.

Genomic footprinting has also revealed that a protein, most likely TBP, constitutively binds to the TATA box centered at position –75, but its interaction with DNA differs radically from that of HSF. Instead of occupying the major groove, the protein tightly interacts with each sugar–phosphate backbone, as revealed by strong protection from hydroxyl radical cleavage between positions –68 and –86 (see Figure 5). In addition, TBP appears to occupy the minor groove between positions –68 and –82, as inferred by the partial protection that it confers to DNase I (Figure 3). TBP–DNA contacts are completely restricted to the minor groove and sugar–phosphate backbone of the TATAAA element *in vitro* (Starr and Hawley, 1991; Lee et al., 1991; Kim et al., 1993a, 1993b).

Marked helical distortions are associated with binding of each transcription factor. HSF binding to HSE1 appears to cause a non-*B*-conformation, based on a local 12 bp periodicity of hydroxyl radical protection and the presence of multiple

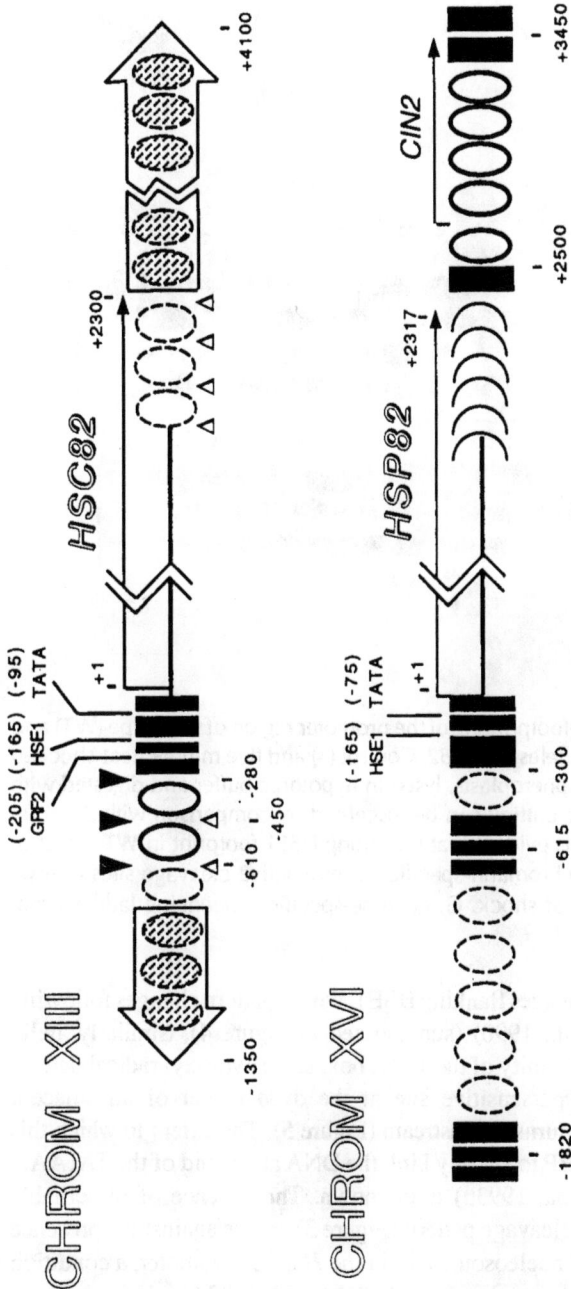

Figure 2. Chromatin structure of the yeast *HSP90* genes ± heat shock. DNase I hypersensitive sites are depicted as solid bars; internal footprints are represented as gaps. Nucleosomes whose positioning has been established are indicated by continuous ovals; those whose positioning is less firmly established are indicated by broken ovals. Regions of regularly spaced nucleosomes are depicted by filled ovals; the boxes signify that positions of the nucleosomes within these regions have not been mapped. Structures cleaved by DNase I at half-nucleosomal intervals are represented by semiovals (from Erkine et al., 1995a).

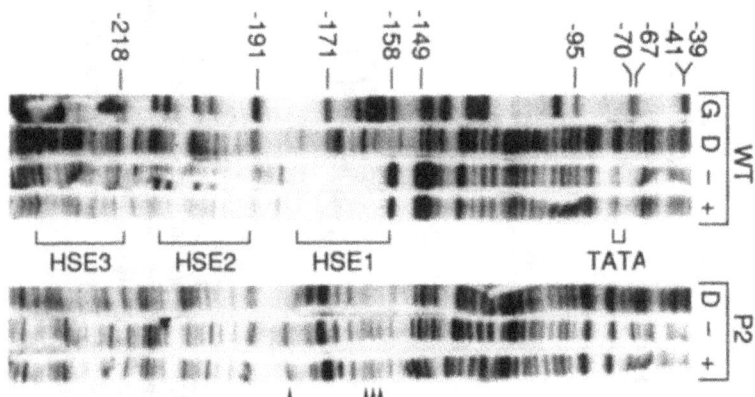

Figure 3. DNase I genomic footprinting of the promoter region of wild-type (**WT**) and double-point-mutated (**P2**) alleles of *hsp82*. Control (–) and five minute heat-shocked (+) cells were converted to spheroplasts, lysed in hypotonic buffer and digested with DNase I. Chromatin-specific cutting can be discerned by comparison with deproteinized DNA controls (D). It is evident that the strong HSE1 footprint in **WT** is lost in **P2**. In its place are several chromatin-specific hyper-reactive cleavage sites (arrowheads) that are present ± heat shock. G, guanine-specific sequencing ladder (from McDaniel et al., 1989).

DNase I hyper-reactive cleavages flanking HSE1, whose pattern changes following heat shock (see Gross et al., 1990) (summarized in Figure 4). Similarly, helix distortion is evident in the vicinity of the TATA box, since hydroxyl radical detects a lower strand-specific hypersensitive site at the dyad center of an adjacent polypurine tract 1.5 helical turns downstream (Figure 5). The extent to which this may reflect the ability of TBP to sharply kink the DNA at the end of the TATAAA sequence (Kim et al., 1993a, 1993b) is unknown. The absence of discernable modulation in the DNase I cleavage pattern (Figure 3) argues against the presence of a specifically positioned nucleosome within the *HSP82*⁺ promoter, a condition that contrasts markedly with certain mutated alleles of *HSP82* (see below). Thus,

```
TACGCTTGTCACATATTGTTCGAACAATTCTGGTTCTTTCGAGTTTCGCAGAACTTTTTG
ATGCGAACAGTGTATAACAAGCTTGTTAAGACCAAGAAAGCTCAAAGCGTCTTGAAAAAC
     -250        -240        -230        -220        -210        -200
                                      HSE3                    HSE2


AATTTTTCTTTTTTTCTAGAACGCCGTGGAAGAAAAACACGCGCATGGTTTTATGAGCG
TTAAAAAGAAAAAAAGATCTTGCGGCACCTTCTTTTTGTGCGCGTACCAAAATACTCGC
  -190        -180        -170        -160        -150        -140
                    HSE1


GTTAATTCTCATCTTAATACCAACCAGGTCCTTCCGCCACCCCCTAAAACATATAAATAT
CAATTAAGAGTAGAATTATGGTTGGTCCAGGAAGGCGGTGGGGGATTTTGTATATTTATA
  -130        -120        -110        -100        -90         -80
                                                              TATA


GCAGCTTATCCCTTCAATTCTTAACATCTGTGACCTCCTCATTTCTTCCCGCTGTATTAG
CGTCGAATAGGGAAGTTAAGAATTGTAGACACTGGAGGAGTAAAGAAGGGCGACATAATC
  -70         -60         -50         -40         -30         -20


AGTTCAAGAAATCATACCTGATAGAAAATAGAGTCCTATAAACAAAAGCACAAACAAACA
TCAAGTTCTTTAGTATGGACTATCTTTTATCTCAGGATATTTGTTTTCGTGTTTGTTTGT
  -10         +1          +10         +20         +30         +40         +50
```

Figure 4. *HSP82* upstream sequence and sites of chemical and enzymatic protection *in vivo*. Numbering is relative to the principal transcription initiation site (+1) (Farrelly and Finkelstein, 1984). Elements bearing similarity to the heat shock consensus sequence (HSCS; see text) are boxed; HSEs 1–3 exhibit 9/12, 9/12, and 8/12 matches, respectively, to conserved nucleotides of the HSCS. Also boxed is the TATA element, which shows a 6/6 match to the conserved TATAAA motif (Chen and Struhl, 1988). (A detailed analysis of *HSP82* upstream regulatory elements is provided in Erkine et al., 1995b.) For the *HSP82*[+] allele, guanines protected from DMS methylation in intact cells are indicated in bold, nucleotides protected from DNase I in spheroplast lysates are indicated by outlined letters, nucleotides protected from hydroxyl radical are delineated by solid bars, and nucleotides hyper-reactive to hydroxyl radical are indicated in bold italics. For the *hsp82*–ΔHSE1 allele, nucleotides specifically cut by DNase I are indicated by filled circles; 5' and 3' extremities of the cleavage ladder are indicated by black triangles while its dyad axis is indicated by a white triangle (from Gross et al., 1993).

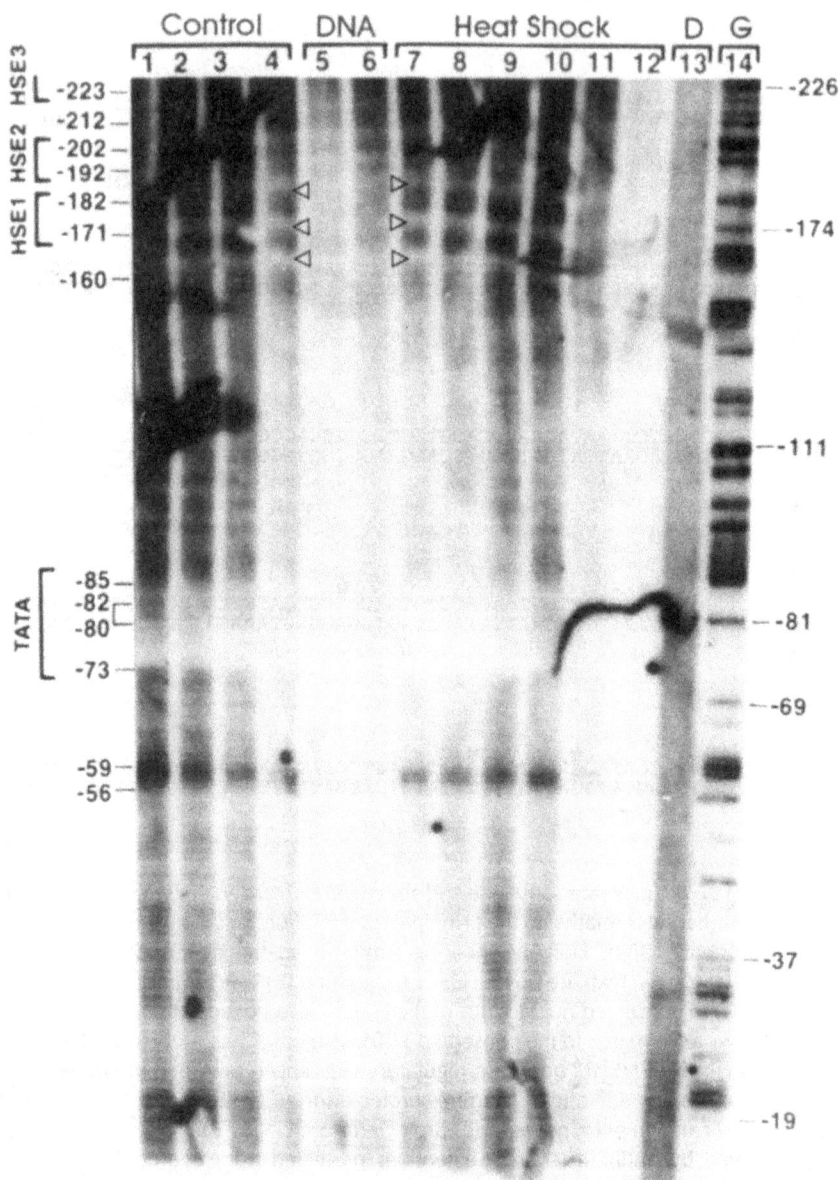

Figure 5. Hydroxyl radical genomic footprinting of the promoter region of the *HSP82⁺* gene in spheroplast lysates. Three regions of constitutive protection/enhancement are evident: (1) a triple footprint over HSE1 between −154 and −183 (arrowheads); (2) an interrupted footprint over the TATA box between −73 and −86; and (3) a hypersensitive site between −56 and −59 that resides at the dyad center of a polypurine tract spanning −49 to −66. DNA (D), naked genomic DNA; G, guanine-specific sequencing ladder (from Gross et al., 1990).

the vacancy at HSEs 2 and 3 is probably not due to their being coiled in a nucleosome in contrast to the vacant HSEs upstream of the *Drosophila hsp26* gene (Thomas and Elgin, 1988); rather, it most likely reflects an intrinsically lower affinity of HSF for these sequences.

1. Biochemical Evidence Implicating HSF and TBP as Factors Binding to HSE1 and TATA

That HSF is in fact bound to HSE1 *in vivo* is supported by several observations. First, the pattern of DNase I cleavage over HSE1 resembles the *in vitro* cutting pattern of two yeast HSF (yHSF)–HSE complexes within the *SSA1* (*HSP70*) promoter, particularly the breadth of the HSE1 footprint, 30 bp, and the presence of flanking hypersensitivity (Wiederrecht et al., 1987). Second, hydroxyl radical footprinting indicates the presence of three sites of enhanced protection on each strand within the HSE1 region, each 5–6 nucleotides in breadth and uniformly spaced in 12 nucleotide intervals over 33 bp, with corresponding upper and lower strand footprints offset by three nucleotides (see Figure 5, arrowheads; data summarized in Figure 4). These results are consistent with a model, presented in Figure 6, Part A, in which the heat shock factor complex contacts one surface of an underwound double helix whose local period is ~12 bp/turn, similar to A–form DNA. The presence of DNase I hypersensitivity is further consistent with helical distortion (due to a return of the helix to the B–form, or perhaps due to compensating overwound helix bookending the underwound footprinted region). Moreover, the presence of six discrete contact sites on one helical face of HSE1 is consistent with evidence indicating that HSF binds DNA as a trimer (Perisic et al., 1989; Sorger and Nelson, 1989). Here, two trimers could bind over the 33 bp region, one between –151 and –167 and the other between –166 and –183. Genetic experiments described below confirm the identification of HSF as the HSE1–bound factor.

With respect to TBP, our genomic footprinting experiments provide strong evidence for the presence of a protein that strongly interacts with the sugar–phosphate backbone of each DNA strand but has no detectable contacts within the major groove. The breadth of DNase I protection, 19 bp, is similar to that mediated by yeast TBP *in vitro* (Buratowski et al., 1988; Hahn et al., 1989). As illustrated in Figure 6, Panel B, the pattern of protected nucleotides readily align to one face of B–form helix (10 bp/turn); the exposed internal segment of each strand being situated to the opposite face. Interestingly, intimate contacts with the sugar-phosphate backbone extend 6–7 nucleotides on each side of the consensus TATAAA element, implying a role for flanking nucleotides in the sequence-specific binding of TBP *in vivo*. This conclusion is consistent with recent X–ray crystallographic evidence indicating that more than the core TATAAA motif is involved in TBP–TATA interactions (Kim et al., 1993a, 1993b). In addition, the 5′ extension of the *in vivo* footprint may reflect the presence of TFIIA, which binds strongly to the TBP/TATA box complex *in vitro* (Buratowski et al., 1989).

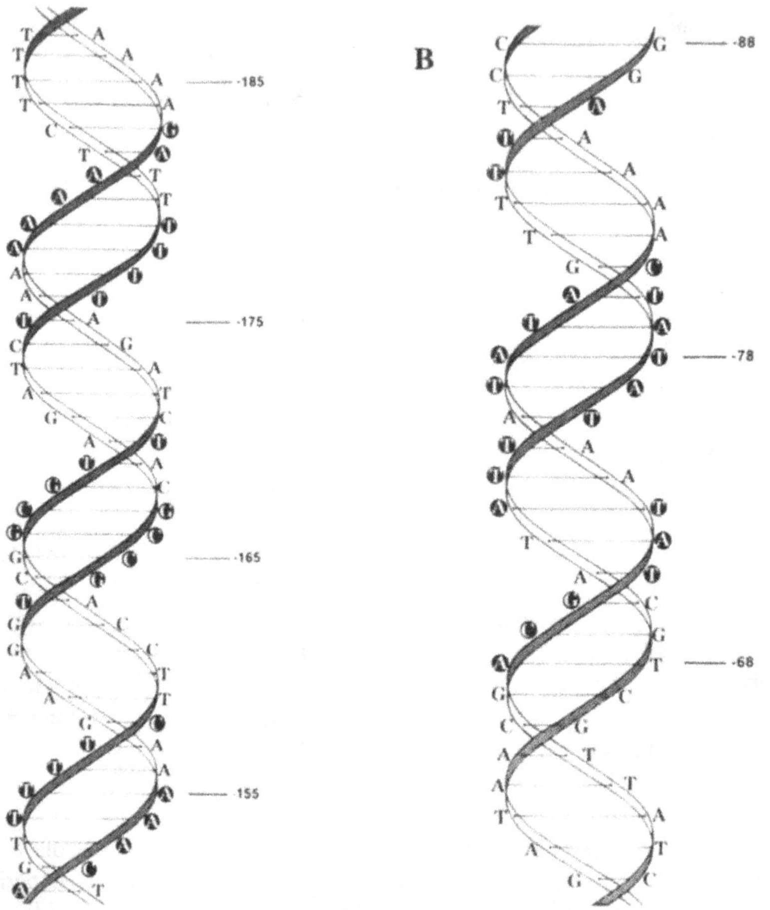

Figure 6. Proposed *in vivo* binding surfaces along the sugar-phosphate backbone of duplex DNA for (**A**) heat shock factor (HSF) and (**B**) TATA-binding protein (TBP). Nucleotides contacted by proteins, as deduced from hydroxyl radical genomic footprinting, are backshadowed (from Gross et al., 1990).

B. Point Mutations within HSE1 Erase the HSF Genomic Footprint and Preferentially Abolish Basal Transcription

To address the role of the HSF/HSE1 complex *in vivo*, we constructed a set of isogenic strains in which the wild-type (WT) *HSP82* gene was replaced by *hsp82* alleles containing 1–5 point mutations within HSE1. Such strains were made by targeting DNA fragments harboring mutated *hsp82* alleles to the gene's native locus by homologous recombination. We employed this approach since it overcomes

A

STRAIN	HSE1 SEQUENCE	RNA LEVEL	
		(−)	(+)
WT	-177 TTTCTAGAACGCCGTGGAAG -158	100	1500
G161	T	10	750
G2	TT	3	600
P2	G C	3	350
P3	G C T	2	300
G5	A T TT C	2	170

B

STRAIN	HSE1 SEQUENCE			RELATIVE RNA LEVELS		
				NHS	HS (11')	HS (25')
HEAT SHOCK CONSENSUS	NTTCNNGAANNTTCNNGAAN					
WILD TYPE	-191 GAATTTTTCTTTT TTTCTAGAACGCCGTGGAAG AAAAACACGCGCA -145			100	1500	1500
ΔHSE1	-191 GAAT Δ AAACACGCGCA -145			≤1	15	15
ΔHSE1•	-191 GAAT GATCTCCTTTAGCTTCTCGACGTGGGCCTTTT AAACACGCGCA -145			≤1	15	25
ΔHSE1•t	-191 GAATTTTTTTTTT TAGCTTATCGACGTGGGCCT AAAAACACGCGCA -145			≤1	170	160

Figure 7. Functional consequences of HSE1 mutations. **(A)** Point mutagenesis. Nucleotides substituted in each isogenic strain are indicated, as is the relative *HSP82* transcript level in control (−) and 15-minute heat-shocked (+) cells. **(B)** Regional mutagenesis. Nucleotides deleted or substituted in each isogenic strain are shown, as are relative *HSP82* transcript levels under nonshocked (NHS) and heat-shocked (HS) conditions. The underlined sequence constitutes HSE1; bold letters represent mutated nucleotides; outlined letters correspond to conserved nucleotides of the HSCS; Δ refers to a 32 bp deletion (−156 to −187).

161

potential artifacts caused by chromosome position effects, flanking foreign DNA sequences, altered gene copy number, and timing of replication during S phase. Furthermore, we could introduce inactivating mutations into the *HSP82* promoter, since *hsp82* null mutants are viable due to the presence of *HSC82*; similarly, *hsc82* null mutants are viable but double null mutants are not (Borkovich et al., 1989).

We anticipated that since the point mutations illustrated in Figure 7, Panel A, altered one or more consensus *n*GAA*n* units, they would severely affect induced transcription. Indeed, each allele harbors mutations in nucleotides strongly protected by HSF from *in vivo* DMS methylation (Gross et al., 1990). However, such mutations had only a mild effect on heat-shock-induced transcription, whereas they virtually abolished basal function. Thus, point mutations within HSE1 convert a gene that is normally 15–fold heat-shock-inducible into one that is induced as much as 200–fold (e.g., strain **G2**). The phenotype of these point-mutated alleles indicates that HSF binding to HSE1 is responsible for triggering basal-level transcription. Such a function resolves the paradox of why in budding yeasts (Sorger et al., 1987; Jakobsen and Pelham, 1988), but not in other eukaryotes (Wu, 1984; Zimarino et al., 1990), HSF binds to HSEs before heat shock induction.

To investigate the impact of point mutations on the chromatin structure of the *hsp82* promoter region, we digested spheroplast lysates with DNase I and visualized *HSP82*-specific cleavage products by indirect end-labeling. As shown in Figure 8, a double point mutation within HSE1, creating a strain termed **P2**, has minimal effect on the pair of broad hypersensitive sites centered at positions –260 and –100 bp. Despite a >30–fold reduction in basal transcription and a >fourfold reduction in induced, the degree of hypersensitivity is not appreciably diminished (compare **WT** and **P2** lanes in Figure 8; see also McDaniel et al., 1989). Notably, however, the region centered between the two hypersensitive sites and mapping to HSE1 is less well protected, suggesting an altered HSF–HSE1 interaction (see below). We have observed similar results for the other point-mutated *hsp82* alleles; for example, even in **G5**, a strain bearing five point mutations within HSE1, promoter-associated DNase I hypersensitivity is retained. These data argue that 1–5 point mutations within HSE1 have little effect on the nucleosome-free environment of the promoter, a conclusion strongly supported by the absence of stable micrococcal nuclease (MNase) cleavage products derived from the promoter region in these strains.

We next examined the effect of HSE1 point mutations on protein/DNA interactions at nucleotide resolution using DMS and DNase I genomic footprinting. Whereas the **WT** allele is strongly protected over HSE1 ± heat shock, the double point mutation in strain **P2** abolishes virtually all detectable protein/DNA interactions within the UAS region (HSEs 1–3), even following heat shock (Figure 3). Likewise, guanine residues within the mutated HSE exhibit no discernable protection from DMS in living cells. These genomic footprinting experiments are consistent with gel shift experiments indicating that the P2 mutation within HSE1 abolishes its ability to bind yHSF (McDaniel et al, 1989). Thus, all detectable

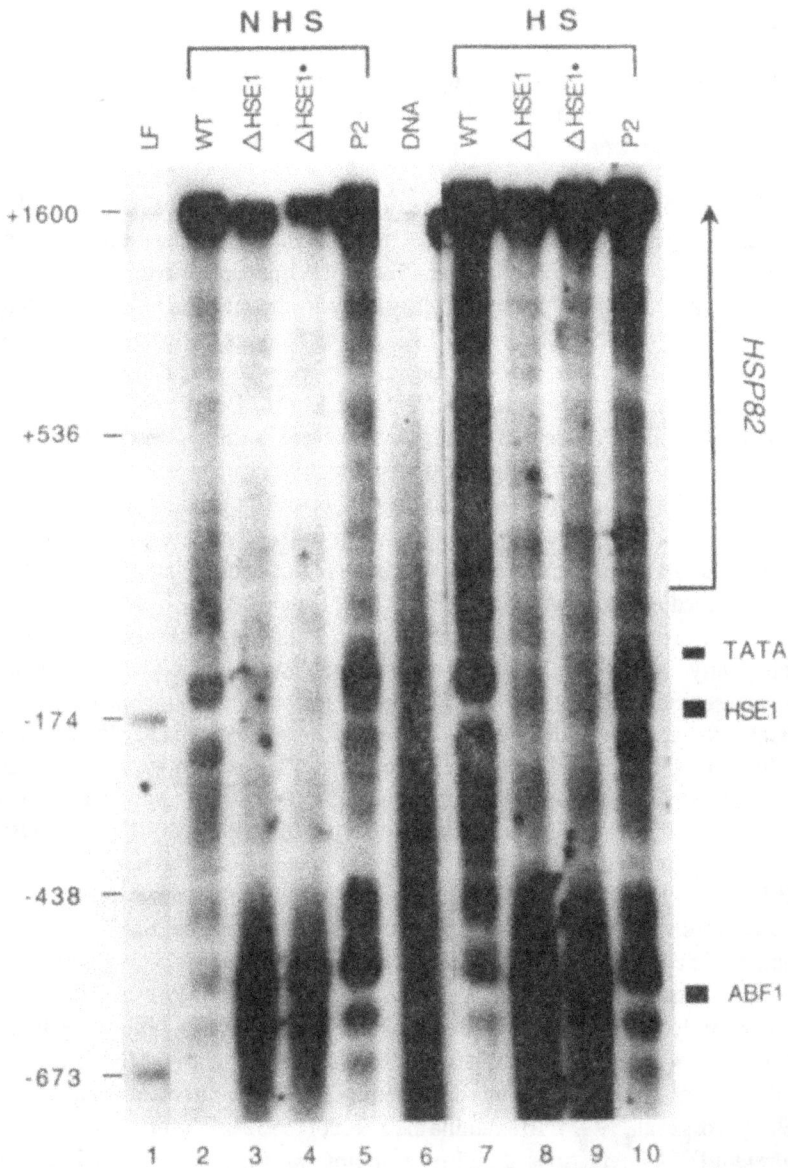

Figure 8. DNase I chromatin footprints of wild-type and mutant alleles of *HSP82*. Spheroplast lysates, derived from nonshocked (NHS) and heat shocked (HS) cells, were digested with DNase I and *HSP82*-specific fragments visualized by indirect end-labeling. Note that promoter-associated DNase I hypersensitive sites, retained in strain **P2** (lanes 5, 10), are abolished in the △**HSE1** strains (lanes 3, 4, 8, 9). LF, landmark restriction fragments; DNA, deproteinized genomic DNA subjected to the same protocol as the lysate samples (from Gross et al., 1993).

163

HSF–HSE1 interactions both *in vivo* and *in vitro* are abrogated by the double point mutation.

C. Deletion of HSE1 and Flanking Nucleotides Abolishes Both Basal and Induced Transcription and Leads to a Dramatic Transition in Promoter Chromatin Structure

The foregoing results would seem to argue against a role for the HSF–HSE1 complex in either directing induced transcription or in establishing the nucleosome-free region over the *HSP82*[+] promoter. To confirm this more directly, we mutated the entire UAS region protected from enzymes and chemicals *in vivo* (Gross et al., 1993). Strikingly, a 32 bp deletion of HSE1 and its flanking nucleotides, generating a strain termed △**HSE1**, results in a considerably more severe phenotype than any of the point mutants: both basal and induced levels of expression are reduced ≥100–fold (Figure 7, Panel B). This more severe phenotype may stem from elements further upstream being functionally compromised by altered spacing. To rule out such a spacing effect, we substituted 32 bp of DNA sequence derived from the *PET56* structural gene—previously shown to lack promoter activity in *Saccharomyces cerevisiae* (Struhl, 1985a)—for the region excised in △**HSE1**, creating a strain termed △**HSE1•**. As shown in Figure 7, Panel B, *hsp82* basal transcription is essentially abolished in the substitution strain. Following heat shock, detectable, albeit greatly reduced, levels of expression are seen. As the phenotype of this mutant closely resembles that of △**HSE1**, it is clear that HSE1 and flanking nucleotides are critically required for both basal and induced expression of *HSP82*. Interestingly, substitution of the 20 bp HSE1 sequence per se results in a phenotype more similar to that of the point-mutated alleles (strain △**HSE1•t** [true]), since strong heat-shock-inducibility is retained (Figure 7, Panel B; see also Figure 11). This result argues for the presence of additional factor(s) bound to the promoter regions of △**HSE1•t** and **P2** whose binding is abolished in the more extensive deletions.

To address the impact of the more extensive mutations on *hsp82* chromatin structure, we conducted a DNase I indirect end-labeling analysis. This assay reveals that the promoter-associated hypersensitivity characteristic of **WT**, and retained in **P2**, is completely abolished in strains △**HSE1** and △**HSE1•** both prior to and following heat shock (Figure 8, lanes 3, 4, 8, and 9). Similarly, MNase analyses confirm the loss of the nucleosome-free promoter region in the deletion/substitution alleles (Gross et al., 1993). To examine the effect of the 32 bp chromosomal deletion on protein/DNA interactions at higher resolution, we performed DNase I genomic footprinting (Figure 9). A striking chromatin-specific modulation in the DNase I cleavage profile is seen in the △**HSE1** samples from both control and heat-shocked cells between positions –120 and +28 (Panel A, lanes 3 and 4). This cleavage ladder, with peaks of DNase I cleavage occurring every 10–11 nucleotides and encompassing a region of ~146 bp, indicates the presence of an extremely stable, rotationally positioned nucleosome centered over the core promoter (nucleosome –**I** of

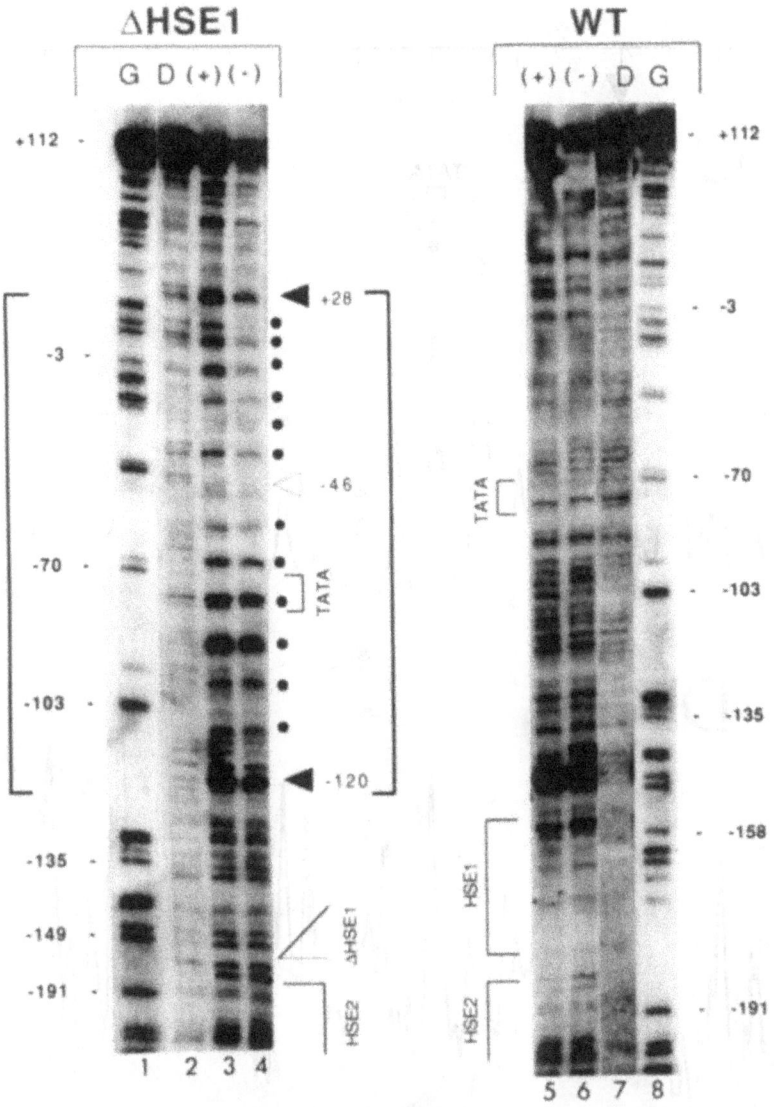

Figure 9. DNase I genomic footprinting of the *HSP82* upstream region in ΔHSE1 and WT strains. **(A)** Spheroplast lysates, generated from control (−) and 15-minute heat-shocked (+) cells, were digested with DNase I, the genomic DNA purified, restricted, and electrophoresed, and the *HSP82*-specific upper strand fragments illuminated by blot-hybridization. The site of the −156 to −187 chromosomal deletion in strain ΔHSE1 is indicated, as is the location of nucleosome −I (large bold brackets; see Figure 10). Chromatin-specific cut sites spaced at 10–11 bp intervals (lanes 3 and 4) are indicated by dots. The extremes of the nucleosome core are delineated by black triangles while its pseudodyad is indicated by a white triangle. D, naked genomic DNA; G, guanine-

(continued)

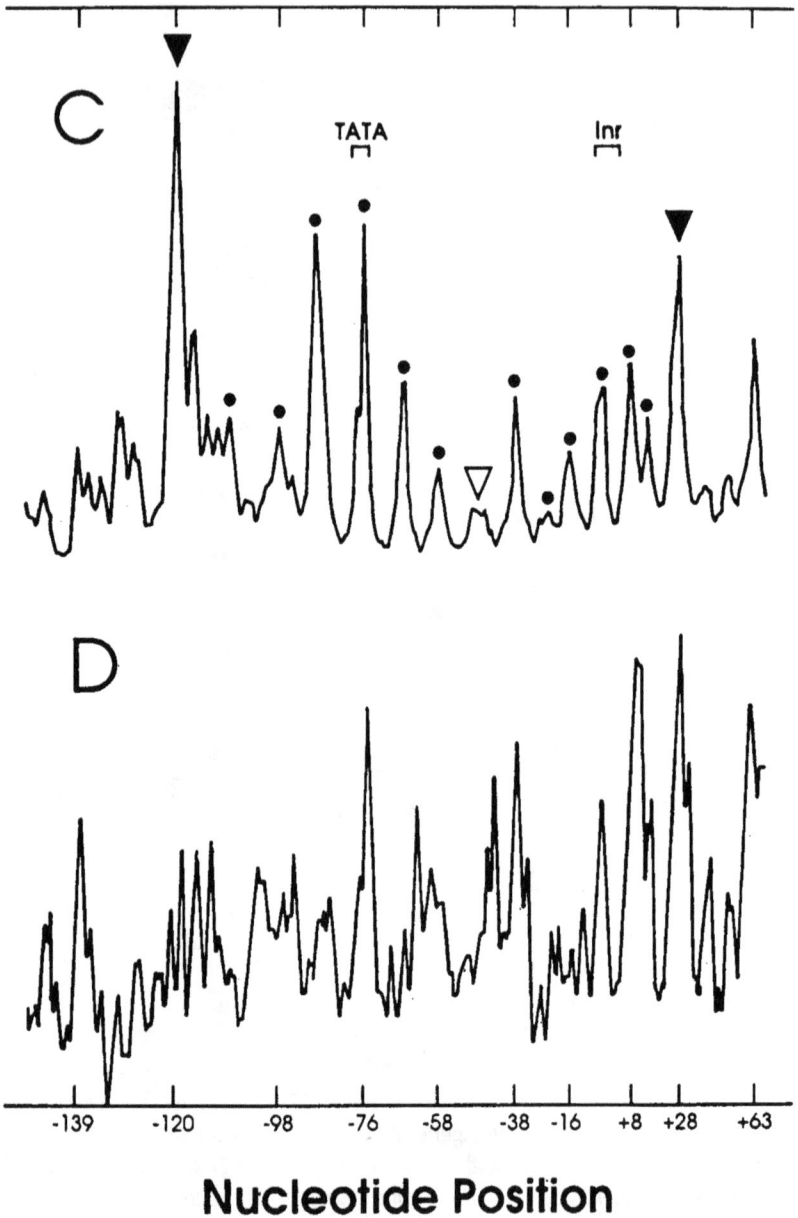

Figure 9. (Continued) specific sequencing ladder. (**B**) Densitometric scans of △**HSE1** chromatin (C) and naked DNA (D) (lanes 3 and 2 of Panel A, respectively). Locations of the TATA box and principal transcription start site (Inr) are indicated. Symbols are as in Panel A (from Gross et al., 1993).

Figure 10). The site of the HSE1 lesion is also assembled into a stable, translationally positioned nucleosome (**-II**) as revealed by DNase I and MNase indirect end-labeling experiments (Gross et al., 1993; see also Figure 12, Panel B).

It is notable that while the pseudodyad of the rotationally positioned nucleosome is relatively inaccessible to DNase I (position –46 [open triangle in Panels A and B of Figure 9]), three of the strongest cleavages map 20–40 bp from the pseudodyad, within the region of the TATA element (at positions –68, –76, and –89; see Panel B, chromatin [C] scan). The intensity and location of these cleavages suggests that while this element is at least partially accessible to its cognate factor, TBP, in chromatin, it is not strongly occupied. Thus, in the absence of the high affinity HSF binding site, a rotationally positioned nucleosome assembles over the core promoter

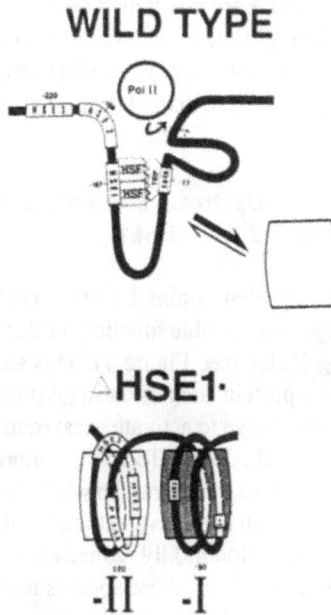

Figure 10. Proposed protein/DNA interactions within the *HSP82* promoter region in wild-type and ∆**HSE1•** strains. Hypothetical nucleoprotein structures are based on genomic footprinting and nucleosome-protected ladder analyses (Gross et al., 1990, 1993). The wild-type promoter is characterized by an open, DNase I hypersensitive chromatin structure. Following a 32 bp substitution of the HSE1 region in strain ∆**HSE1•**, two translationally positioned nucleosomes, designated **–I** and **–II**, replace the DNase I hypersensitive structure. The nucleosome assembled over the core promoter in this strain is extremely stable, being rotationally positioned with respect to the underlying DNA helix (indicated by shading). Interestingly, an unstable nucleosome may exist over the core promoter in the wild-type strain (depicted here as a histone octamer in quasiequilibrium with downstream factors). Note that the 32 bp deletion strain (∆**HSE1**) exhibits an identical mutant phenotype to ∆**HSE1•** (from Gross et al., 1993).

of *HSP82*, potentially blocking TBP–TATA interactions and impeding the formation of the preinitiation complex.

These results indicate that the open chromatin structure characteristic of the wild-type promoter undergoes a dramatic transition upon *in situ* deletion or substitution of the high-affinity HSF site, as suggested by the structural models presented in Figure 10. These illustrate two important points. First, there appears to be a metastable nucleosomal structure over the transcription initiation region even in the wild-type promoter, based on the presence of an MNase-protected fragment of mononucleosomal length. This structure is not a positioned nucleosome, since its presence is not detected by either DNase I or MNase indirect end-labeling (Gross et al., 1993). Second, chromatin mapping experiments indicate that within the △HSE1 promoter, HSEs 2 and 3 are positioned very close to the dyad (within 20 bp) of nucleosome **–II**, the region least accessible to *trans*-acting factors *in vivo* (Simpson, 1990). Hence, if HSF were to bind to either (or both) of these low-affinity sites, it would almost certainly have to disrupt the underlying nucleosomal structure (see below).

D. Functionally Redundant Upstream Elements Underlie the Phenotypic Difference between Strains P2 and △HSE1

Comparison of the phenotype of strains **P2** and △**HSE1•t** with that of strains △**HSE1** and △**HSE1•** suggests a possible functional role for the poly(dT) sequence upstream and overlapping HSE1 (see Figure 7). This sequence could potentially serve as a binding site for a protein such as datin (Winter and Varshavsky, 1989). Similar sequences have been shown to activate yeast transcription both *in vitro* and *in vivo* (Struhl, 1985b; Lue et al., 1989). To address more directly the role of this sequence *in vivo*, we created an isogenic strain in which the poly(dT) tract spanning –178 to –188 was substituted with an equivalent length of inert DNA sequence. As indicated in Figure 11, this mutation mildly reduced basal transcription, but it had no effect on induced expression. Thus, there appears to be functional redundancy between HSE1 and the adjacent poly(dT) element.

A second example of functional redundancy has been seen when *in situ* deletions of *HSP82* far upstream sequence are combined with point mutations in HSE1. Two types of *HSP82* alleles have been generated: (1) those with a 483 bp deletion extending 5′ of the poly(dT) tract (position –190), and (2) those with a 408 bp deletion extending 5′ of HSE3 (position –265). The upstream extremity of both deletions, a *Cla* I site at position –673, was selected since insertions there as large as 5 kb have no effect on *HSP82* promoter function (Lee and Gross, 1993). Strains harboring only one upstream deletion or the other exhibit an essentially normal phenotype (Figure 11, strains **–190** and **–265**). In contrast, when the –190 deletion is coupled with the P2 mutation, basal transcription is virtually abolished while induced transcription is reduced nearly 20-fold (strain **–190/P2**). Importantly, the –265 deletion coupled with the P2 mutation (strain **–265/P2**) has a phenotype

Strain

Strain	NHS	HS	DNaseI HSS	Promoter-Associated Nucleosomes
WT	+++	+++	+++	–
P2	±	++	+++	–
G5	±	+	+	–
-265	+++	+++	ND	ND
-265/P2	±	++	ND	ND
-190	++	++	ND	ND
-190/P2	–	+	ND	ND
ΔdT	+	+++	++	ND
ΔHSE1•t	–	+	++	ND
ΔHSE1•	–	–	–	++
ΔHSE1•2•3	–	–	ND	ND

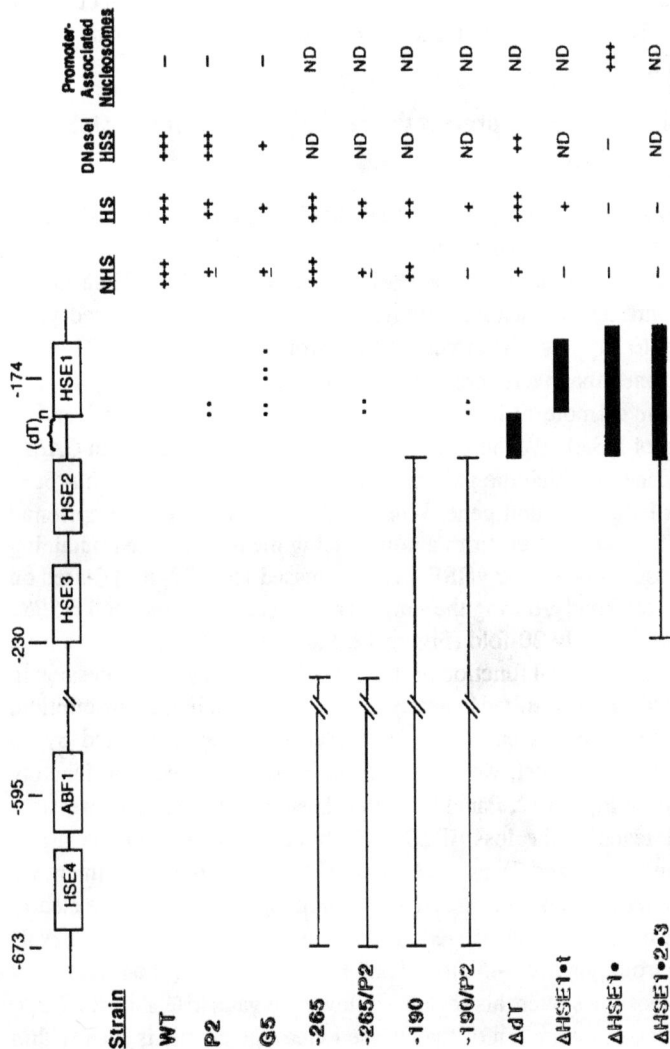

Figure 11. Phenotypes of *hsp82* promoter mutants. The nature and location of mutations introduced into each isogenic strain are schematically depicted, with dots representing point substitutions, filled bars representing regional substitutions, and horizontal lines representing regional deletions. Four phenotypes are summarized: *HSP82* transcript levels in nonshocked (NHS) and heat-shocked (HS) cells, the presence of promoter-associated DNase I hypersensitive sites (HSS), and the presence of stable nucleosomes over the core and proximal promoter regions. N.D., not determined

identical to strain **P2** (Figure 11), indicating that the crucial compensatory function is localized between –190 and –265 (implicating HSE2 and/or HSE3). Therefore, despite the absence of a discernable DNase I genomic footprint over the UAS region of the *hsp82*–P2 allele, these experiments provide genetic evidence for the cooperative binding of factors—presumably HSF trimers—to the HSE1 and HSE2/HSE3 regions. While such cooperative interactions probably also exist within the **WT** allele, they are minimally important to promoter function. In contrast, they are abolished in **–190/P2** and △**HSE1**, which may account for the null phenotype of these strains.

E. Overexpression of HSF Suppresses the Null Phenotype of the △HSE1 Strains

The above analyses conclusively demonstrate that HSE1 is an essential determinant of the nucleosome-free state of the *HSP82* promoter. To investigate more directly the role of HSF in activating *HSP82* transcription and establishing the nucleosome-free chromatin structure over its promoter, we overexpressed yeast HSF in an attempt to suppress the effects of the chromosomal deletion (Gross et al., 1993). We reasoned that overexpression of the protein might result in its stable association with one or more of the low affinity HSF sites (e.g., HSEs 2 and 3) disposed upstream of HSE1. We therefore transformed strain △**HSE1** with *GAL1– HSF*, a chimeric gene in which the galactose-inducible *GAL1* promoter has been fused to the yeast HSF structural gene. When △**HSE1** cells bearing an episomal copy of *GAL1–HSF* were shifted from a noninducing medium to one containing 0.5 percent galactose, intracellular yHSF levels increased 15 to 30–fold (based on Western analysis). Strikingly, during the same period, heat-shock-induced *HSP82* RNA levels increased nearly 20-fold (Figure 12, Panel A). While there was little detectable suppression of basal function in strain △**HSE1**, yHSF overexpression in HSE1 point-mutated strains resulted in nearly full restoration of basal transcription.

To ask if the suppression seen at the functional level was paralleled by an alteration at the structural level, we performed a DNase I indirect end-labeling analysis. As shown in Figure 12, Panel B, in the absence of HSF overexpression, deletion of HSE1 leads to the loss of DNase I hypersensivity over the *hsp82* promoter (compare lanes 1 and 7), as seen above. However, in this experiment we electrophoresed DNA samples less extensively than in Figure 8, allowing a clearly resolved ladder of bands to be visualized. These bands map at ~170 bp intervals and span the gene from position –450 to at least +1400, indicating the presence of 11 positioned nucleosomes over this region. Following a galactose shift for 3.5 to 7.5 hours, a progressive disruption of the DNase I cleavage profile is seen within the *hsp82* promoter, particularly in the heat-shock induced samples. Thus, high levels of yHSF appear to disrupt the stable nucleosomal structure over the mutant *hsp82* promoter region. To confirm yHSF-mediated nucleosomal disruption more directly, we performed an MNase nucleosome-protected ladder analysis (Figure 12,

Panel C). While the **WT** promoter is cleaved into a heterodisperse array of fragments (lanes 1 and 2), a discrete nucleosome-protected ladder is evident in the △**HSE1** samples in the absence of HSF overexpression (lanes 3 and 5). However, cells transformed with pGAL1HSF show a marked galactose-dependent disruption of the nucleosomal ladder both prior to and following a 15-minute heat shock (most prominently seen in the 7.5 hour samples; lanes 8 and 10). Therefore, high levels of yHSF disrupt the stable nucleosomal structure over the mutant *hsp82* promoter region. Such alterations are more marked following transcriptional derepression, and are accompanied by a re-establishment of DNase I hypersensitivity strongly resembling the hypersensitivity of the wild-type allele.

To rule out a nonspecific global effect mediated by overexpressed yHSF, we constructed a strain in which the 42 bp immediately upstream of the HSE1 mutation in △**HSE1•** were deleted (Figure 11). This strain, lacking all three promoter-proximal heat shock elements and termed △**HSE1•2•3**, was transformed with *GAL1–HSF* and subjected to galactose shift as above. As shown in Figure 13, *hsp82* transcript levels in the triple HSE mutant are dramatically reduced relative to those seen in △**HSE1•**, indicating that the sequence lying immediately upstream of HSE1—and encompassing HSEs 2 and 3—mediates the effects of overexpressed yHSF. As HSEs 2 and 3 map very close to the dyad of nucleosome **–II** (see Figure 10), our results are consistent with the possibility that yHSF can bind to a stably positioned nucleosome, and disrupt (or displace) it. The MNase result in particular argues that binding and disruption of the *underlying* nucleosome can take place under noninducing conditions (Figure 12, Panel C, compare lanes 4 versus 3, and 8 versus 7). The DNase I result indicates that additional disruption, probably of nucleosome **–I**, occurs following heat shock. Thus, it is likely that suppression of the null phenotype of the induced *hsp82–*△HSE1 gene involves a two-step process:

Figure 12. (overleaf) Overexpression of yHSF suppresses the null phenotype of the *hsp82–*△HSE1 allele and reinstates the nucleosome-free, DNase I hypersensitive environment over the mutated promoter. **(A)** Northern analysis of pGAL1HSF-transformed △**HSE1** cells subjected to a 15-minute, 30°–39° C heat shock. Cells were cultivated in synthetic medium containing 2 percent raffinose, shifted to 0.5 percent galactose/1.5 percent raffinose for the indicated times, then heat-shocked. Quantifications of relative *HSP82* RNA levels, internally normalized to those of *ACT1*, are provided. **(B)** DNase I chromatin footprints of pGAL1HSF-transformed △**HSE1** cells in control (–) and heat shocked (+) cells. Galactose shift for 0, 3.5, and 7.5 hours was as above; DNA isolation and analysis were as in Figure 8. The DNase I cleavage profile of wild-type (**WT**) chromatin is provided in lane 7. **(C)** Nucleosome-protected ladder analysis of pGAL1HSF-(+) or YCp50-transformed (–) △**HSE1•** cells ± heat shock following galactose shift for 3.5 or 7.5 hours. Nuclei were isolated, digested with micrococcal nuclease, and the purified DNA eletrophoresed and blot-hybridized with a probe spanning the gene's UAS region (from Gross et al., 1993).

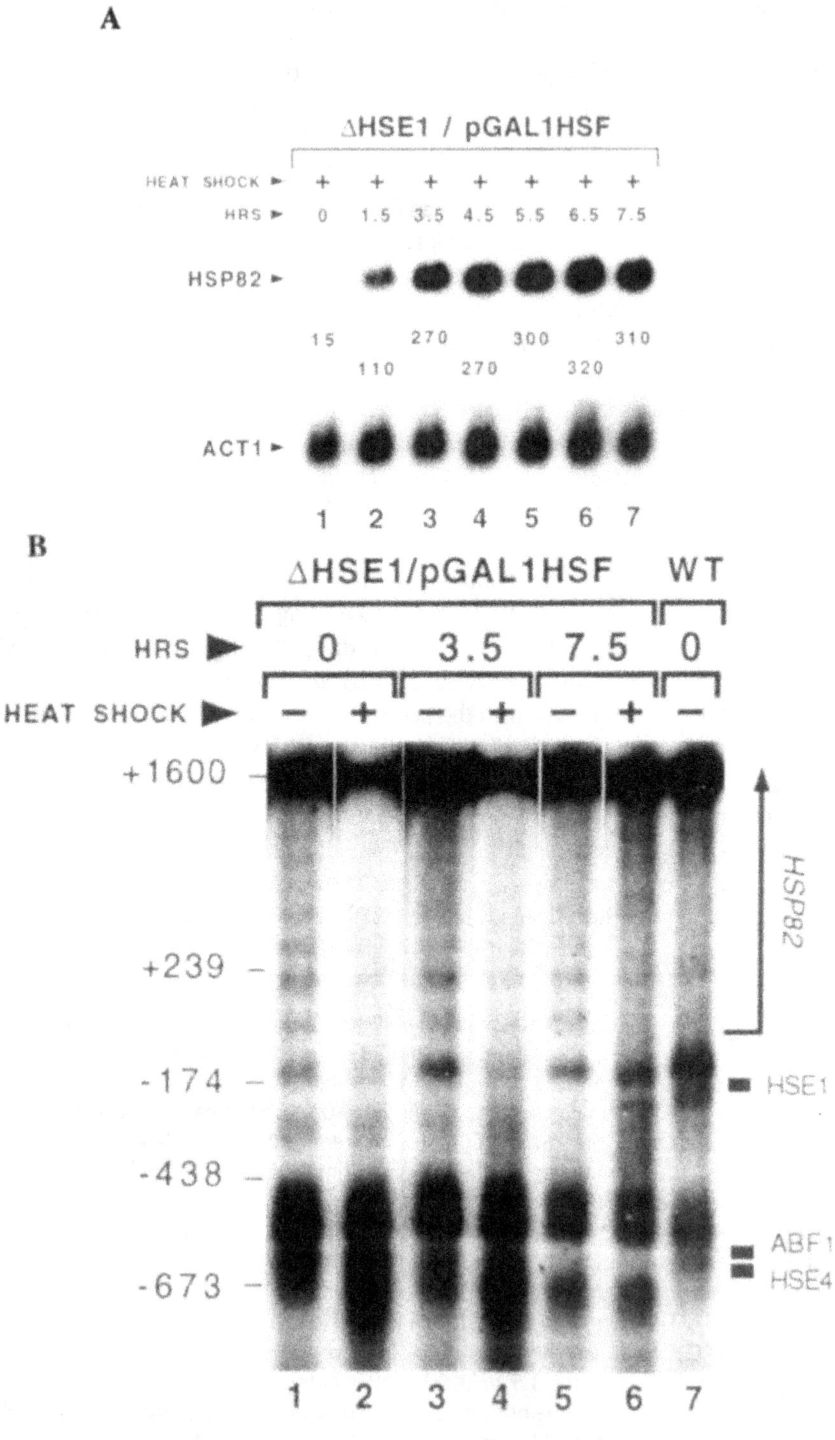

Figure 12. (Continued)

172

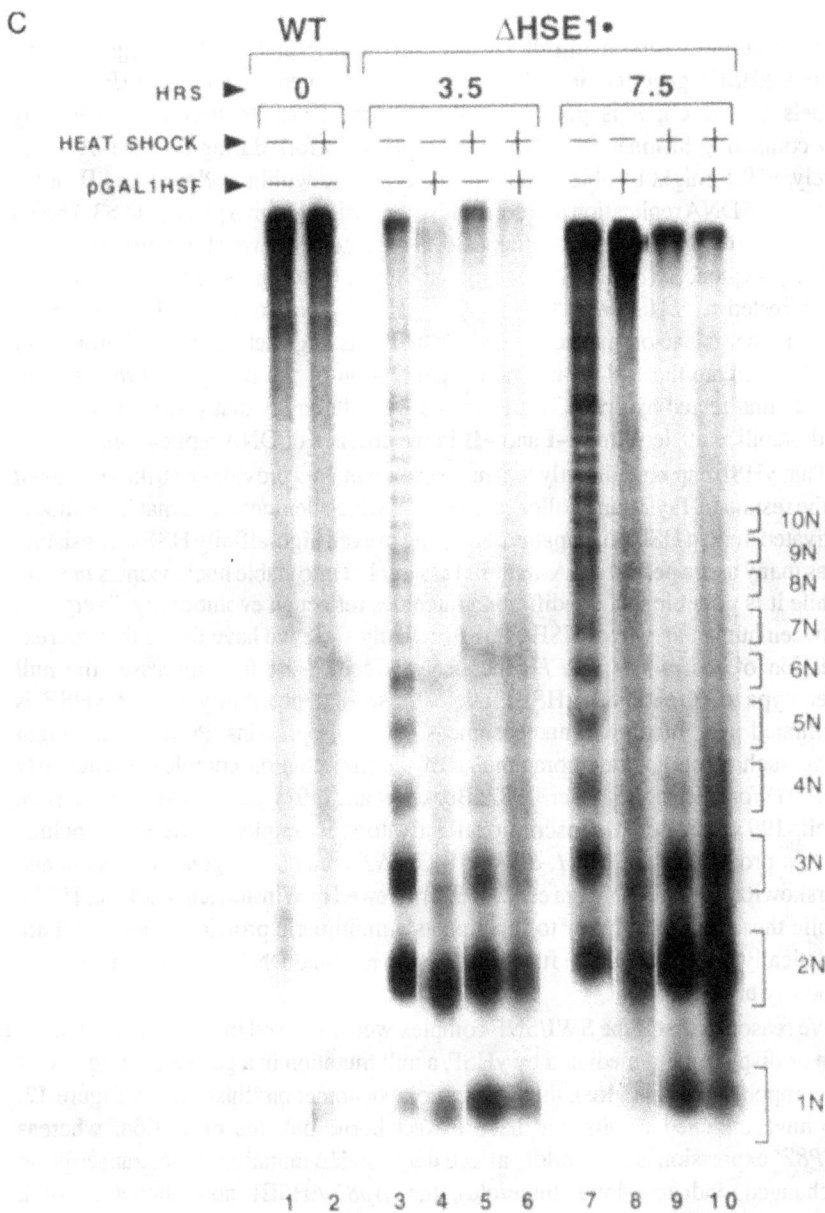

Figure 12. (Continued)

173

(1) HSF binding and disruption of the UAS nucleosome **-II**; and (2) heat-shock-dependent disruption or displacement of the TATA-Inr nucleosome **-I**.

F. HSF–mediated Suppression Can Occur in the Absence of DNA Replication But Is Blocked by an Extragenic Mutation in the *SWI1* Gene

Since definitive chromatin alterations over the UAS and TATA-Inr regions of the *hsp82*–ΔHSE1 gene require >3.5 hour shift in 0.5 percent galactose (Figure 12, Panels B and C), it is possible that yHSF mediates its structural effects by out-competing histones for DNA at the replication fork during S phase. Alternatively, yHSF might be able to bind to its target site within nucleosome **-II** in the absence of DNA replication and concomitant transit through S phase, but >3.5 hours are required to accumulate a sufficient level of protein to permit binding to the lower affinity sequences (HSE2/HSE3). To distinguish between these two possibilities, we arrested a *GAL1–HSF* transformed **ΔHSE1** derivative in G1 with mating pheromone (α–factor) and subjected arrested cells to galactose shift as before. Our results indicate that yHSF–mediated suppression of the null phenotype is completely unaffected by prior G1 arrest. These results imply that yHSF can displace or destabilize nucleosomes **-I** and **-II** in the absence of DNA replication.

That yHSF can bind directly to a nucleosome *in vivo* provides a striking contrast to the results of Taylor and colleagues (1991), which demonstrated that heat-shock-activated human HSF is incapable of binding to even high-affinity HSEs (consisting of as many as six perfect *n*GAA*n* units) assembled into stable nucleosomes *in vitro*. While it is possible that the different outcomes reflect an evolutionary divergence between human and yeast HSFs, this is unlikely since we have found that overexpression of the mouse *HSF1* gene in yeast cells partially suppresses the null phenotype of the *hsp82*–ΔHSE1 allele. A second possibility is that yHSF is facilitated in its binding to nucleosome **-II** by other proteins. Proteins that might serve such a role include components of the preinitiation complex, particularly TFIID (Workman and Roeder, 1987; Becker et al., 1991); acetylated histones (Lee et al., 1993); or global transcriptional activators. Examples of the latter include protein products of the *SWI1*, *SWI2*, *SWI3*, *SNF5*, and *SNF6* genes (Peterson and Herskowitz, 1992; Yoshinaga et al., 1992; reviewed in Winston and Carlson, 1992). While these proteins appear to function as a multimeric protein complex and are genetically linked to histone function, they do not bind DNA and their *in vivo* role is poorly understood.

We reasoned that if the SWI/SNF complex were involved in nucleosome disruption or displacement mediated by yHSF, a null mutation in a gene encoding one of its components might block the suppression phenomenon illustrated in Figure 12. We have obtained results that have in fact borne out this prediction: whereas *HSP82*[+] expression is only mildly affected by a *swi1*Δ mutation (basal transcription unchanged; induced down threefold), the *hsp82*-ΔHSE1 null phenotype of a *GAL1-HSF* transformed *swi1*Δ strain is not detectably suppressed following galac-

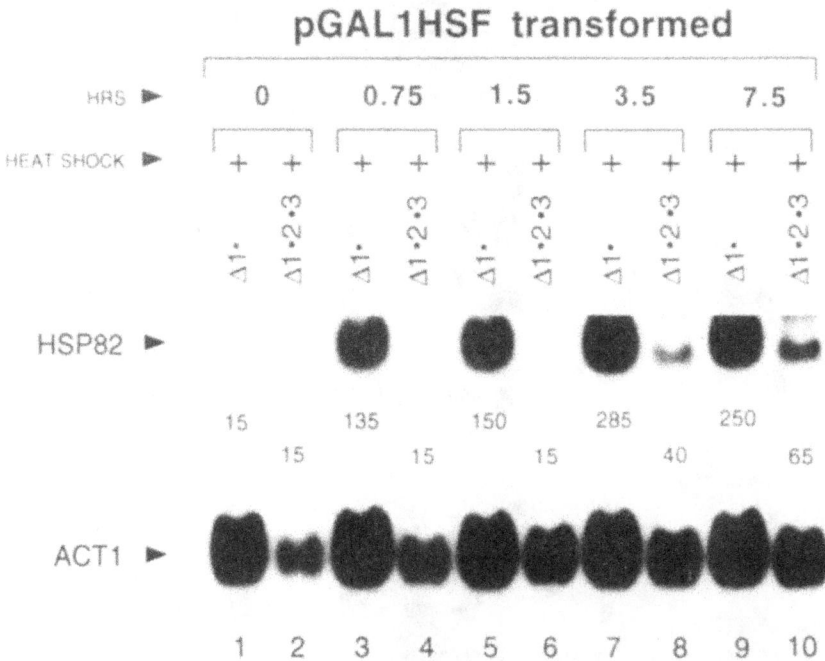

Figure 13. Deletion of heat shock elements 2 and 3 virtually abolishes yHSF-mediated suppression of the *hsp82–*ΔHSE1• allele. Shown is a Northern analysis of pGAL1HSF-transformed Δ**HSE1•** and Δ**HSE1•2•3** cells subjected to a 15-minute, 30°–39° C heat shock following shift to galactose-containing medium for the indicated times (from Gross et al., 1993).

tose shift. This result indicates that HSF requires a functional *SWI1* product to suppress the null phenotype of strain Δ**HSE1**. Whether the *swi1*Δ mutation blocks HSF binding to nucleosome –**II**, the subsequent, heat-shock-induced displacement of nucleosome –**I**, or a step further downstream in the transcription initiation pathway, is currently under investigation.

III. STUDIES ON THE *HSC82* GENE

Similar to *HSP82*, the promoter region of *HSC82* is characterized by the presence of a broad DNase I hypersensitive site (Figure 14; see Figure 2 for a schematic representation of the gene's chromatin structure). Within this hypersensitive domain are internal footprints reflecting the presence of sequence-specific regulatory proteins, whose sites of interaction map to the TATA box at position –95, the promoter–*distal* heat shock element (HSE1) at position –165, and an upstream GRF2 site at position -205, as shown by DNase I genomic footprinting (Erkine et al., 1995a). Interestingly, a second heat shock element, termed HSE0, remains vacant under both transcriptional states despite exhibiting an equally strong match

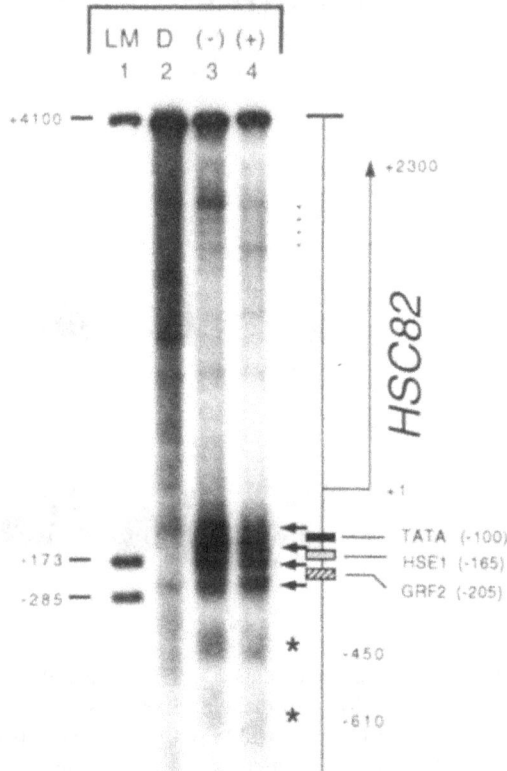

Figure 14. DNase I chromatin footprints of the *HSC82* locus ± heat shock. Spheroplasts were lysed in a hypotonic buffer, digested with DNase I, and the purified DNA restricted, electrophoresed, blotted, and indirectly end-labeled with an *HSC82*-specific hybridization probe. A single hypersensitive domain, mapping to the promoter region of *HSC82*, can be seen in both control (–) and heat-shocked (+) samples. Three internal footprints corresponding to the TATA element, the promoter-*distal* heat shock element (HSE1), and a consensus GRF2 site, are indicated (see Figure 15). LM, landmark restriction fragments; D, naked genomic DNA (from Adams et al., 1995).

(10/12) to conserved nucleotides of the HSCS. Preferential occupancy of the heat shock element centered at position –165 parallels the situation at *HSP82* and supports the notion that additional nucleotides besides the conserved TTC/GAA core sequences are important in heat shock element function (Amin et al., 1988; Fernandes et al., 1994). Indeed, within each HSE1 is the octameric sequence TTCTAGAA, an upstream motif characteristic of most, if not all, heat shock genes in *Saccharomyces cerevisiae* (Tuite et al., 1988). The upstream sequence of *HSC82* and sites protected from DNase I cleavage in spheroplast lysates are provided in Figure 15.

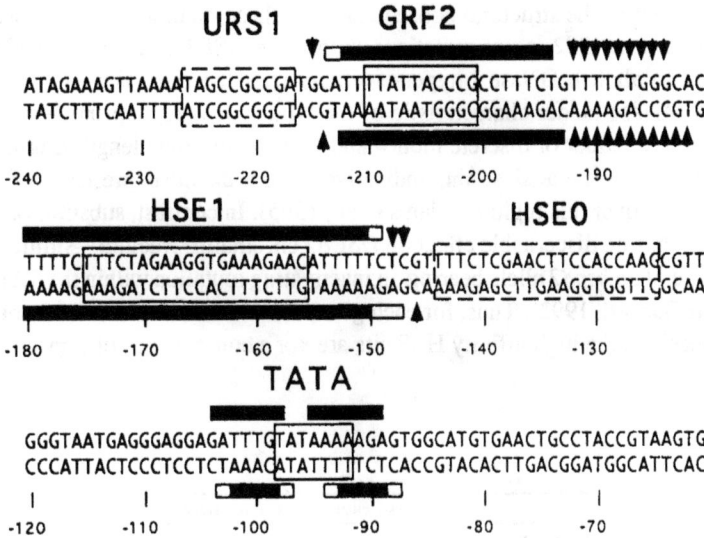

Figure 15. DNA sequence of the *HSC82* promoter region, which is numbered relative to the principal transcription start site (Borkovich et al., 1989). The three continuously boxed elements serve as binding sites for sequence-specific proteins, as deduced from DNase I genomic footprinting and mutational analyses (see text and Figures 14 and 16). Dashed lines enclose sequences with strong homology to either upstream repressor or heat shock consensus sequences (URS1 and HSE0, respectively), but which engage in only weak interactions with proteins *in vivo*. The extent of strand-specific protection from DNase I is indicated by solid bars; nucleotides hypersensitive to DNase I cleavage are indicated by arrowheads (from Erkine et al., 1995a). A detailed analysis of upstream regulatory motifs is provided in Erkine et al., 1995b.

A. Point Mutations within the HSE1 Sequence of *HSC82* Abolish Promoter Function and Promoter-associated DNase I Hypersensitivity

To address the functional role of *HSC82* promoter elements protected from enzymatic cleavage in chromatin, we constructed isogenic strains harboring mutations in the TATA box, HSE0, HSE1, GRF2, or URS1 (Figure 16, Panel A). As illustrated in Figure 16, Panel B, when point mutations were introduced into each of the three conserved *n*GAA*n* modules of HSE1, *HSC82*-basal transcription was reduced ~50–fold while heat-shock induced expression was down nearly 200–fold. A 6 bp substitution of the GRF2 site diminished promoter activity nearly threefold ± heat shock whereas a 6 bp substitution of the TATAAA motif reduced basal expression nearly one order of magnitude in the nonstressed state, and nearly fourfold following a 15-minute heat shock. By comparison, mutations in either URS1 or HSE0 elements had virtually no effect on gene function (Adams et al., 1995).

Assessment of the structural consequences of these mutations has revealed that in contrast to *HSP82*, point mutations within the HSE1 element of *HSC82* are sufficient to abolish DNase I hypersensitivity over the promoter. Moreover, the smeared MNase ladder characteristic of the wild-type *HSC82* promoter is transformed into a ladder of discrete mono- and oligonucleosome length fragments in the HSE1 point-mutated strain, indicative of the *de novo* presence of stable nucleosomes over this region (Adams et al., 1995). In contrast, substitution of the TATA box has no discernable effect on *HSC82* chromatin structure—similar to the phenotype of an *hsp82* allele bearing an inactivating mutation within the TATA box (Lee and Garrard, 1992). Thus, for each gene, the structural effects accompanying mutagenesis of the high affinity HSF site are *not* a consequence of transcriptional

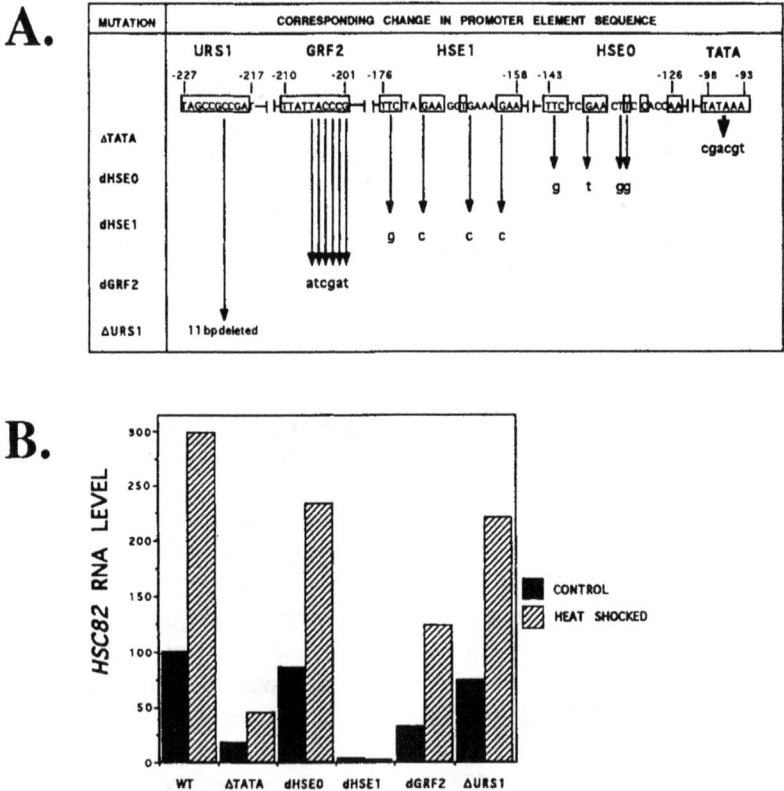

Figure 16. Phenotypes of *hsc82* promoter mutants. (A) Nucleotide substitutions and deletions introduced *in situ* within the promoter region of *HSC82*. (B) Relative *HSC82* transcript levels in control and 15-minute heat-shocked cells for isogenic strains bearing the indicated mutations (from Adams et al., 1995).

inactivation. Interestingly, despite having a negligible effect on promoter function, deletion of URS1 causes a striking alteration in the positioning of the nucleosomes abutting the 5′ DNase I hypersensitive domain (Adams et al., 1995). Thus, both HSF and the URS1-binding factor (URSF) play important roles in generating the chromatin structure over the *HSC82* upstream region. HSF's structural role appears relegated to the crucial proximal and core promoter regions, while that of URSF appears restricted to the 5′ intergenic region. It is striking that while point mutations within the high affinity HSF site upstream of *HSP82* have little effect on the nucleosome-free phenotype of that gene's promoter, analogous point mutations within *HSC82* have a dramatic effect. This difference between the two genes may derive from HSF's ability to cooperatively bind to multiple heat shock elements upstream of one gene (*HSP82*) but not of the other (*HSC82*).

IV. CONCLUSION

Biochemical and genetic studies of the yeast *HSP90* genes have revealed a critical role for yHSF in establishing nucleosome-free, DNase I hypersensitive chromatin structures over two heat shock promoters in *Saccharomyces cerevisiae*. Moreover, yHSF appears capable of directly binding to a positioned nucleosome *in vivo* and mediating its subsequent disruption or displacement. That yHSF may collaborate with other factors in binding and/or disrupting nucleosomes is implied by the observation that the *SWI1* gene product is critically required for yHSF-mediated suppression of the *hsp82*–ΔHSE1 allele. At that gene's promoter, yHSF not only releases the underlying nucleosome (**–II**) in noninduced cells, but also disrupts the adjacent, stably bound nucleosome centered over the core promoter (**–I**) following heat shock induction. We suggest that DNA-binding proteins which serve dual structural and functional roles—such as yHSF, GAL4, PHO4, and steriod hormone receptors—be designated *POWER* (*P*romoter *O*pen *W*indow *E*ntry *R*egulator) factors.

Our results indicating that yHSF can bind to nucleosomal DNA is consistent with a recent report that a high-affinity HSE assembled into a positioned nucleosome can activate heat-shock induced transcription of a reporter gene in *Saccharomyces cerevisiae* (Pederson and Fidrych, 1994). In contrast to our results, however, those of Pederson and Fidrych suggest that yHSF can form a stable ternary complex with DNA and the histone octamer. While the basis for the difference between these two studies is not entirely clear, it is possible that the translational location of the target HSE within the nucleosome dictates whether bound yHSF can coexist with the underlying nucleosome or release it. In the study by Pederson and Fidrych (1994), the HSF binding site is positioned close to the edge of the nucleosome, whereas HSEs 2 and 3 map very close to the dyad of nucleosome **–II** (see Figure 10). Thus, yHSF may form a stable ternary complex with an underlying nucleosome in certain sequence contexts but not in others.

Yeast HSF therefore presents a striking contrast to that of fly or vertebrate HSF, which binds DNA only in response to heat shock (Wu, 1984; Sorger et al., 1987; Zimarino et al., 1990; Abravaya et al., 1991), and, at least in the case of fly HSF, appears to play virtually no role in establishing the chromatin structure of heat shock genes (Lu et al., 1993). In the *Drosophila HSP26* gene, for example, $(CT)_n$ elements are necessary for establishing the DNase I hypersensitive region (Lu et al., 1992). Recent *in vitro* reconstitution experiments of the chromatin structure of the *Drosophila HSP70* promoter confirm a critical role for the $(CT)_n$–binding protein, GAGA factor, in mediating ATP-dependent nucleosome disruption (Tsukiyama et al., 1994). It thus appears that yeast HSF embodies functions of both GAGA factor and fly HSF. Whether this capability is intrinsic to yHSF per se, or is a property of complexes it may form with other proteins *in vivo* (for example, the SWI/SNF complex), is currently under investigation.

ACKNOWLEDGMENTS

I thank all members of my laboratory, past and present, for their many technical contributions, insights, and ideas. Persons who have contributed to the work described here include Chris Adams, Kate Bethea, Kerry Collins, Tuba Diken, Karen English, Alex Erkine, Mingxing Gao, Seewoo Lee, Serena Simmons, and Bruce Stentz. I am grateful to Tuba Diken for her assistance in preparing the figures for this chapter. Work described here was funded by grants awarded to the author from the National Institutes of Health, the American Cancer Society, the Louisiana Education Quality Trust Fund, the March of Dimes Birth Defects Foundation, and the Center for Excellence in Cancer Research at Louisiana State University Medical Center–Shreveport.

REFERENCES

Abravaya, K., Phillips, B., & Morimoto, R.I. (1991). Heat shock-induced interactions of heat shock transcription factor and the human hsp70 promoter examined by *in vivo* footprinting. Mol. Cell. Biol. 11, 586–592.

Adams, C.C., & Gross, D.S. (1991). The yeast heat shock response is induced by conversion of cells to spheroplasts and by potent transcriptional inhibitors. J. Bacteriol. 173, 7429–7435.

Adams, C.C., Erkine, A.M., Diken, T., & Gross, D.S. (1995). Distinct functional and structural roles for heat shock transcription factor, GRF/REB1, URS-binding factor and TATA-binding protein at the yeast *HSC82* heat shock promoter. Submitted.

Amin, J., Anathan, J., & Voellmy, R. (1988). Key features of heat shock regulatory elements. Mol. Cell. Biol. 8, 3761–3769.

Becker, P.J., Rabindran, S.K., & Wu, C. (1991). Heat shock-regulated transcription *in vitro* from a reconstituted chromatin template. Proc. Natl. Acad. Sci. USA 88, 4109–4113.

Buratowski, S., Hahn, S., Sharp, P.A., & Guarente, L. (1988). Function of a yeast TATA element-binding protein in a mammalian transcription system. Nature 334, 37–42.

Buratowski, S., Hahn, S., Guarente, L., & Sharp, P.A. (1989). Five intermediate complexes in transcription initiation by RNA polymerase II. Cell 56, 549–561.

Borkovich, K.A., Farrelly, F.W., Finkelstein, D.B., Taulien, J., & Lindquist, S. (1989). hsp82 is an essential protein that is required in higher concentrations for growth of cells at higher temperatures. Mol. Cell. Biol. 9, 3919–3930.

Chen, W., & Struhl, K. (1988). Saturation mutagenesis of a yeast *his3* TATA element: Genetic evidence for a specific TATA-binding protein. Proc. Natl. Acad. Sci. USA 85, 2691–2695.

Elgin, S.C.R. (1988). The formation and function of DNase I hypersensitive sites in the process of gene activation. J. Biol. Chem. 263, 19259–19262.

Erkine, A.M., Adams, C.C., Gao, M., & Gross, D.S. (1995a). Multiple protein-DNA interactions over the yeast *HSC82* heat shock gene promoter. Nucleic Acids Res. 23, 1822–1829.

Erkine, A.M., Szent-Gyorgyi, C., Simmons, S.F., & Gross, D.S. (1995b). The upstream sequences of the *HSP82* and *HSC82* genes of *Saccharomyces cerevisiae*: Regulatory elements and nucleosome positioning motifs. Yeast 11, in press.

Farrelly, F.W., & Finkelstein, D.B. (1984). Complete sequence of the heat shock-inducible *HSP90* gene of *Saccharomyces cerevisiae*. J. Biol. Chem. 259, 5745–5751.

Felsenfeld, G. (1992). Chromatin as an essential part of the transcription mechanism. Nature 355, 219–224.

Fernandes, M., Xiao, H., & Lis, J.T. (1994). Fine structure analyses of the *Drosophila* and *Saccharomyces* heat shock factor—heat shock element interactions. Nucleic Acids Res. 22, 167–173.

Gross, D.S., & Garrard, W.T. (1988). Nuclease hypersensitive sites in chromatin. Annu. Rev. Biochem. 57, 159–197.

Gross, D.S., Szent-Gyorgyi, C., & Garrard, W.T. (1986). Yeast as a model system to dissect the relationship between chromatin structure and gene expression. UCLA Symp. Mol. Cell. Biol. New Ser. 33, 345–366.

Gross, D.S., English, K.E., Collins, K.W., & Lee, S. (1990). Genomic footprinting of the yeast *HSP82* promoter reveals marked distortion of the DNA helix and constitutive occupancy of heat shock and TATA elements. J. Mol. Biol. 216, 611–631.

Gross, D.S., Adams, C.C., Lee, S., & Stentz, B. (1993). A critical role for heat shock transcription factor in establishing a nucleosome-free region over the TATA-initiation site of the yeast *HSP82* heat shock gene. EMBO J. 12, 3931–3945.

Grunstein, M. (1990a). Histone function in transcription. Annu. Rev. Cell Biol. 6, 643–678.

Grunstein, M. (1990b). Nucleosomes: Regulators of transcription. Trends Genet. 6, 395–400.

Hahn, S., Buratowski, S., Sharp, P.A., & Guarente, L. (1989). Yeast TATA-binding protein TFIID binds to TATA elements with both consensus and nonconsensus DNA sequences. Proc. Natl. Acad. Sci. USA 86, 5718–5722.

Jakobsen, B.K., & Pelham, H.R.B. (1988). Constitutive binding of yeast heat shock factor to DNA *in vivo*. Mol. Cell. Biol. 8, 5040–5042.

Kim, Y., Geiger, J.H., Hahn, S., & Sigler, P.B. (1993a). Crystal structure of a yeast TBP/TATA-box complex. Nature 365, 512–520.

Kim, J.L., Nikolov, D.B., & Burley, S.K. (1993b). Co-crystal structure of TBP recognizing the minor groove of a TATA element. Nature 365, 520–527.

Kornberg, R.D., & Lorch, Y. (1991). Irresistible force meets immovable object: Transcription and the nucleosome. Cell 67, 833–836.

Lee, M.-S., & Garrard, W.T. (1992). Uncoupling gene activity from chromatin structure: Promoter mutations can inactivate transcription of the yeast *HSP82* gene without eliminating nucleosome-free regions. Proc. Natl. Acad. Sci. USA 89, 9166–9170.

Lee, S., & Gross, D.S. (1993). Conditional silencing: The *HMRE* mating-type silencer exerts a rapidly reversible position effect on the yeast *HSP82* heat shock gene. Mol. Cell. Biol. 13, 727–738.

Lee, D.K., Horikoshi, M., & Roeder, R.G. (1991). Interaction of TFIID in the minor groove of the TATA element. Cell 67, 1241–1250.

Lee, D.Y., Hayes, J.J., Pruss, D., & Wolffe, A.P. (1993). A positive role for histone acetylation in transcription factor access to nucleosomal DNA. Cell 72, 73–84.

Lu, Q., Wallrath, L.L., Allan, B.D., Glaser, R.L., Lis, J.T., & Elgin, S.C.R. (1992). Promoter sequence containing $(CT)_n \bullet (GA)_n$ repeats is critical for the formation of the DNase I hypersensitive sites in the *Drosophila HSP26* gene. J. Mol. Biol. 225, 985–998.

Lu, Q., Wallrath, L.L., Granok, H., & Elgin, S.C.R. (1993). $(CT)_n \bullet (GA)_n$ repeats and heat shock elements have distinct roles in chromatin structure and transcriptional activation of the *Drosophila HSP26* gene. Mol. Cell. Biol. 13, 2802–2814.

Lue, N.F., Buchman, A.R., & Kornberg, R.D. (1989). Activation of yeast RNA polymerase II transcription by a thymidine-rich upstream element *in vitro*. Proc. Natl. Acad. Sci. USA 86, 486–490.

McDaniel, D., Caplan A.J., Lee, M.S., Adams, C.C., Fishel, B.R., Gross, D.S., & Garrard, W.T. (1989). Basal-level expression of the yeast *HSP82* gene requires a heat shock regulatory element. Mol. Cell. Biol. 9, 4789–4798.

Pederson, D.S., & Fidrych, T. (1994). Heat shock factor can activate transcription while bound to nucleosomal DNA in *Saccharomyces cerevisiae*. Mol. Cell. Biol. 14, 189–199.

Perisic, O., Xiao, H., & Lis, J.T. (1989). Stable binding of *Drosophila* heat shock factor to head-to-head and tail-to-tail repeats of a conserved 5 bp recognition unit. Cell 59, 797–806.

Peterson, C.L., & Herskowitz, I. (1992). Characterization of the yeast *SWI1*, *SWI2*, and *SWI3* genes, which encode a global activator of transcription. Cell 68, 573–583.

Simpson, R.T. (1990). Nucleosome positioning can affect the function of a *cis*-acting DNA element *in vivo*. Nature 343, 387–390.

Sorger, P.K., & Nelson, H.C.M. (1989). Trimerization of a yeast transcriptional activator via a coiled-coil motif. Cell 59, 807–813.

Sorger, P.K., Lewis, M.J., & Pelham, H.R.B. (1987). Heat shock factor is regulated differently in yeast and HeLa cells. Nature 329, 81–84.

Starr, D.B., & Hawley, D.K. (1991). TFIID binds in the minor groove of the TATA box. Cell 67, 1231–1240.

Struhl, K. (1985a). Negative control at a distance mediates catabolite repression in yeast. Nature 317, 822–824.

Struhl, K. (1985b). Naturally occurring poly (dA-dT) sequences are upstream promoter elements for constitutive transcription in yeast. Proc. Natl. Acad. Sci. 82, 8419–8423.

Szent-Gyorgyi, C., Finkelstein, D.B., & Garrard, W.T. (1987). Sharp boundaries demarcate the chromatin structure of a yeast heat-shock gene. J. Mol. Biol. 193, 71–80.

Taylor, I.C.A., Workman, J.L., Schuetz, T.J., & Kingston, R.E. (1991). Facilitated binding of GAL4 and heat shock factor to nucleosomal templates: Differential function of DNA binding domains. Genes & Dev. 5, 1285–1298.

Thomas, G.H., & Elgin, S.C.R. (1988). Protein/DNA architecture of the DNase I hypersensitive region of the *Drosophila* hsp26 promoter. EMBO. J. 7, 2191–2201.

Tsukiyama, T., Becker, P.B., & Wu, C. (1994). ATP-dependent nucleosome disruption at a heat-shock promoter mediated by binding of GAGA transcription factor. Nature 367, 525–532.

Tuite, M.F., Bossier, P., & Fitch, I.T. (1988). A highly conserved sequence in yeast heat shock gene promoters. Nucl. Acids Res. 16, 11845.

Wiederrecht, G., Shuey, D.J., Kibbe, W.A., & Parker, C.S. (1987). The *Saccharomyces* and *Drosophila* heat shock transcription factors are identical in size and DNA binding properties. Cell 48, 507–515.

Winter, E., & Varshavsky, A. (1989). A DNA binding protein that recognizes oligo(dA)•oligo(dT) tracts. EMBO.J. 8, 1867–1877.

Winston, F., & Carlson, M. (1992). Yeast SNF/SWI transcriptional activators and the SPT/SIN chromatin connection. Trends Genet. 8, 387–391.

Wolffe, A.P. (1994). Transcription: In tune with the histones. Cell 77, 13–16.

Workman, J.L., & Buchman, A.R. (1993). Multiple functions of nucleosomes and regulatory factors in transcription. Trends Biochem. Sci. 18, 90–95.

Workman, J.L., & Roeder, R.G. (1987). Binding of transcription factor TFIID to the major late promoter during *in vitro* nucleosome assembly potentiates subsequent initiation by RNA polymerase II. Cell 51, 613–622.

Wu, C. (1980). The 5' ends of *Drosophila* heat shock genes in chromatin are hypersensitive to DNase I. Nature 286, 854–860.

Wu, C. (1984). Two protein-binding sites in chromatin implicated in the activation of heat-shock genes. Nature 309, 229–234.

Xiao, H., & Lis, J.T. (1988). Germline transformation used to define key features of heat-shock response elements. Science 239, 1139–1142.

Yoshinaga, S.K., Peterson, C.L., Herskowitz, I., & Yamamoto, K.R. (1992). Roles of SWI1, SWI2, and SWI3 proteins for transcriptional enhancement by steroid receptors. Science 258, 1598–1604.

Zimmarino, V., Tsai, C., & Wu, C. (1990). Complex modes of heat shock factor activation. Mol. Cell. Biol. 10, 752–759.

Part IV

Molecular Machines that Interact with Nucleosomes

Chapter 8

The SWI/SNF Protein Machine:

Helping Transcription Factors Contend With Chromatin-Mediated Repression

CRAIG L. PETERSON

The Nucleus
Volume 1, pages 185–206.
Copyright © 1995 by JAI Press Inc.
All rights of reproduction in any form reserved.
ISBN: 1-55938-940-0

ABSTRACT

The SWI/SNF protein complex plays a crucial role in the regulation of eukaryotic gene expression. This large complex (about 2 MDa) is required for transcriptional induction of a large number of yeast genes and is also required for many transcriptional activators to enhance transcription. Although originally identified in the budding yeast, *Saccharomyces cerevisiae*, putative homologs for SWI/SNF subunits have been identified in *Drosophila*, mice, and humans. A current hypothesis is that the SWI/SNF complex facilitates activator function by contending with chromatin-mediated repression of transcription. This proposal is based on genetic studies in which chromatin components were identified as mutations that alleviated the requirement for the SWI/SNF complex in activator function. Recently the yeast SWI/SNF complex has been purified to near homogeneity. The purified SWI/SNF complex antagonizes chromatin function—the complex uses the energy of ATP hydrolysis to stimulate the binding of an activator to nucleosomal DNA. I propose a model for SWI/SNF action in which the primary role of the complex is to facilitate the binding of transcription factors to nucleosomal DNA.

I. INTRODUCTION

Much that we know about the regulatory mechanisms that control transcription of protein-encoding genes derives from biochemical studies using partially purified or purified transcription factors and DNA templates. These studies have shown that building a functional transcription initiation complex requires a myriad of proteins that communicate regulatory information through both protein–DNA and protein–protein interactions. *In vivo*, however, the scenario is even more complex, because transcription factors must contend with the packaging of DNA into nucleosomes. *In vitro* and *in vivo* studies over the past few years have indicated that nucleosomes are indeed potent inhibitors of transcription that compete with transcription factors for occupancy of DNA binding sites (for reviews see Kornberg and Lorch, 1991; Felsenfeld, 1992; Grunstein, 1992; Workman and Buchman, 1993). *In vitro*, nucleosomes inhibit transcription initiation by either blocking access of the general transcription machinery to promoter sequences (Lorch et al., 1987, 1992; Laybourn and Kadonaga, 1991) or by hindering the binding of upstream activator proteins (for a review, see Adams and Workman, 1993). Other nonhistone components of chromatin, such as histone H1, can enhance the ability of nucleosomes to repress transcription (Laybourn and Kadonaga, 1991). The lowered affinity of transcription factors for nucleosomal DNA raises the important question as to whether additional accessory factors assist in the effective binding of these factors to nucleosomes *in vivo*.

This paper will focus on a large protein complex, called SWI/SNF, that functions *in vivo* and *in vitro* to facilitate the activity of transcription factors on chromatin templates. Although subunits of this complex were initially identified in the budding yeast, *Saccharomyces cerevisae*, it is now apparent that at least one subunit of this complex has been highly conserved throughout the evolution of multicellular eukaryotes. In the first section, I will review *in vivo* studies in yeast which led to the identification of SWI/SNF subunits and the connection between SWI/SNF action and chromatin. In a second part I will discuss the identification and characterization of putative *Drosophila* and human homologs of the SWI2 subunit; and in the final section I will review recent biochemical advances that have led to a new working model for the function of the SWI/SNF complex in transcription. For an earlier review on SWI/SNF and chromatin see Winston and Carlson (1992).

II. IDENTIFICATION OF GENES ENCODING SWI/SNF SUBUNITS

Over the past several years genetic analyses of two yeast genes, *HO* and *SUC2*, have led to the identification of five genes that encode subunits of the SWI/SNF complex. The *SWI1*, *SWI2*, and *SWI3* genes were originally identified by the Herskowitz laboratory as positive regulators of the *HO* gene which encodes an endonuclease that initiates mating type switching (Stern et al., 1984). In the absence of any one of these three gene products, *HO* mRNA levels are reduced 50 to 100-fold, and consequently these mutants are defective in mating type switching (hence the name **SWI**tch genes). Parallel studies in the laboratory of Dr. Marian Carlson have also led to the identification of three genes, *SNF2*, *SNF5*, and *SNF6* that were identified as positive regulators of the *SUC2* gene which encodes the enzyme invertase (invertase activity is required for yeast to utilize sucrose as a carbon source; hence, *snf* mutants are **S**ucrose **N**on**F**ermentors; Neigeborn and Carlson, 1984). In addition to being defective for switching or sucrose utilization, these five mutants grow very slowly on rich medium, are unable to grow on media-containing glycerol or ethanol (nonfermentable carbon sources), and homozygous *swi⁻* or *snf⁻* diploids cannot sporulate (progress through meiosis). These diverse phenotypes suggested that *SWI* and *SNF* gene products were unlikely to function as gene-specific activators but might in fact be more general components of the transcription machinery. Indeed, it is now known that *SWI1*, *SWI2*, and *SWI3* are required for expression of a large number of genes, including *INO1*, *ADH1*, *ADH2*, *SUC2*, *GAL1*, and *GAL10* (Peterson and Herskowitz, 1992). Likewise, *SNF2*, *SNF5*, and *SNF6* are required for expression of many genes, including *HO*, *TY* elements, and *PHO5* (Abrams et al., 1986; Estruch and Carlson, 1990; Happel et al., 1991; Peterson and Herskowitz, 1992). In addition, *SNF5* appears to be required for expression of two cell-type specific genes, *BAR1* and *MFa1* (Laurent et al., 1990), and Yamashita's group found that the *SNF2* and *SWI1* genes were required for expression of the *STA1* gene (Yoshimoto and Yamashita, 1991). What is the relationship among these *SWI* and *SNF* genes? DNA sequencing of the *SNF2* gene by the Carlson laboratory (Laurent et al., 1991) and of the *SWI2* gene by the Nasmyth laboratory has provided the key link between these two sets of genes— *SWI2* and *SNF2* are two names for the same gene (from here on I will refer to *SWI2* as *SWI2/SNF2*). *SNF5* and *SNF6*, however, **are not** equivalent to *SWI1* and *SWI3*. Thus, five gene products—*SWI1*, *SWI2/SNF2*, *SWI3*, *SNF5*, and *SNF6*—are required for expression of a large set of genes (presumably the same gene set).

The observation that all five *SWI* and *SNF* genes are required for expression of many genes was one piece of evidence that led to the hypothesis that all five *SWI* and *SNF* products might function together as components of a large multisubunit complex. Other pieces of indirect evidence were also consistent with this view; for example, the phenotype of multiply-defective *swi⁻* or *snf⁻* mutants is identical to that of a single *swi⁻* or *snf⁻* mutant (Peterson and Herskowitz, 1992); the stability of SWI3 protein is reduced in the absence of SWI1 and SWI2 (Peterson and Herskowitz, 1992); the SWI3 protein can associate with the mammalian glucocorticoid receptor *in vitro* and this interaction requires the SWI1 and SWI2 proteins (Yoshinaga et al., 1992; see below); and analyses of LexA–SWI and LexA–SNF fusion proteins indicated a functional interdependence among SWI and SNF proteins for transcriptional activation (Laurent et al., 1992; Laurent and Carlson, 1992). Recently we reported that SWI1, SWI2, SWI3, SNF5, and SNF6 all coelute from a gel filtration column as a large complex with an apparent molecular weight of approximately 2 MDa (Peterson et al., 1994). When extracts were prepared from individual *swi* or *snf* mutants, the elution of the other SWI/SNF polypeptides were altered, indicating that all five SWI/SNF products are subunits of the same protein complex.

What role might *SWI* and *SNF* products play in the regulation of transcription? Studies from the Young, Herskowitz, and Carlson laboratories have shown that all five SWI and SNF polypeptides are localized to the nucleus (O'Hara et al., 1988; Estruch and Carlson, 1990; Laurent et al., 1990; Laurent et al., 1991; Peterson and Herskowitz, 1992); and furthermore, LexA-SWI and LexA-SNF fusion proteins can activate transcription from LexA binding sites linked to a test promoter (Laurent et al., 1991; Laurent and Carlson, 1992). These results indicated that SWI/SNF products are likely to play a direct role in the transcription process. Can we learn anything from the types of genes whose expression requires SWI/SNF function? It is intriguing that the majority of these SWI/SNF-dependent genes are highly regulated, inducible genes that are known to require other gene-specific activators; for example, induction of *GAL* genes requires the GAL4 activator. This observation led me to suggest that SWI/SNF products might function by facilitating transcriptional enhancement by many different gene-specific activators (Peterson and Herskowitz, 1992). Subsequent studies with simple reporter genes have validated this view. A typical reporter gene contains one or more binding sites for a particular activator, such as GAL4, placed upstream of a TATA box and a *LacZ* gene. A plasmid containing the reporter is introduced into different yeast strains, and the capacity of the activator protein to enhance transcription is measured by monitoring β-galactosidase activity in protein extracts. Many different activator proteins have now been assessed for *SWI* or *SNF* dependence using simplified reporters; for example, transcriptional activation by the yeast activators, GAL4 (Peterson and Herskowitz,

1992), INO2/INO4 (Peterson, unpublished), Heat Shock Factor (Peterson and Yoshinaga, unpublished), and GCN4 (Peterson, unpublished) require one or more SWI or SNF proteins; heterologous activators, *Drosophila* ftz (Peterson and Herskowitz, 1992) and mammalian steroid receptors (Yoshinaga et al., 1992)), also require *SWI* products to activate transcription in yeast; and furthermore, LexA–Bicoid and LexA–GAL4 fusion proteins require SNF proteins to activate transcription from LexA binding sites fused to a yeast promoter (Laurent and Carlson, 1992). In each of these cases, enhancement of transcription is reduced at least 10 to 50-fold in the absence of SWI or SNF polypeptides.

How do *SWI* and *SNF* products facilitate activator function? One can imagine many models in which *SWI/SNF* products influence any of several steps during the stimulation of transcription by an activator. *SWI/SNF* products could regulate the nuclear localization of activators, the binding of an activator to its recognition sequence, or they could regulate the ability of an activator to communicate regulatory information to the general transcription machinery. In the latter two models, *SWI/SNF* products may play a specific role in helping an activator bind its sequence when it is assembled onto a histone octamer, or in helping an activator remove or perturb nucleosomes that block promoter elements. Below I describe genetic studies that suggest a role for *SWI/SNF* products in helping activators contend with chromatin-mediated repression.

III. A GENETIC LINK BETWEEN SWI/SNF FUNCTION AND CHROMATIN

SIN Genes

When the *SWI1*, *SWI2*, and *SWI3* gene products were initially identified as positive regulators of *HO* expression, two models were put forth: (1) SWI1, SWI2, and SWI3 act in a direct fashion to stimulate transcription, for instance, like GAL4; or (2) these *SWI* products activate transcription by removing or antagonizing a repressor of *HO* transcription. These models were, of course, based on paradigms of that time—regulation of *lac* genes in bacteria or *GAL* genes in yeast, both of which require activators and/or repressors. In 1987 Sternberg and colleagues hypothesized that if SWI1, SWI2, and SWI3 functioned by antagonizing a repressor, then removal of the repressor by mutation would make *HO* expression independent of *SWI1*, *SWI2*, and *SWI3* (Sternberg et al., 1987). The genetic screen that they devised to test this theory was quite simple—they screened for mutations in other genes that would allow *HO* transcription in the absence of *SWI1*. Mutations in two genes, *SIN1* and *SIN2* (**SWI-IN**dependent), were identified that restored transcription of a *HO–lacZ* fusion gene to nearly wild-type levels even in the absence of the *SWI1*, *SWI2*, or *SWI3* genes. Formally, *SIN1* and *SIN2* encode repressors of transcription. One interpretation of these functional interactions is that

SWI1, SWI2, and SWI3 facilitate transcription by antagonizing the inhibitory effects of *SIN* gene products.

SWI1, SWI2, SWI3 ------- | *SIN1, SIN2* ------- | Transcription

How might *SIN* products repress transcription? One simple possibility is that *SIN1* and *SIN2* encode sequence-specific DNA binding proteins that repress *HO* transcription. Several lines of evidence suggested, however, that SIN1 and SIN2 might repress transcription by a much more global mechanism. First, these SIN repressors were not specific to the regulation of *HO* expression, as *sin* mutations also alleviated the defects in growth, in transcription of the *INO1* gene, and in GR activity caused by mutations in either *SWI1, SWI2,* or *SWI3* (Peterson et al., 1991; Yoshinaga et al., 1992). Furthermore, *sin* mutants alleviated the defects in growth and in transcription of *INO1* due to a truncation of the conserved carboxy-terminal domain of the largest subunit of RNA polymerase II (Peterson et al., 1991; Kruger et al., 1994).

With the cloning of both *SIN* genes, their role as global repressors became more obvious—both *SIN* genes appear to encode components of chromatin. The *SIN1* gene encodes a very highly charged, nuclear protein that shows some similarity to a mammalian nonhistone chromatin protein, HMG1 (Kruger and Herskowitz, 1991). Although SIN1 does not contain a perfect "signature" sequence to be a true HMG1 relative, the observed sequence similarity led Kruger and Herskowitz (1991) to hypothesize that *SIN1* might encode a nonhistone component of chromatin. This hypothesis was supported by biochemical data where it was shown that SIN1 protein binds nonspecifically to DNA *in vitro* with an affinity in the range of 10^5 to 10^7 M^{-1}, which is similar to that observed for HMG proteins (reviewed in van Holde, 1989).

The sequence of the *SIN2* gene left no doubts that it encoded a bona fide chromatin component. *SIN2* is identical to the *HHT2* gene which is one of the two yeast genes that encodes histone H3 (Smith and Andresson, 1983). Of the five independent *sin2* alleles, all result in single amino acid changes within a 15 amino acid region of histone H3 (Figure 1). These altered residues are evolutionarily invariant and the 15 amino acid region lies within a central, globular domain of histone H3. This globular domain is believed to be involved in histone–histone interactions that drive assembly of the histone octamer. These missense mutations are not recessive, but semidominant, suggesting that altered histone H3 proteins interfere with the function of the wild-type protein. Consistent with this view, removal of one of the two copies of the histone H3 gene does not lead to a Sin phenotype. Recently, Prelich and Winston (1993) also isolated one of these *sin* alleles (R117H; *bur5–1* in this study) as a suppressor of a partial deletion of the *SUC2* upstream regulatory region.

The histone octamer is composed of two copies of each of the four histone proteins, H3, H4, H2A, and H2B. The octamer is organized as one tetramer of

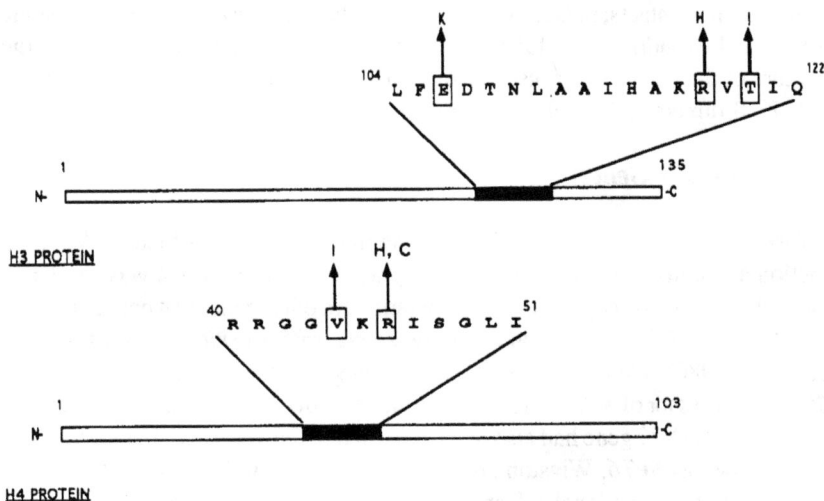

Figure 1. Sin mutations in histones H3 and H4.

histones H3 and H4 and two dimers of histones H2A and H2B. Based on the physical association of H3 and H4, Kruger and colleagues (1994) asked whether *sin* alleles in the gene that encodes histone H4 could also be identified. The *HHF2* gene, which is one of the two yeast genes that encodes histone H4, was mutagenized *in vitro* with hydroxylamine and *sin* alleles were identified *in vivo* in yeast (they restored expression of an *HO–lacZ* fusion gene in a *swi1* mutant). Ten independent mutants were identified and each mutated histone H4 gene was sequenced. Surprisingly, all ten mutations resulted in changes at only one of two amino acids, V44 or R46 (Figure 1). As is the case for histone H3 residues, these positions are conserved in all sequenced histone genes and they are located within a central domain known to be crucial for nucleosome assembly.

The three-dimensional crystal structure of the chicken histone octamer has been determined at 3.1 Å resolution (Arents et al., 1991). Kruger and colleagues (1994) took advantage of the extreme sequence conservation of the histone proteins to place the yeast histone sequence onto the chicken octamer structure. Remarkably, H3 and H4 residues that were altered in *sin* mutants appear to comprise a single, contiguous domain of the octamer. Thus, the clustering of residues on the primary histone sequence was dramatically reflected in the three-dimensional structure as well. Unfortunately, the chicken octamer structure is not yet at the resolution such that amino acid side chains can be oriented, thus it is unclear how *sin* mutations might disrupt octamer structure and function. The tight clustering of residues, however, indicates that all mutations, both in H3 and H4, might disrupt the structure or function via a common mechanism. One possibility is that Sin mutations disrupt

DNA–histone contacts; in fact, Arg46 of H4 is believed to be a DNA contact residue (Arents and Moudrianakis, 1993). Alternatively, Sin mutations may disrupt the structure of the octamer, for instance by destabilizing the interactions between the H2A–H2B dimers and the H3–H4 tetramer.

1. SSN and SPT Genes

Work from the Carlson and Winston laboratories has also linked SWI/SNF function and chromatin. The Carlson group used an approach that was similar to that of Herskowitz—they screened for mutations in other genes that could alleviate the defect in *SUC2* transcription due to *swi2/snf2*, *snf5*, or *snf6* mutations (Neigeborn et al., 1986). This mutant hunt yielded a large number of alleles of one gene, *SSN20* (Suppressor of *SNF2*). To date, it is unclear how *SSN20* relates to chromatin; however, the *SSN20* gene had also been identified in the Winston laboratory where it was known as *SPT6*. Winston and colleagues had identified *SSN20/SPT6* as a suppressor of transcriptional defects caused by insertions of a transposable element, Ty . Seventeen different *SPT* (SuPpressor of Ty) genes have been placed into two functional categories based upon the ability of each *spt* mutant to suppress the phenotypes of different Ty insertions (reviewed in Winston and Carlson, 1992). One group is believed to encode components of the general transcription machinery since one member of this group, *SPT15*, encodes the TATA box binding protein (TBP). The second group of *SPT* genes are believed to encode components of chromatin, since two members, *HTA1* and *HTB1*, encode histones H2A and H2B, respectively. This group also includes *SSN20/SPT6*. And, like mutations in *SSN20/SPT6*, a deletion of the *HTA1–HTB1* locus, which is believed to cause a depletion of histones H2A and H2B from cellular chromatin, also alleviates the transcriptional defects caused by *swi* and *snf* mutations (it has a Sin phenotype; Hirschhorn et al., 1992). One possibility is that all *sin* mutations in chromatin components alter chromatin structure by directly or indirectly affecting the assembly or stability of the H2A/H2B dimers (this point is discussed in detail in Hirschhorn et al., 1992; see later in this paper).

Recently Hirschhorn and colleagues (1992) provided the first functional evidence that SWI/SNF proteins might function in transcription by perturbing chromatin structure. These investigators analyzed the chromatin structure surrounding the *SUC2* gene in *SWI⁺* and *swi/snf* cells grown under repressed and transcriptionally induced states. In nuclei prepared from wild-type cells (*SWI⁺*) grown under repressing conditions (+ glucose), sequences surrounding the *SUC2* TATA box were inaccessible to micrococcal nuclease digestion, which is consistent with the presence of a nucleosome. In contrast, when nuclei were prepared from *SWI⁺* cells grown under inducing conditions (– glucose), these sequences were accessible to digestion by the nuclease. These results were consistent with a nucleosome blocking the TATA sequence in the repressed state and subsequent perturbation or removal of that nucleosome when *SUC2* transcription was induced. When nuclei were prepared from *swi2/snf2⁻* and *snf5⁻* yeast cells, sequences surrounding the TATA box were inaccessible to nuclease digestion even when cells were grown under inducing conditions. Furthermore, when chromatin was altered in these swi/snf mutants by deletion of the *HTA1–HTB1* locus (encodes histones H2A and H2B), the changes in chromatin structure associated with induction were observed and *SUC2* transcription was restored. These results indicate that SWI/SNF products are required for the observed changes in chromatin structure upon induction of *SUC2* transcription, and that chromatin represses transcription in the absence of SWI/SNF action.

Do these changes in chromatin structure reflect a discrete step in gene activation or are they simply the result of the transcription process? To address this question, Hirschhorn and colleagues (1992) analyzed the chromatin structure of an altered *SUC2* promoter that contained a deletion of the TATA box. This deletion eliminated transcription of the *SUC2* gene, however, under inducing conditions the changes in nuclease accessibility were still observed. This result strongly argues that perturbation of chromatin is a discrete step in the process of transcriptional activation and that this step occurs independently of (and presumably prior to) the formation of a preinitiation complex at the promoter. Furthermore, these studies support the hypothesis that the SWI/SNF complex functions in transcription to disrupt nucleosomes positioned on the TATA sequence; alternatively, the results are also consistent with SWI/SNF complex functioning at an earlier step, for example, the complex may facilitate the binding of an activator protein to the *SUC2* upstream regulatory region.

IV. SWI2/SNF2 ENCODES A HIGHLY CONSERVED DNA-STIMULATED ATPase

A. SWI2/SNF2 Encodes a DNA-Stimulated ATPase

SWI2/SNF2 encodes a protein of almost 200 kDa (Laurent et al., 1992). After the sequence of *SWI2/SNF2* was reported, Davies and colleagues (1992) and Laurent and colleagues (1991) reported that SWI2/SNF2 contained seven sequence motifs within a central 600 amino acid region that were characteristic of nucleic acid-stimulated ATPases (Figure 2). Since a small subset of these nucleic acid-stimulated ATPases are known to have either RNA or DNA helicase activity, a provocative model was put forth by Travers (1992). In this model it was suggested that the SWI/SNF complex might perturb chromatin by acting as a DNA helicase machine that could move processively along the DNA, removing nucleosomes via a DNA-unwinding mechanism. As of yet no biochemical or genetic data has lent support for this model (see later in this paper).

Laurent and colleagues (1992) proposed that SWI2/SNF2 may define a new subfamily of DNA-stimulated ATPases or helicases, as a large number of proteins are now known to contain ATPase sequence motifs which are more similar to

```
                    I                               Ia
SWI2        GILADEMGLGKTIQTI            LVIVPLSTLS
brm         GILADEMGLGKTIQSI            LIIVPLSTLP
BRG1        GILADEMGLGKTIQTI            LIIVPLSTLS
CONS          +      G GKT    +           ++  O
                              S


                    II                              III
SWI2        SKVKWVHMIIDEGHRMKN         RLILTGTPLQNNLPELWALLNFV
brm         AKTRWKYMIIDEGHRMKN         RLLLTGTPLQNNLPELWALLNFL
BRG1        AKTRWKYMIVDEGHRMKN         RLLLTGTPLQNKLPELWALLNFL
CONS          ++++DE H                   +  +TAT   O     +  O
                      D                          SGS


                    IV                              V
SWI2        PFLLRRLKKEVEK              FILSTRAGGLGLNLQTADTVIIFDTDWNP
brm         RFLLRRLKKVVEH              FLLSTRAGGLGLNLQTADTVVIFDSDWNP
BRG1        PFLLRRLLKEVEA              FLLSTRAGGLGLNLQSADTVIIFDSDWNP
CONS          F   S O O                 +++  TO+    G  O+    O ++
              Y   T                          S        S


                    VI
SWI2        QAQDRAHRIGQKNEVR
brm         QAQDRAHRIGQRNEVR
BRG1        QAQDRAHRIGQRNEVR
CONS          O   O   H   GR   R
                      Q
```

Figure 2. The seven sequence motifs characteristic of nucleic acid-stimulated AT-Pases within yeast SWI2, *Drosophila brm*, and the human BRG1 protein. Consensus sequences (CONS) are taken from Gorbalenya and colleagues (1989). Hydrophobic residues (I, L, V, M, F, Y, W) and polar residues (S, T, D, E, N, Q, K, R) are designated by + and O, respectively.

SWI2/SNF2 than to other nucleic acid-stimulated ATPase families. These genes include putative homologs of *SWI2/SNF2* (see later in this paper) as well as other genes, such as yeast *STH1, MOT1, RAD54, RAD5, RAD16, FUN30, Drosophila lodestar* and *ISWI*, and human *hSNF2L* and *ERCC6*. Furthermore, a new analysis of the sequence motifs that define the SWI2/SNF2 family indicates that they are not highly related to motifs from proteins that are known to be helicases, in fact, they appear to define a distinct protein family that includes two well-characterized DNA-stimulated ATPases encoded by polio virus (Henikoff, 1993).

It is now clear that these ATPase motifs are crucial for function of SWI2/SNF2. Laurent and colleagues (1993) and Khavari and colleagues (1993) reported that single amino acid changes within motif I (contains the GxGKT motif that defines the putative ATP binding loop; Figure 2) eliminates SWI2/SNF2 function *in vivo*. Mutations in several other motifs also eliminate SWI2/SNF2 activity (Richmond and Peterson, unpublished). Khavari and colleagues (1993) found that changing the conserved lysine residue to an alanine in motif I exhibited a dominant negative phenotype *in vivo*—even in the presence of wild-type, this mutant SWI2/SNF2 caused a 50 percent reduction in transcriptional activation by the rat glucocorticoid receptor. This result is consistent with a model in which the mutant SWI2/SNF2 subunit is assembled into a SWI/SNF complex and that this defective complex interferes with the function of a wild-type complex. Recent biochemical results have in fact shown that this mutant subunit is assembled into a SWI/SNF complex (Peterson et al., 1994).

Laurent and colleagues (1993) showed that *SWI2/SNF2* encodes a DNA-stimulated ATPase. They expressed a C–terminal fragment of SWI2/SNF2 in bacteria as a fusion protein with maltose binding protein (the fusion lacked sequences N–terminal to the ATPase motif region). The purified fusion protein had measurable ATPase activity that was stimulated more than fivefold by double-stranded DNA. A fusion protein carrying a single amino acid change in motif I eliminated the ATPase activity, proving that the activity was due to SWI2/SNF2. The bacterially-expressed SWI2/SNF2 fusion protein, however, lacked associated DNA helicase activity. Similar results have also been found with the purified SWI/SNF complex (Côté et al., 1994; see later in this paper).

B. Identification of Putative Homologs for SWI/SNF Subunits

Our observation that several heterologous activators from multicellular eukaryotes require the SWI/SNF machinery to enhance transcription in yeast suggests that SWI/SNF components might have been conserved during the evolution of multicellular eukaryotes. Below I describe several studies that indicate that at least some of the SWI/SNF subunits have been conserved in more complex eukaryotes and that a functional homolog of the SWI/SNF complex appears to function in mammals.

1. The Drosophila brm Gene Encodes a Putative SWI2/SNF2 Homolog

Tamkun and colleagues (1992) reported the cloning and sequencing of a *Drosophila* gene called *brahma*, or *brm*, that encodes a protein with 40 percent overall sequence identity to the yeast SWI2/SNF2 subunit, and is also very similar in size (about 200 kD). Furthermore, the DNA-stimulated ATPase motifs are nearly identical between yeast SWI2/SNF2 and *Drosophila brm* (Figure 2). Protein chimeras between *brm* and SWI2/SNF2, in which the 600 amino acid motif region of SWI2/SNF2 was replaced by the corresponding region of *brm*, retains function in yeast, which suggests that *brm* may in fact be a functional homolog of *SWI2/SNF2* (Elfring et al., 1994).

Genetic studies of *brm* function in *Drosophila* indicate an amazing parallel to the genetics of *SWI2/SNF2*, suggesting that *brm* may play a similar role in transcription. *Brm* encodes an activator of multiple homeotic genes, and *brm* mutations strongly suppress mutations in Polycomb, a repressor of homeotic genes (Kennison and Tamkun, 1988). Based on similarities between Polycomb and a *Drosophila* heterochromatin protein, HP1, it has been suggested that Polycomb represses homeotic gene transcription by influencing chromatin structure (Paro and Hogness, 1991). Thus, in flies, a *SWI2/SNF2* homolog, *brm*, has the genetic properties of a SWI/SNF subunit—it acts as a positive regulator of transcription that functions antagonistically to proteins that are thought to influence chromatin structure! Unfortunately, the relationship of Polycomb function to chromatin is purely speculative at this point—none of the Polycomb group genes, for instance, encode known chromatin components. But the hypothesis that Polycomb functions as a chromatin modifier was well established **prior** to the identification of the *brm* gene as a homolog of *SWI2/SNF2*; thus the genetic relationships between *SWI* and chromatin, both in *Drosophila* and yeast, were established independently.

Once the *brm* gene was identified as a putative homolog of *SWI2/SNF2*, it seemed likely that other SWI/SNF subunits might also be present in *Drosophila*. We found that antibodies to the yeast SWI1 and SWI3 polypeptides cross-reacted with proteins in *Drosophila* extracts that were of similar molecular weight to the bona fide SWI proteins (Yoshinaga et al., 1992). Furthermore, *in vitro* studies that used *Drosophila* embryo extracts also provided evidence for a *SWI3* homolog (Yoshinaga et al., 1992). In these studies, a polyclonal antibody to the yeast SWI3 protein was added to *Drosophila in vitro* transcription reactions. These antibodies had no effect on the basal, unstimulated level of transcription, but they decreased the ability of the rat glucocorticoid receptor (GR) to enhance transcription. Unrelated antibodies had no affect on GR activation. Thus, these antibody inhibition studies indicated that a protein that was antigenically-related to SWI3 was present in *Drosophila* extracts and appeared to play a crucial role in mediating activation of transcription by the glucocorticoid receptor. It remains to be determined, however, whether this protein is in fact a functional homolog of yeast *SWI3*.

2. Identification of Mouse and Human Homologs of SWI2/SNF2

Several groups have recently described the isolation of putative mouse and human homologs of *SWI2/SNF2* (Randazzo et al., 1993; Khavari et al., 1993; Muchardt and Yaniv, 1993). These genes were cloned by cross hybridization to either *Drosophila brm* (Khavari et al., 1993; Muchardt and Yaniv, 1993) or human *BRG1* (Randazzo et al., 1994) sequences, and as one might expect, they are all closely related—mouse and human homologs on average are about 52 percent identical to *Drosophila brm* and all genes encode proteins of about 200 kDa. In humans there appears to be at least two different putative *SWI2/SNF2* homologs. The Crabtree laboratory isolated one gene, called *BRG1* (Brm Related Gene 1), that is 52 percent identical to brm and is also highly related in sequence to SWI2/SNF2 (Khavari et al., 1993; Figure 2). Yaniv's group has also isolated a putative *SWI2/SNF2* homolog (Muchardt and Yaniv, 1993). The sequence of this gene, called *hbrm*, is 56 percent identical to brm, but it is not identical to the *BRG1* gene. Although the central, motif region of hbrm and BRG1 are nearly identical (97%

identity over 600 residues), the N–termini are divergent, suggesting that they may carry out distinct functions or associate with different polypeptides. Khavari and colleagues (1993) reported that a protein chimera between SWI2/SNF2 and BRG1 (the central 600 amino acids motif region of SWI2/SNF2 was swapped with BRG1 sequences) retains function in yeast, suggesting that *BRG1* may encode a functional homolog of *SWI2/SNF2*. Furthermore, BRG1 (like SWI2/SNF2) appears to be a component of a very large protein complex which elutes from gel filtration with an apparent molecular weight of about 2 MDa (Khavari et al., 1993).

Several lines of evidence indicates that BRG1 and hbrm function as transcription factors in human cells. First, a dominant negative BRG1 protein was constructed based upon the dominant negative phenotype of a yeast *SWI2/SNF2* mutant (K798A; see above). This mutant BRG1 protein was overexpressed in mammalian cells and found to inhibit transcription of a subset of cotransfected reporter genes (Khavari et al., 1993). For example, transcription from the *EF–1–alpha* promoter was reduced to about 30 percent of the wild-type level upon overexpression of the BRG1 mutant, whereas transcription mediated by the AP–1 transcription factor was not affected. Two results reported by Muchardt and Yaniv (1993) suggest that hbrm also functions as a transcription factor. First, they found that hbrm could activate transcription *in vivo* if tethered to DNA via the DNA binding domain of the BPV–1 E2 protein. Similarly, SWI2/SNF2 can also activate transcription if tethered to DNA via a LexA DNA binding domain (Laurent et al., 1991). Unlike the case with SWI2/SNF2, however, a mutation in the putative ATP binding loop of the E2–hbrm fusion had no affect on transcriptional activation. The second experiment indicated that hbrm facilitated the function of the glucocorticoid (GR) receptor. In these studies an hbrm cDNA was cotransfected into cervical carcinoma cells with glucocorticoid receptor expression and reporter plasmids. These cells lacked endo-genous hbrm, so the presence or absence of hbrm on GR function could be investigated. They found that hbrm facilitated transcriptional activation by the GR–expression of hbrm produced nearly a 10-fold increase in GR–mediated enhancement of transcription. Hbrm function was not restricted to the GR; a threefold stimulation of RAR–α function was also observed, although hbrm did not facilitate the activity of HNF–1 or AP–2. As is the case for SWI2/SNF2, the ability of hbrm to facilitate activator function requires an intact ATPase domain. These results indicate that human homologs of SWI2/SNF2 may function as transcription factors in mammalian cells, and suggest that they might function in the context of a functionally homologous SWI/SNF complex.

V. BIOCHEMICAL TESTS OF SWI/SNF FUNCTION IN TRANSCRIPTION

The *in vivo* and *in vitro* studies described above suggested a model in which the SWI/SNF complex facilitated the function of sequence-specific activators by antagonizing chromatin-mediated repression. But the link to chromatin was based primarily on genetic suppression studies; thus to test this rather speculative model for SWI/SNF function, we recently purified the SWI/SNF complex and have initiated the first biochemical tests of SWI/SNF function in transcriptional enhancement.

A. Purification of the SWI/SNF Complex

Previous studies in crude whole cell extracts indicated that the SWI/SNF complex might be very low in abundance; therefore, a yeast strain was constructed that allowed an affinity purification step. A *SWI2/SNF2* fusion gene (*SWI2–HA–6HIS*) was engineered that contained at the C–terminus of the protein sequences encoding an epitope for a monoclonal antibody (α-hemagglutinin [HA]) and six tandem histidines (Peterson et al., 1994). The six tandem histidines provided an affinity purification tag for chromatography on Ni^{+2}–NTA agarose (Hochuli et al., 1986). The *SWI2–HA–6HIS* fusion gene was stably introduced into a *swi2* deletion strain to generate a *SWI*$^+$ strain that was used for purification of the SWI/SNF complex.

The SWI/SNF complex was purified nearly 100,000-fold by three sequential chromatography steps—Ni^{+2}–NTA agarose affinity, anion exchange, and gel filtration (Côté et al., 1994). The purified SWI/SNF complex contains a total of ten different subunits: SWI1, SWI2/SNF2, SWI3, SNF5, SNF6, and five subunits (78, 68, 50, 47, and 25 kDa) of unknown identity. These additional polypeptides not only copurify with SWI/SNF polypeptides, but they also coimmunoprecipitate with the SWI2/SNF2 subunit. The purified SWI/SNF complex, as observed previously for the isolated SWI2/SNF2 subunit (Laurent et al., 1993), has ATPase activity that is stimulated >30-fold by double-stranded or single-stranded DNA (the specific activity was about 60 pmoles of Pi released per minute per microgram of SWI/SNF complex). Côté and colleagues (1994) also purified a SWI/SNF complex that contained a mutant SWI2/SNF2 subunit (K798A, see above). The ATPase activity of this mutant complex was at least eightfold lower than that of the wild-type complex. The SWI/SNF complex was rigorously tested for DNA helicase activity, but no helicase activity was detected. Thus, this result further weakens the central hypothesis of the helicase model put forth by Travers (1992; see earlier in paper).

Why were all ten SWI/SNF subunits not identified through mutant analysis? One possibility is that all ten subunits are not required for the function of the SWI/SNF complex, or, alternatively, that a subset of the subunits is required for specialized activities. It is more likely, however, that the mutant hunts that have been carried out to date have not been saturated. In fact, in all of the published *swi* and *snf* mutant screens, each hunt identified only a subset of the five identified SWI/SNF subunits.

B. SWI/SNF Complex Facilitates the Binding of an Activator to Nucleosomal DNA

The genetic studies described above suggested two possible roles for the SWI/SNF complex in antagonizing chromatin-mediated repression. In the first scenario, the SWI/SNF complex might facilitate the binding of activators to nucleosomal sites; and in the second model, the SWI/SNF complex might function at a later step to disrupt nucleosomes that block promoter elements. Côté and colleagues (1994) tested the first possibility by asking if the SWI/SNF complex could facilitate the binding of GAL4 derivatives to a binding site that was assembled onto a nucleosome.

Previously, the Workman and Kingston laboratories showed that small derivatives of the yeast GAL4 protein can recognize their binding sites even when they are assembled onto a histone octamer (reviewed in Adams and Workman, 1993). These derivatives contain only the DNA binding and dimerization domains of GAL4 and contain (GAL4–AH or GAL4–VP16) or lack (GAL4 [1–94]) a transcriptional activation domain. If a single GAL4 binding site is assembled onto a nucleosome, the affinity of these GAL4 derivatives is reduced at least 100-fold compared to their binding to unassembled DNA sites. The binding of GAL4 derivatives to nucleosomes generates a ternary complex that contains the GAL4 derivative, a complete octamer of histones, and the DNA. Côté and colleagues (1994) added purified SWI/SNF complex to DNA binding reactions which contained GAL4 derivatives and a nucleosomal DNA probe. They observed that the SWI/SNF complex increased the affinity of GAL4 (1–94), GAL4–AH, and GAL4–VP16 for nucleosomal DNA by 10 to 30-fold and that this reaction required ATP hydrolysis. These results were also confirmed by DNase I footprinting. SWI/SNF action did not appear to lead to complete displacement of the histones, since the SWI/SNF complex appeared to facilitate the formation of a GAL4 derivative/histone/DNA ternary complex. The SWI2/SNF2 polypeptide was responsible for the ATP hydrolysis, as addition of the SWI/SNF complex that contained the mutant SWI2/SNF2 subunit did not stimulate the binding of GAL4 derivatives.

What do we know about the molecular mechanism of SWI/SNF function on nucleosomal DNA? Côté and colleagues (1994) observed that the SWI/SNF complex had a low affinity for nucleosomal DNA which resulted in a retardation of the mobility of the nucleosomal probe in an acrylamide gel. Furthermore, the SWI/SNF complex, in the absence of the GAL4 derivative, was found to disrupt the helical repeat of DNA on the surface of the nucleosome core in an ATP-dependent reaction. These results suggest that the SWI/SNF complex has an inherent affinity for nucleosomal DNA.

One additional clue to the mechanism of SWI/SNF action has come from experiments with the histone chaperone, nucleoplasmin. Nucleoplasmin is a protein purified from *Xenopus laevis* which binds all four histone proteins *in vitro*, although it preferentially binds the histone H2A–H2B dimers. *In vivo*, nucleoplasmin is believed to play a crucial role in nucleosome assembly and disassembly by binding these histone dimers (Kleinschmidt et al., 1985, 1990; Dilworth et al., 1987). Chen and colleagues (1994) found that nucleoplasmin could also stimulate the binding of GAL4 derivatives to nucleosomal DNA; stimulation by nucleoplasmin, however, did not require ATP hydrolysis. Chen and colleagues (1994) analyzed the protein composition of the nucleoplasmin-stimulated GAL4 ternary complex and found that it contained only the GAL4 derivative and histones H3 and H4—histones H2A and H2B had been depleted by nucleoplasmin. Thus, GAL4 derivatives appear to have a much higher affinity for a tetramer of histones H3 and H4, and the H2A–H2B dimers are responsible for inhibiting GAL4 binding. RNA polymerase II and an RNA polymerase III transcription factor, TFIIIA, are also known to have a much higher affinity for H3–H4 tetramers than for complete octamers (including the H2A–H2B dimers; Baer and Rhodes, 1983; Tremethick et al., 1990; Almouzni et al., 1991; Hayes and Wolffe, 1992).

Côté and colleagues (1994) hypothesized that if the SWI/SNF complex also facilitated binding by disrupting the H2A–H2B dimers, then perhaps small concentrations of nucleoplasmin might stimulate the reaction by acting as a histone "sink." Consistent with this model, addition of a small amount of the histone chaperone did stimulate the SWI/SNF effect, leading to a further enhancement of GAL4 binding. Upon addition of higher concentrations of nucleoplasmin, which stimulated binding by itself, there was no further stimulation of SWI/SNF action above that observed with the low concentration. These results suggested that the SWI/SNF complex and nucleoplasmin might facilitate the binding of GAL4 derivatives by related mechanisms. SWI/SNF function is distinct, however, from that of nucleoplasm as SWI/SNF action requires ATP hydrolysis whereas nucleoplasmin does not. Furthermore, it is likely that SWI/SNF acts by interacting primarily with the nucleosomal DNA and the activator (see later in paper), while nucleoplasmin functions through interactions with the histones.

C. Association of the SWI/SNF Complex with an Activator

The SWI/SNF complex may stimulate activator binding by interacting primarily with the nucleosomal template, or it may also contact the activator. Studies with the mammalian glucocorticoid receptor support the latter possibility. As mentioned earlier, the mammalian glucocorticoid receptor (GR) is a potent activator of transcription in yeast, but its activity is reduced at least 10-fold in the absence of either *SWI1*, *SWI2*, or *SWI3* (Yoshinaga et al., 1992). Yoshinaga and colleagues (1992) tested the possibility that SWI/SNF function might involve protein–protein interactions among SWI/SNF subunits and the glucocorticoid receptor. Yeast whole cell extracts were prepared from *SWI*[+] cells, a purified derivative of the GR was added, and then GR and associated proteins were immunoprecipitated with a monoclonal antibody directed against the GR. Western blot analysis of these immunoprecipitates showed that the SWI3 subunit was present. SWI3 was not precipitated by the GR monoclonal if GR protein was excluded from the reaction. The interaction of the GR with SWI3 appeared to be of low affinity since only about five percent of the total SWI3 protein could be driven into a complex with the GR even when GR was in vast excess. To determine whether this interaction required the intact SWI/SNF complex, extracts were prepared from *swi1*[−] or *swi2*[−] cells.

When GR was immunoprecipitated from these mutant extracts, no SWI3 protein was found associated with the GR, even if the concentration of SWI3 in the extracts was increased more than 10-fold. Thus, it appears that the GR does not interact with the isolated SWI3 subunit, but rather it interacts with the intact SWI/SNF complex. Presently it is not known which residues of GR mediate the interactions with the SWI/SNF complex or which subunit is contacted by GR.

Since the SWI/SNF complex interacts with one activator, the GR, Yoshinaga and colleagues (1992) proposed that the SWI/SNF complex might associate with all activators that require SWI/SNF function. How the complex could interact with so many different activators is an intriguing, and so far, unanswered question; in the case of the GR, a 150 amino acid fragment that encompasses the Zn-finger DNA binding domain was sufficient *in vitro* to associate with the SWI3 protein. This GR derivative does not activate transcription in yeast, suggesting that SWI3 association is not dependent upon a functional transcriptional activation domain.

VI. SUMMARY AND MODEL

In vivo studies from many groups have provided convincing evidence that the SWI/SNF complex plays a crucial role in transcriptional activation. Most transcriptional activator proteins appear to require this SWI/SNF machinery to enhance transcription in yeast, and likewise, transcription of most inducible yeast genes is dependent upon a functional SWI/SNF complex. Subunits of the SWI/SNF complex have been conserved during the evolution of more complex eukaryotes, and studies in *Drosophila* and mammalian cells indicate that the function of this complex may have been conserved as well. The primary role of the SWI/SNF complex appears to be the antagonism of chromatin-mediated repression of transcription. This hypothesis is based primarily on genetic suppression studies, where chromatin components were isolated as mutations which alleviate the requirement for a SWI/SNF complex. *In vitro*, the SWI/SNF complex also functions to antagonize chromatin—the complex facilitates the binding of GAL4 derivatives to nucleosomal DNA.

What is the mechanism of SWI/SNF action *in vivo*? *In vitro* studies with the purified SWI/SNF complex suggest that the complex might function by facilitating the binding of transcriptional activators to a chromatin template. Consistent with this view, *in vivo* the GAL4 activator only requires the SWI/SNF machinery when forced to bind to low affinity binding sites. If provided with sites of higher affinity, GAL4 activity *in vivo* is independent of the presence or absence of the SWI/SNF complex (Peterson and Herskowitz, unpublished).

Figure 3 illustrates our current, speculative model for SWI/SNF function in transcription. Panel A addresses the following question: How is the SWI/SNF complex targeted to specific chromosomal locations? We know that the SWI/SNF complex is too low in abundance to be a general component of chromatin (only 50–150 copies per cell), and work from Hirschhorn and colleagues (1993) suggests

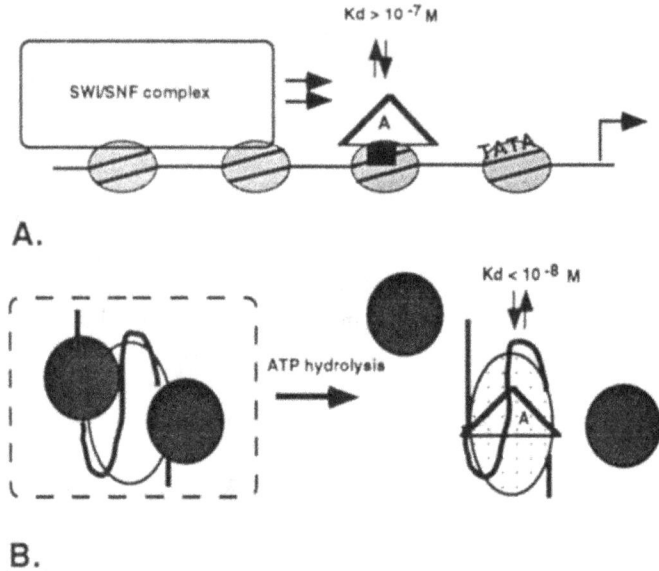

Figure 3. Speculative Model for SWI/SNF function. Panel A. Targeting of the SWI/SNF complex. SWI/SNF complex is shown "scanning" the chromosome for an activator (triangle) bound weakly to a nucleosomal binding site.

Panel B. SWI/SNF complex (dashed box) is shown bound to a nucleosome; activator is presumed to be bound as well, but it is omitted for clarity. SWI/SNF complex uses the energy of ATP hydrolysis to destabilize the DNA helix within the core particle, resulting in removal of the H2A–H2B dimers and dissociation of the SWI/SNF complex. SWI/SNF action results in a stable complex of the activator bound to a tetramer of H3 and H4. The indicated dissociation constants for activator binding is based upon the affinity of GAL4 derivatives for nucleosomal DNA (Côté et al., 1994).

that SWI/SNF action only affects one or two nucleosomes within the SUC2 upstream regulatory region. These observations indicate that SWI/SNF function must be targeted to the correct chromosomal position. One possibility is that the SWI/SNF complex interacts with activator proteins off the DNA and that the activator targets the complex to the correct chromosomal position via its binding sites. This model would be consistent with the association of the SWI/SNF complex and the GR, described above. Alternatively, the SWI/SNF complex may recognize a transcription factor that is bound weakly to a nucleosomal binding site (Figure 3, Panel A). In this model, the SWI/SNF complex could utilize its weak affinity for nucleosomal DNA to scan the chromosome for its target. This model also predicts that the SWI/SNF complex should have a low affinity for the activator, and is therefore consistent with the GR results. This model has the advantage that it does not require the assembly of the SWI/SNF complex into a large number of stable activator:SWI/SNF complexes, but suggests that the SWI/SNF complex plays a catalytic role to "lock-down" transcriptional activators that are bound weakly to nucleosomal sites.

Once the SWI/SNF complex has been brought to the correct chromosomal position, how does it stimulate binding? I propose that the SWI/SNF complex uses ATP hydrolysis to remove the histone H2A–H2B dimers, resulting in an activator bound with high affinity to a tetramer of histones H3 and H4 (Figure 3, Panel B). This model is based on the *in vitro* stimulation of SWI/SNF action by nucleoplasmin, and by the genetic suppression studies that indicated that depletion of dimers *in vivo* alleviates the requirement for the SWI/SNF complex (see preceding text). How does ATP hydrolysis disrupt the dimers? McMurray and van Holde (1986) found that unwinding of nucleosomal DNA by intercalation of EtBr led initially to loss of the H2A–H2B dimers. Likewise, the complex may use the energy of ATP hydrolysis to destabilize or unwind a small region of DNA duplex. Destabilized dimers would be removed from the SWI/SNF:GAL4 derivative:nucleosome complex by transfer to a histone chaperone, such as a nucleoplasmin-like protein, transfer to the SWI/SNF complex (the acidic N-terminus of the SWI3 subunit? See Peterson and Herskowitz, 1992), or by exchange of these dimers into bulk chromatin (discussed in Hansen and Ausio, 1992). Once the SWI/SNF complex has loaded the activator onto an H3–H4 tetramer, it might dissociate from the activator:tetramer complex and thus be free to "scan" for new activator:nucleosome complexes.

Obviously the biochemical testing of this model is still in its infancy. With purified SWI/SNF complex in hand, however, each aspect of this model should be tested within the next few years. The relative ease of purification of SWI/SNF complexes that contain specific mutations or lack individual subunits will also facilitate a structure-function analysis of the SWI/SNF complex. Previous biochemical studies have made exclusive use of human histones; we hope that in the future yeast histones (and mutant versions as well) will be reconstituted onto DNA probes for *in vitro* analyses. Thus, by coupling the isolation of mutants and their subsequent biochemical analysis, we can dissect the molecular mechanisms of chromatin-mediated repression as well as the machinery that has evolved to counteract it.

ACKNOWLEDGMENTS

I would like to thank J. Tamkun for helpful comments on the manuscript, and Masayori Inouye for suggesting the "nucleosome scanning" model for targeting the SWI/SNF complex. Studies in the Peterson laboratory are funded by grants from the National Institutes of Health and the March of Dimes Birth Defects Foundation. The author is a Leukemia Society of America Scholar.

REFERENCES

Abrams, E., Neigeborn, L., & Carlson, M. (1986). Molecular analysis of *SNF2* and *SNF5*, genes required for expression of glucose-repressible genes in *Saccharomyces cerevisiae*. Mol. Cell. Biol. 6, 3643–3651.

Adams, C.C., & Workman J.L. (1993) Nucleosome displacement in transcription. Cell 72, 305–308.

Almouzni, G., Mechali, M., & Wolffe, A.P. (1991). Transcription complex disruption caused by a transition in chromatin structure. Mol. Cell. Biol. 11, 655–665.

Arents, G., Burlingame, R.W., Wang, B-C., Love, W.E., & Moudrianakis, E.N. (1991). The nucleosomal core histone octamer at 3.1 resolution: A tripartite protein assembly and a left-handed superhelix. Proc. Natl. Acad. Sci. USA 88, 10148–10152.

Arents, G., & Moudrianakis, E.N. (1993). Topography of the histone octamer surface: Repeating structural motifs utilized in the docking of nucleosomal DNA. Proc. Natl. Acad. Sci. USA 90, 10489–10493.

Baer, B.W., & Rhodes, D. (1983). Eukaryotic RNA polymerase II binds to nucleosome cores from transcribed genes. Nature 301, 482–488.

Chen, H., Li, B., & Workman, J.L. (1994). A histone-binding protein, nucleoplasmin, stimulates transcription factor binding to nucleosomes and factor-induced nucleosome disassembly. EMBO J. 13, 380–390.

Côté, J., Workman, J., & Peterson, C.L. (1994). The yeast SWI/SNF complex facilitates the binding of GAL4 derivatives to nucleosomal DNA. Science 265, 53–60.

Davis, J.L., Kunisawa, R., & Thorner, J. (1992). A presumptive helicase (*MOT1* gene product) affects gene expression and is required for viability in the yeast *Saccharomyces cerevisae*. Mol. Cell. Biol. 12, 1879–1892.

Dilworth, S.M., Black, S.J., & Laskey, R.A. (1987). Two complexes that contain histones are required for nucleosome assembly *in vitro*: Role of nucleoplasmin and N1 in *Xenopus* egg extracts. Cell 51, 1009–1018.

Estruch, F., & Carlson, M. (1990). SNF6 encodes a nuclear protein that is required for expression of many genes in *Saccharomyces cerevisiae*. Mol. Cell. Biol. 10, 2544–2553.

Elfring, L.K., Deuring, R., McCallum, C.M., Peterson, C.L., & Tamkun, J.W. (1994). Identification and characterization of *Drosophila* relatives of the yeast transcriptional activator *SNF2/SWI2*, Mol. Cell. Biol. 14, 2225–2234.

Felsenfeld, G. (1992). Chromatin as an essential part of the transcriptional mechanism. Nature (London) 355, 219–224.

Gorbalenya, A.E., Koonin, E.V., Donchenko, A.P., & Blinov, V.M. (1989). Two related superfamilies of putative helicases involved in replication, recombination, repair and expression of DNA and RNA genomes. Nucleic Acids Res. 17, 4713–4730.

Grunstein, M. (1992). Histones as regulators of genes. Scientific Amer. 267, 68–74B.

Hansen, J.C., & Ausio, J. (1992). Chromatin dynamics and the modulation of genetic activity. TIBS 15, 187–191.

Happel, A.M., Swanson, M.S., & Winston, F. (1991). The *SNF2*, *SNF5*, and *SNF6* genes are required for Ty transcription in *Saccharomyces cerevisiae*. Genetics 128, 69–77.

Hayes, J.J., & Wolffe, A.P. (1992). Histones H2A/H2B inhibit the interaction of transcription factor IIIA with the *Xenopus borealis* somatic *5S RNA* gene in a nucleosome. Proc. Natl. Acad. Sci. USA 89, 1229–1233.

Hirschhorn, J.N., Brown, S.A., Clark, C.D., & Winston, F. (1992). Evidence that *SNF2/SWI2* and *SNF5* activate transcription in yeast by altering chromatin structure. Genes and Dev. 6, 2288–2298.

Henikoff, S. (1993). Transcriptional activator components and poxvirus DNA-dependent ATPases comprise a single family. TIBS 18, 291–292.

Hochuli, E., Bannwarth, W., Dobeli, H., Gentz, R., & Stuber, D. (1988). Genetic approach to facilitate purification of recombinant proteins with a novel metal chelate adsorbent. Bio/Technology 6, 1321–1325.

Kennison, J.A., & Tamkun, J.W. (1988). Dosage-dependent modifiers of Polycomb and Antennapedia mutations in *Drosophila*. Proc. Natl. Acad. Sci. USA 85, 8136–8140.

Khavari, P.A., Peterson, C.L., Tamkun, J.W., & Crabtree, G.R. (1993). BRG1 contains a conserved domain of the SWI2/SNF2 family necessary for normal mitotic growth and transcription. Nature, 366, 170–174.

Kleinschmidt, J.A., Fortklamp, E., Krohne, G., Zentgraf, H., & Franke, W.W. (1985). Coexistence of two different types of soluble histone complexes in nuclei of *Xenopus laevis* oocytes. J. Biol. Chem. 260, 1166–1176.

Kleinschmidt, J.A., Seiter, A., & Zentgraf, H. (1990). Nucleosome assembly in vitro: Separate histone transfer and synergistic interaction of native histone complexes purified from nuclei of *Xenopus laevis* oocytes. EMBO J. 9, 1309–1318.

Kornberg, R.D., & Lorch, Y. (1991). Irresistible force meets immovable object: transcription and the nucleosome. Cell 67, 833–836.

Kruger, W., & Herskowitz, I. (1991). A negative regulator of HO transcription, SIN1 (SPT2), is a nonspecific DNA-binding protein related to HMG1. Mol. Cell. Biol. 11, 4135–4146.

Kruger, W., Peterson, C.L., Sil, A., Coburn, C., & Herskowitz, I. (1994). Residues in the globular domains of histones H3 and H4 are necessary for proper transcriptional regulation. Submitted.

Laurent, B.C., & Carlson, M. (1992). Yeast SNF2/SWI2, SNF5, and SNF6 proteins function coordinately with the gene-specific transcriptional activators GAL4 and Bicoid. Genes and Dev. 6, 1707–1715.

Laurent, B.C., Treitel, M.A., & Carlson, M. (1991). Functional interdependence of the yeast SNF2, SNF5, and SNF6 proteins in transcriptional activation. Proc. Natl. Acad. Sci. USA 88, 2687–2691.

Laurent, B.C., Treitel, M.A., & Carlson, M. (1990). The SNF5 protein of *Saccharomyces cerevisiae* is a glutamine- and proline-rich transcriptional activator that affects expression of a broad spectrum of genes. Mol. Cell. Biol. 10, 5616–5625.

Laurent, B.C., Treich, I., & Carlson, M. (1993). The yeast SNF2/SWI2 protein has DNA-stimulated ATPase activity required for transcriptional activation. Genes and Dev. 7, 583–591.

Laurent, B.C., Yang, X., & Carlson, M. (1992). An essential *Saccharomyces cerevisiae* gene homologous to *SNF2* encodes a helicase-related protein in a new family. Mol. Cell. Biol. 12, 1893–1902.

Laybourn, P.J., & Kadonaga, J.T. (1991). The role of nucleosomal cores and histone H1 in regulation of transcription by RNA polymerase II. Science 254, 238–245.

Lorch, Y., LaPointe, J.W., & Kornberg, R.D. (1987). Nucleosomes inhibit the initiation of transcription but allow chain elongation with the displacement of histones. Cell 49, 203–210.

Lorch, Y., LaPointe, J.W., & Kornberg, R.D. (1992). Initiation on chromatin templates in a yeast RNA polymerae II transcription system. Genes and Dev. 6, 2282–2287.

McMurray, C.T., & van Holde, K.E. (1986). Binding of ethidium bromide causes dissociation of the nucleosome core particle. Proc. Natl. Acad. Sci. USA 83, 8472–8476.

Muchardt, C., & Yaniv, M. (1993). A human homologue of *Saccharomyces cerevisiae* SNF2/SWI2 and *Drosophila* brm genes potentiates transcriptional activation by the glucocorticoid receptor. EMBO J. 12, 4279–4290.

Neigeborn, L., & Carlson, M. (1984). Genes affecting the regulation of *SUC2* gene expression by glucose repression in *Saccharomyces cerevisiae*. Genetics 108, 845–858.

Neigeborn, L., Rubin, K., & Carlson, M. (1986). Suppressors of snf2 mutations restore invertase derepression and cause temperature-sensitive lethality in yeast. Genetics 112, 741–753.

O'Hara, P.J., Horowitz, H., Eichinger, G., & Young, E.T. (1988) The yeast *ADR6* gene encodes homopolymeric amino acid sequences and a potential metal-binding domain. Nuc. Acids Res. 16, 10153–10169.

Paro, R., & Hogness, D.S. (1991). The Polycomb protein shares a homologous domain with a heterochromatin-associated protein of *Drosophila*. Proc. Natl. Acad. Sci. USA 88, 263–267.

Peterson, C.L., Dingwall, A., & Scott, M.P. (1994). Five SWI/SNF gene products are components of a large multi-subunit complex required for transcriptional enhancement, Proc. Natl. Acad. Sci. USA 91, 2905–2908.

Peterson, C.L., Kruger, W., & Herskowitz, I. (1991). A functional interaction between the C-terminal domain of RNA polymerase II and the negative regulator SIN1. Cell 64, 1135–1143.

Peterson, C.L., & Herskowitz, I. (1992). Characterization of the yeast *SWI1*, *SWI2*, and *SWI3* genes, which encode a global activator of transcription. Cell 68, 573–583.

Prelich, G., & Winston, F. (1993). Mutations that suppress the deletion of an upstream activating sequence in yeast: Involvement of a protein kinase and histone H3 in repressing transcription *in vivo*. Genetics 135, 665–676.

Randazzo, F.M., Khavari, P., Crabtree, G., Tamkun, J., & Rossant, J. (1994). Brg1: A putative murine homologue of the *Drosophila brahma* gene, a homeotic gene regulator. Dev. Biol. 161, 229–242.

Smith, M.M., & Andresson, O.S. (1983). DNA sequences of yeast H3 and H4 histone genes from two non-allelic gene sets encode identical H3 and H4 proteins. J. Mol. Biol. 169, 663–690.

Stern, M., Jensen, R., & Herskowitz, I. (1984). Five SWI genes are required for expression of the *HO* gene in yeast. J. Mol. Biol. 178, 853–868.

Sternberg, P.W., Stern, M.J., Clark, I., & Herskowitz, I. (1987). Activation of the yeast *HO* gene by release from multiple negative controls. Cell 48, 567–577.

Tamkun, J.W., Deuring, R., Scott, M.P., Kissinger, M., Pattatucci, A.M., Kaufman, T.C., & Kennison, J.A. (1992). Brahma: A regulator of *Drosophila* homeotic genes structurally related to the yeast transcriptional activator SNF2/SWI2. Cell 68, 561–572.

Travers, A.A. (1992). The reprogramming of transcriptional competence. Cell 69, 573–575.

Tremethick, D., Zucker, K., & Worcel, A. (1990). The transcription complex of the 5S RNA gene, but not transcription factor IIIA alone, prevents nucleosomal repression of transcription. J. Biol. Chem. 265, 5014–5023.

van Holde, K.E. (1989). Chromatin. Springer, Berlin.

Winston, F., & Carlson, M. (1992). Yeast SNF/SWI transcriptional activators and the SPT/SIN chromatin connection. TIG 8, 387–391.

Workman, J.L., & Buchman, A.R. (1993). Multiple functions of nucleosomes and regulatory factors in transcription. TIBS 18, 90–95.

Yoshimoto, H., & Yamashita, I. (1991). The *GAM1/SNF2* gene of *Saccharomyces cerevisiae* encodes a highly charged nuclear protein required for transcription of the *STA1* gene. Mol Gen. Genet. 228, 270–280.

Yoshinaga, S.K., Peterson C.L., Herskowitz I., & Yamamoto, K.R. (1992). Roles of SWI1, SWI2, and SWI3 proteins for transcriptional enhancement by steroid receptors. Science 258, 1598–1604.

Chapter 9

Transcription Through the Nucleosome

DAVID J. CLARK

The Nucleus
Volume 1, pages 207–239.

ABSTRACT

The mechanism of transcription through nucleosomes *in vivo* and *in vitro* is addressed. The nucleosome core is a highly compact structure containing nearly two coils of DNA tightly wrapped around a central core histone octamer. It is sterically impossible for a bulky RNA polymerase molecule to transcribe through such a structure. Two classes of model have been proposed to circumvent this problem: nucleosome unfolding and nucleosome displacement/transfer. The experimental evidence which must be accounted for by the models is critically discussed. *In vivo*, transcriptionally competent genes are packaged in domains of chromatin which are decondensed relative to domains containing permanently inactive genes, but nevertheless probably exist in the form of 30 nm chromatin filament. Domains of transcriptionally competent chromatin are probably enriched in acetylated core histones and perhaps in HMG proteins 14/17. Transcribed regions are organized into nucleosomes, at least some of which contain histone H1. Electron micrographs suggest that transcription is accompanied by a further decondensation of the chromatin filament containing the gene to an extent that depends primarily on the polymerase density (i.e., gene activity). The passage of RNA polymerase disrupts the nucleosomal organization of a gene. There is evidence both for displacement/transfer of nucleosomes and for transient or stable unfolding of nucleosomes as a result of transcription *in vivo*. *In vitro*, prokaryotic RNA polymerases have been used extensively as models for transcription through the nucleosome core. Most of the evidence is consistent with the displacement/ transfer model. Experiments with eukaryotic RNA polymerases also favor the displacement model. Finally, a hypothetical model for the dynamic nucleosomal organization of a eukaryotic gene is presented.

I. INTRODUCTION

The compact nature of the nucleosome is obviously compatible with its fundamental role in packaging DNA into chromosomes, but how is this packaged information made accessible to the enzymes involved in DNA metabolism? How do processes such as DNA replication, recombination, repair, and transcription occur in their natural chromatin environment? To what extent are nucleosomes obstacles to these processes? Some answers to these questions are now beginning to emerge. This paper is strictly focused on transcription in chromatin and, in particular, on the question of how RNA polymerase transcribes through nucleosomes. The mechanism of transcript initiation in chromatin has been reviewed by Felsenfeld (1992), Kornberg and Lorch (1992), and Workman and Buchman (1993) and will not be discussed here. In addition, there are several excellent reviews on active chromatin and transcription through nucleosomes (Gross and Garrard, 1987; Kornberg and Lorch, 1991, 1992; Wolffe, 1992: Morse, 1992; Hansen and Ausio, 1992; van Holde et al., 1992). In this paper, a brief discussion of nucleosome structure is followed by a description of the various models which have been proposed for transcription through the nucleosome. The next section is a critical review of experiments designed to investigate the nature of active chromatin *in vivo*. This is followed by a critical review of experiments aimed at elucidating the mechanism of transcription through nucleosomes *in vitro*. The paper concludes with an overview and a tentative model for the dynamic nucleosomal organization of active genes.

II. THE STRUCTURE OF THE NUCLEOSOME CORE

Chromatin is usually prepared by digestion of cell nuclei with micrococcal nuclease (MNase), followed by lysis at low salt concentration. In the electron microscope (see Thoma et al., 1979), fragments of chromatin resemble regularly spaced beads on a string—each bead is a nucleosome core, the string is linker DNA connecting one nucleosome core to the next and the structural repeat unit (the core plus linker) constitutes the nucleosome. The protein components of the nucleosome core are the core histones (H2A, H2B, H3, and H4); the linker histone, H1, is less tightly bound on the outside of the core and to the linker DNA. If salt is added to 15 mM or less, the chromatin fragments contract to form more regular zig-zag filaments about 10 nm in diameter ("10 nm filaments"). Addition of more salt induces further contraction of the filaments, which become shorter and thicker, almost certainly through helical coiling of the 10 nm filament (reviewed by Felsenfeld & McGhee, 1986). The limit products of the folding process are very compact filaments about 30 nm in diameter which are formed at salt concentrations near 0.1 M, close to physiological ionic strength. These filaments resemble those observed in nuclei. Histone H1 is required for maximal, ordered compaction of the 30 nm filament, probably mediated through its effects on linker DNA (see Clark and Kimura, 1990 for a discussion).

If isolated chromatin fragments are stripped of H1 and digested further with MNase, the linker DNA is destroyed and nucleosome core particles are obtained. The structure of the core particle has been solved by X-ray crystallography to 7 Å resolution (Richmond et al., 1984). It contains 146 bp of DNA wrapped in 1.75 negative (left-handed) superhelical turns around a core histone octamer. This negative superhelicity is partly offset by a positive change in DNA twist and, by direct measurement of supercoiling, it was determined that almost exactly one negative supercoil is stabilized within the nucleosome core (Simpson et al., 1985). The DNA is not smoothly bent but kinked at several points on the surface of the core. The histone octamer is stable in the absence of DNA in 2 M NaCl (Thomas and Kornberg, 1975), but at lower salt concentrations it dissociates into an $(H3/H4)_2$ tetramer and two H2A/H2B dimers, indicating that dimer-tetramer interactions are weak relative to the interactions within the dimer and within the tetramer (Eickbush and Moudrianakis, 1978). The structure of the octamer has been solved to 3.1 Å resolution (Arents and Moudrianakis, 1991, 1993), revealing a central tetramer flanked by dimers (Figure 1). Inspection of the octamer surface revealed a left-handed histone superhelix with an array of positively charged residues which could

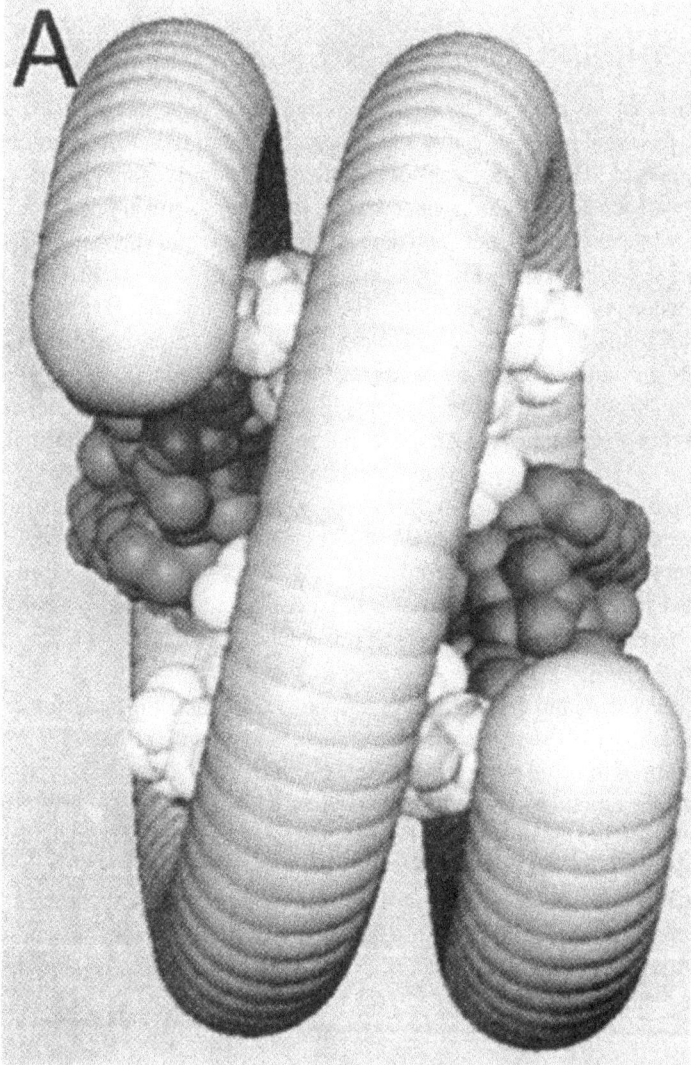

Figure 1. The structure of the nucleosome core. A scale model for the structure of the nucleosome core particle. DNA represented as a 20 Å tube was coiled around a model of the histone octamer (derived from the crystal structure). The two H2A/H2B dimers (shown in black) flank the central (H3/H4)$_2$ tetramer (shown in white) (from Arents & Moudrianakis, 1993).

organize the DNA around the octamer, predicting a structure for the core particle that is very similar to that described by Richmond and colleagues (1984).

III. MODELS FOR TRANSCRIPTION THROUGH THE NUCLEOSOME

The nucleosome core is a highly compact structure in which the DNA coils are very close together. Furthermore, it is an extremely stable structure, at least *in vitro*. A eukaryotic RNA polymerase is about five times larger than the histone octamer. To continue synthesizing a transcript, it must separate the DNA strands and rotate once with respect to the DNA helix every 10 bp. This task is clearly stereochemically impossible unless something dramatic happens to the nucleosome core. Two classes of model have been proposed to resolve this difficulty (Figure 2): (1) Unfolding: Transcribing RNA polymerases induce a dramatic conformational change in the nucleosome to facilitate traversal of nucleosomal DNA, or nucleosomes on a gene adopt a stable unfolded conformation prior to transcription; and (2) Displacement/ transfer: RNA polymerase causes the octamer to be displaced into solution or its transfer to another location, thus removing the obstacle altogether. A variant of this is "progressive displacement" in which RNA polymerase displaces the first H2A/H2B dimer, transcribes through the H3/H4 tetramer by uncoiling its DNA (but leaves it in place) and then displaces the second H2A/H2B dimer (van Holde et al., 1992).

A. Transcription-Induced DNA Supercoiling

Some of the experiments discussed below were motivated by the conclusions of a theoretical analysis of the mechanics of transcription by Liu and Wang (1987). They pointed out that viscous drag is likely to offer considerable resistance to the rotation of a transcribing polymerase with its associated transcript around the DNA helix (it must rotate once around the DNA helix for every 10 bp transcribed). They suggested that the DNA would rotate instead, that is, become supercoiled, such that DNA ahead of the advancing polymerase would become positively supercoiled and the DNA behind would become negatively supercoiled (Figure 2, Part a). If the DNA is free to rotate about its own axis then these supercoils would be rapidly dissipated, and in a circular DNA the supercoils would eventually cancel one another as they diffuse through the helix toward each other. They went on to demonstrate transcription-induced supercoiling experimentally *in vitro* and *in vivo* (e.g., Tsao et al., 1989). These ideas inspired many in the chromatin field because positive supercoils preceding a transcribing polymerase might destabilize a nucleosome core in its path (because it contains one negative supercoil).

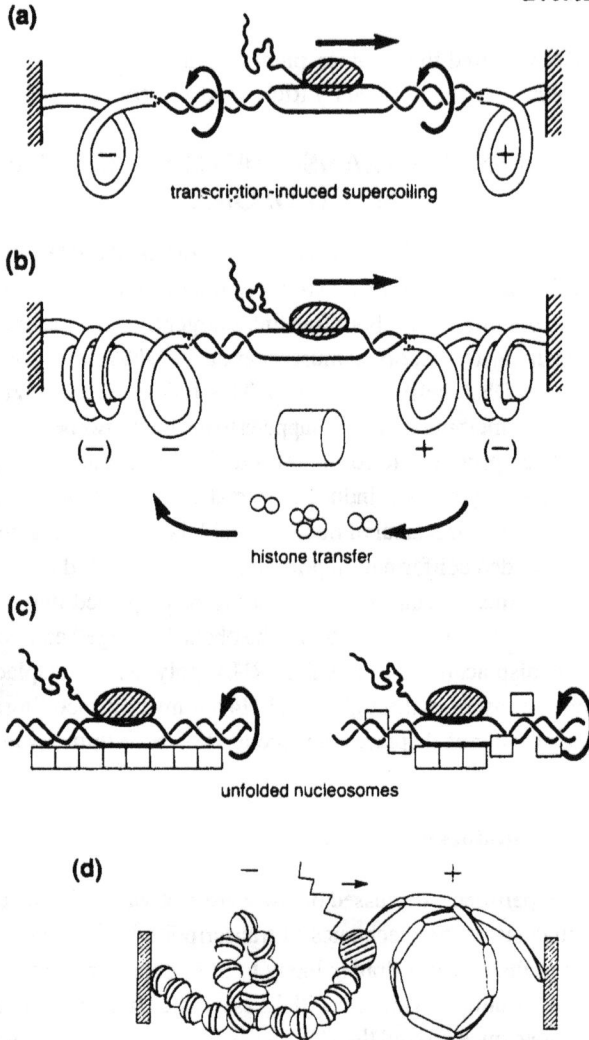

Figure 2. Transcription-induced supercoiling and models for transcription through the nucleosome. (a) The transcription-induced supercoiling model of Liu and Wang (1987). The transcribing polymerase does not drag the transcript and associated protein once around the DNA helix every 10 bp; instead, the DNA rotates and because it is tethered to a support at both ends (perhaps the nuclear matrix), the superhelical stress cannot be relieved. Consequently, the DNA ahead of the polymerase becomes positively supercoiled and the DNA behind becomes negatively supercoiled, as indicated. (b) The histone displacement/transfer model for transcription through the nucleosome (Clark and Felsenfeld, 1991). Positive supercoils ahead of the transcribing polymerase cause the displacement of the histone octamer (and perhaps its dissociation into tetramer and dimers, as shown) which is then recaptured on the negatively supercoiled DNA behind the polymerase. This is an octamer release/recapture model;

(*continued*)

Figure 2. (Continued) an alternative (not shown) is that the octamer is transferred directly from in front of the polymerase to behind it, without release into solution. (c) The unfolding model for transcription through the nucleosome. The nucleosome is linearized to facilitate passage of the polymerase. This might be caused by positive supercoils preceding the polymerase or it might occur before transcription. The nucleosomal DNA coils are removed but the core histones remain bound as a result of a major conformational change in the octamer. The unfolded conformation would have to be such that the histone-DNA contacts do not impede the passage of the polymerase and this may require contacts on different DNA strands as suggested in the figure, or the histone-DNA contacts might be transiently ruptured. (d) The nucleosome unfolding model of Lee and Garrard (1991a). Positive supercoils ahead of the polymerase cause nucleosome splitting to facilitate passage of the polymerase and negative supercoils in the wake of the polymerase cause the nucleosomes to refold. Parts (a), (b), and (c) are from Thoma (1991) and (d) is from Lee and Garrard (1991a).

IV. TRANSCRIPTIONALLY ACTIVE CHROMATIN

For transcription to occur, various transcription factors must locate their binding sites at the promoter (and at regulatory elements such as enhancers) before an RNA polymerase molecule is recruited at the promoter and traverses the gene to synthesize a transcript. When occupied, these binding sites are hypersensitive to DNase I, reflecting early events in gene activation which will not be discussed here because they are not directly relevant to transcript elongation. In this section, the structure of active chromatin is examined to establish the nature of the template for RNA polymerase II (Pol II) *in vivo*. This is followed by a discussion of experiments which cast light on the mechanism of transcription through nucleosomes *in vivo*.

A. The Ultrastructure of Transcribing Genes

What does chromatin active in transcription look like? The most direct method of analysis is electron microscopy. The numerous early studies of transcriptionally active "lampbrush" and polytene chromosomes have been discussed in the excellent book by van Holde (1988). This section focuses on recent work on polytene chromosomes; similar observations have been made with lampbrush chromosomes (Scheer, 1987).

Insect polytene chromosomes have been studied intensively because actively transcribing regions appear as chromosome "puffs" and can therefore be recognized with ease (e.g., the Balbiani rings). When polytene chromosomes are treated with certain stains, they develop a reproducible and specific banding pattern; each band corresponds to a chromatin domain containing a gene or co-coordinately regulated set of genes. In response to environmental stimuli such as heat shock or drugs, specific bands puff out as a result of heavy transcription. In early micrographs ("Miller" spreads, e.g., McKnight et al., 1978), these highly active genes resembled

"Xmas trees": each "trunk" is a gene, to which multiple transcribing RNA polymerase molecules are attached, each associated with a "branch" corresponding to a nascent transcript. Beaded structures identified as nucleosomes were observed, and the number of nucleosomes depended inversely on the number of transcribing polymerases on the gene. However, these dramatic micrographs were obtained by subjecting chromosomes to "stretching" conditions by fixation at very low salt concentration, making it likely that they do not represent the native state. Despite this concern, it was established that transcriptionally active genes are much less condensed than inactive chromatin and that they are not devoid of nucleosome cores.

More recently, relatively intact chromosomes have been obtained for analysis by gentle disruption of cells and fixation of individual chromosomes in buffers approximating physiological salt concentrations. Stained sections of isolated chromosomes resembled chromosomes observed *in situ* (Björkroth et al., 1988). Inactive genes are in highly condensed chromatin thought to contain tightly packed loops of 30 nm filament (Andersson et al., 1984). When the Balbiani ring genes are induced, specific bands puff out as the chromatin containing these genes undergoes a dramatic decondensation. Each puff consists of many "transcription loops" emanating from more condensed chromatin, and each loop contains a highly decondensed gene under heavy transcription (Ericsson et al., 1989). Transcribing RNA polymerases were visualized on the gene through associated densely staining ribonucleoprotein particles (i.e., transcripts) resembling granules on stalks which increase in size as the nascent transcript is synthesized, making it possible to establish the direction of transcription unambiguously. Presumably this is the native form of the Xmas tree (Figure 3). The frequency of intact nucleosome cores on the gene depends on the distance between transcribing polymerases. Polymerase molecules very close together (less than 150 bp) are separated by a nucleofilament about 5 nm wide, which could be naked DNA or an unfolded nucleosome. Short stretches of slightly coiled nucleosomal 10 nm filament are observed between polymerase molecules further apart and, occasionally, even short regions of 30 nm filament are observed between transcribing polymerases (Björkroth et al., 1988). The promoter region is a thin filament about 5 nm wide and about 500 bp long, but the DNA immediately upstream is in 30 nm filament. The region downstream of the last polymerase (about 3 kb) is in a loosely coiled 10 nm filament, which eventually coils up into 30 nm filament (Ericsson et al., 1989). When reinitiation of transcription of active Balbiani ring genes was inhibited, the genes were rapidly condensed into 30 nm filament, indicating that *transcription* is responsible for the decondensation of 30 nm filament containing the gene. The highly condensed chromatin band characteristic of the uninduced inactive gene was not formed, suggesting that cessation of transcription was not sufficient to cause complete condensation of the chromatin (Andersson et al., 1984), and that potentially active genes might differ from permanently inactive genes at a level of chromatin organization above that of the 30 nm filament. In summary, transcription induces

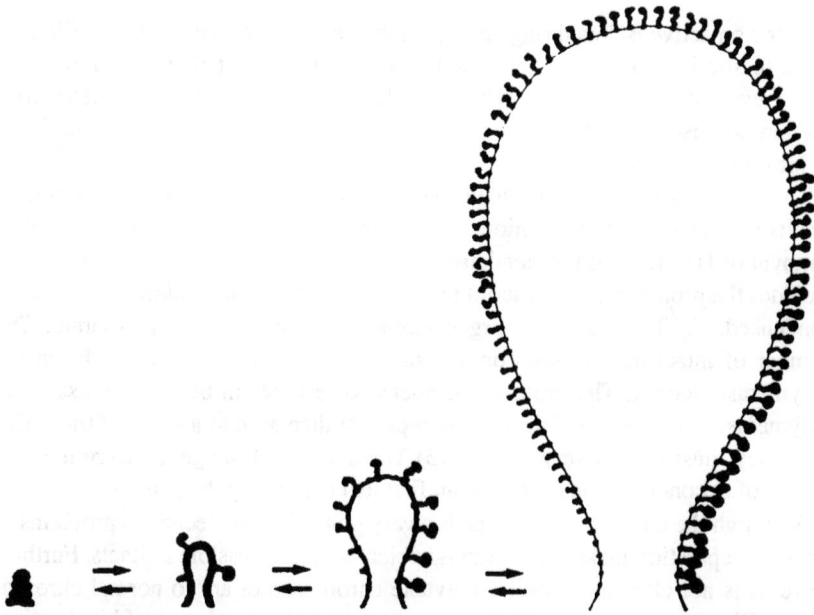

Figure 3. Schematic representation of the structure of a chromatin loop at different levels of activity. A condensed, transcriptionally inactive chromatin loop (left) becomes moderately decondensed (center left) as a result of a low level of transcription (transcribing RNA polymerases are indicated by their associated ribonucleoprotein, which resemble granules on stalks) with most of the chromatin in the form of 30 nm filament. Increased activity results in further decondensation of the loop to yield 10 nm nucleosomal filament (center right). The extremely active gene is almost fully decondensed (right) with transcribing polymerases very close together, separated by a 5 nm filament (which may represent unfolded nucleosomes or free DNA). This decondensation is reversed on addition of inhibitors of transcript initiation (from Björkroth et al. 1988).

the decondensation of 30 nm filament containing the transcribed region and limited flanking sequences only, to a degree determined by the activity of the gene.

These observations are consistent with MNase digestion studies on Balbiani ring genes: the inactive gene gave a strong MNase repeat indicating a regular nucleosomal organization, but on the moderately active gene the repeat was faint, and on the extremely active gene (on which the polymerase density was very high) the repeat was absent, indicating a major disruption of the nucleosomal array (Widmer et al., 1984). Immuno-electron microscopy studies indicate that chromosomal puffs contain H1, perhaps in reduced amounts (Hill et al., 1989; Ericsson et al., 1990) and also an HMG14-like protein (Westermann and Grossbach, 1984) which was not detected in the condensed bands. Both unacetylated and acetylated H4 were detected on active Balbiani ring genes, but the latter was concentrated in "islands" close to the boundaries of the decondensed region rather than in transcribing chromatin (Turner *et al.*, 1990). However, these studies should be interpreted with caution because it is difficult to control for differences in accessibility to antibody or for buried epitopes.

These studies led to several major conclusions: (1) Induction of transcription results in the local decondensation of the gene, which is not the result of complete removal of H1. This local decondensation is confined to the body of the gene and includes the promoter and transcript termination region: The flanking chromatin is condensed. (2) The chromatin organization of an active gene is dynamic: The number of intact nucleosomes on a transcribing gene depends inversely on the polymerase density. The presence of nucleosome cores in between transcribing polymerases implies that if they are disrupted or displaced as a result of transcription, they must reform very rapidly. (3) The activity of the gene determines the degree of decondensation of the 30 nm filament containing the gene.

Although the electron microscope is a very powerful tool, technical problems in sample preparation make micrographs subject to many possible artifacts. Furthermore, it is not clear how similar polytene chromosomes are to normal chromosomes. These conclusions must therefore be viewed in the context of biochemical studies of active genes.

B. The Chromatin Structure of the Active and Inactive β-Globin Genes

Many papers have addressed the biochemical properties of active chromatin and it is beyond the scope of this paper to review all of these. Most of them have been discussed by van Holde (1988). The chromatin structure of the active and transcriptionally competent β-globin genes has been chosen as an example.

Active genes and potentially active genes (those which might be expressed at some time in a particular cell type) are located in regions of chromatin which are more sensitive to digestion with DNase I than the bulk of chromatin ("general" sensitivity) (see van Holde, 1988). The structural basis of general sensitivity is directly relevant to this discussion because it is likely to reflect important changes in the nature of the chromatin substrate for RNA polymerase. The β-globin gene cluster, which has served as an important model system for studying the chromatin structure of active and inactive genes, exhibits general sensitivity in erythroid cells (which express or have the potential to express one or more of the genes in the cluster), but not in other cells (in which these genes are permanently inactive). General sensitivity is not confined to transcribed regions but also includes many kilobases of flanking DNA. Thus, general sensitivity is clearly not the result of transcription per se.

Several groups have examined the nucleosomal conformation of the active and inactive β-globin gene using nucleases. There are two types of assay (Wu et al.,

1979): (1) The characteristic nucleosomal repeat pattern obtained by digesting nuclei with MNase is probed with a radio-labeled DNA fragment containing the region of interest to determine whether it is organized into nucleosomes and the nucleosomal spacing in this region. (2) Very limited digestion with nuclease followed by digestion of the purified DNA with a restriction enzyme. A blot is probed to determine whether nucleosomes are phased (i.e., adopt unique positions with respect to DNA sequence) in the region of interest. If nucleosomes are phased, patterns of protection (nucleosomal DNA) and digestion (linker DNA) are observed, but if they are not phased, a smear is observed (this does *not* necessarily indicate loss or disruption of nucleosomes).

All studies of the β-globin genes confirm the presence of nucleosomes irrespective of transcriptional activity. However, when the transcriptionally competent gene was compared with the inactive gene, there was some evidence for atypical nucleosomes on the chicken gene (Sun et al., 1986), for partial depletion of nucleosomes from the mouse gene which is more extensive on induction of transcription (Cohen and Sheffery, 1985), and for an array of phased nucleosomes on the mouse gene which is disrupted only over the 5' half of the gene (Benezra et al., 1986). Smith and colleagues (1983) observed an increase in the nucleosomal repeat length, Villeponteau and colleagues (1992) observed a shortened repeat, and Sun and colleagues (1986) reported no change! Most of these studies did not detect significant changes on the transcriptionally active gene, perhaps reflecting the relatively modest transcription rates of the globin genes, and indicating that none of these effects is the result of transcription through nucleosomes.

Within the β-globin gene cluster, transcriptionally active genes are more sensitive to DNase I than inactive genes (Smith et al., 1984). This differential sensitivity was not observed in nuclei digested at very low salt concentration (to induce extensive chromatin unfolding) or if H1 was removed from the nuclei before digestion, suggesting that the chromatin containing the active gene is less condensed, perhaps due to H1 depletion.

The degree of condensation of the transcriptionally competent chicken β^A-globin gene has been compared with that of the permanently inactive gene (Kimura et al., 1983; Fisher and Felsenfeld, 1986; Caplan et al., 1987). The degree of unfolding of a 6.2 kb chromatin fragment containing the gene was determined by sedimenting chromatin fragments obtained by digestion of nuclei with *Eco*RI. This fragment contains the entire gene and some flanking sequences, and was previously shown to exhibit general sensitivity to DNase I in erythrocytes (Wood and Felsenfeld, 1982). The fragment was detected by blot hybridization of DNA from gradient fractions. When prepared from erythrocytes (in which it is transcriptionally competent but not active), it sedimented somewhat more slowly than bulk chromatin fragments of the same size, but when prepared from oviduct or other tissues in which the gene is permanently inactive, it sedimented with bulk chromatin fragments of the same size. Equivalent results were obtained for a fragment containing the ovalbumin gene prepared from erythrocytes (where it is inactive) or oviduct (where it is active). The difference in sedimentation rate was attributed to a modest impairment of the ability of H1 to condense the filament (e.g., perhaps due to modification of the nucleosome core) and/or to the presence of two open regions (corresponding to DNase I hypersensitive sites) at the promoter and the enhancer, rather than a general decondensation of the chromatin containing the gene.

Using UV cross-linking, Kamakaka and Thomas (1990) demonstrated that H1 and H5 (an erythrocyte-specific variant of H1) were both partially depleted from chromatin containing the transcriptionally competent β^A-globin gene *in vivo*. The chromatin containing the *active* β^A-globin gene is also partially depleted of H1 and H5, and is modestly enriched in HMGs 14/17 (Postnikov et al., 1991). Taken together, these experiments imply that the modest depletion of H1 observed is not the result of transcription.

The association of core histone acetylation with transcriptional activity is controversial (see van Holde, 1988). However, convincing evidence is now available for a direct correlation between acetylation and general sensitivity in the chicken β-globin gene cluster (Hebbes et al., 1988, 1994). The correlation between increased levels of the nucleosome core binding proteins HMGs 14/17 and transcriptional activity is far more controversial (see van Holde, 1988). Nucleosome cores containing acetylated core histones appear to be closely associated with nucleosome cores containing HMGs14/17 (Malik et al., 1984). The separate effects of hyperacetylation (McGhee et al., 1983) and HMGs 14/17 (McGhee et al., 1982) on the folding of the 30 nm filament *in vitro* are very subtle; only depletion of H1 results in significant unfolding (Thoma et al., 1979). These factors might work in concert, and there is some evidence that histone acetylation reduces the capacity of H1 to condense chromatin (Ridsdale et al., 1990). However, these factors correlate with the establishment of the transcriptionally competent state rather than with transcription per se.

In conclusion, the active, potentially active, and permanently inactive forms of the gene are all organized into nucleosomes. The chromatin containing the transcriptionally competent β-globin gene is slightly less condensed than that containing the permanently inactive gene, but it is not in an extended conformation. Whether this is true of the active gene is not known. Other factors which have been linked with active chromatin include the distribution of histone variants (reviewed by Wolffe, 1991), other histone modifications such as phosphorylation and ubiquitination (Davie and Murphy, 1990), domain supercoiling and DNA methylation (see van Holde, 1988), but the evidence is not sufficiently compelling to warrant discussion here. Although there are many contradictions in this field, it may be concluded that the nucleosomes which must be transcribed through by RNA polymerase II (Pol II) probably contain acetylated core histones, are probably somewhat depleted of H1, and might be enriched in HMGs 14/17.

C. Transcription-dependent Changes in the Nucleosomal Organization of Yeast Genes

Studies of the nucleosomal organization of several inducible yeast genes have provided the most interesting examples of changes in chromatin structure as a result of transcription *in vivo*. The nucleosomal organization of the heat-shock inducible *HSP82* gene was analyzed by Szent–Györgyi and colleagues (1987) using DNase I mapping. The flanking regions of the gene are organized into arrays of phased nucleosomes with several open regions including the promoter, defined by DNase I hypersensitive sites. The gene was more sensitive to DNase I than the flanking regions, but sensitivity to MNase was not enhanced over the gene. Most of the gene has an indeterminate nucleosomal organization (i.e., there are no phased nucleosomes), but the 3' end exhibited a striking cleavage periodicity of about 85 bp, about half the nucleosomal repeat length in yeast. Many of the bands coincided with preferential cleavage sites in free DNA, suggesting that this repeat might reflect the sequence preference of DNase I rather than an altered chromatin structure (Szent–Györgyi et al., 1987). However, Lee and Garrard (1991a) presented evidence that the half-nucleosome repeat represented phased, conformationally altered ("split") nucleosomes (presumably with a DNase I sensitive site at the dyad axis). They showed that mutations which abolished transcription of the *HSP82* gene also abolished the half-nucleosome repeat, and they ruled out the possibility that the half-nucleosome repeat reflected two overlapping sets of phased nucleosomes. Once transcription had been induced, the half-nucleosome repeat was stable even if subsequent transcription was inhibited.

Lee and Garrard (1991b) addressed the possibility that transcription-induced supercoiling is important in generating the altered chromatin structure of active genes by expressing *E. coli* topoisomerase I (topo I), which relaxes only negative supercoils, in yeast cells carrying mutated topoisomerase genes. The *REP2* gene on the endogenous 2 μm plasmid became sensitive to DNase I when the plasmid was positively supercoiled as a result of transcription, but a half-nucleosome repeat was not observed. The increased sensitivity persisted even after removal of the positive supercoils by relaxation of the isolated minichromosome *in vitro*, but there were a few less nucleosomes in the minichromosomes that were initially positively supercoiled. These results do not distinguish between models for transcription through the nucleosome, but they do indicate that positive supercoiling can induce an altered chromatin structure with increased sensitivity to DNase I.

These observations led to an unfolding model for transcription through the nucleosome *in vivo* (Lee and Garrard, 1991a, 1991b; reviewed by Thoma, 1991). Positive supercoils preceding the first RNA polymerase to traverse the gene after induction of transcription cause the nucleosomes to split, facilitating the passage of the polymerase (Figure 2, Part d), and nucleosomes refold in the wake of the polymerase as a result of negatively supercoiling. However, nucleosome refolding appears to be at odds with the apparent stability of split nucleosomes. To facilitate passage of the polymerase, the split nucleosome would have to be significantly unfolded, but the observed change in topological linking number is consistent with only a very modest uncoiling of nucleosomal DNA (Lee and Garrard, 1991b). Moreover, Pederson and Morse (1990) were unable to detect a transcription-dependent change in the linking number of a plasmid carrying the *HSP26* gene in yeast cells on induction by heat shock. They concluded that any alterations in nucleosome topology which might accompany polymerase passage must be rapidly reversed (i.e., if they occurred, they were not stable). The absence of a half-nucleosome repeat over the 5′ half of the *HSP82* gene is presumably due to lack of phasing, which might also explain why the half-nucleosome repeat has been observed on very few genes. The *Drosophila hsp26* gene exhibits a clear half-nucleosomal repeat on heat shock (Cartwright and Elgin, 1986), but in this case also, the naked DNA exhibits a periodic DNase I cleavage pattern.

Other possible explanations for the half-nucleosome repeat include: (1) A queue of paused RNA polymerases at the 3′ end of the gene awaiting termination (Szent–Györgyi et al., 1987), although it is not apparent why the polymerases would pause. (2) An array of phased $(H3/H4)_2$ tetramer complexes (van Holde et al., 1992). These protect about 80 bp from MNase digestion (Camerini-Otero et al., 1976) and are intermediates in nucleosome assembly *in vitro* (Fotedar and Roberts, 1989). However, the region exhibiting the half-nucleosome repeat with DNase I yielded a normal MNase repeat (Lee and Garrard, 1991a), indicating that there are at least some octamers present in this region. (3) Partial depletion of nucleosomes from the 3′ region of the gene as a result of transcription; this would tend to reveal the DNase I pattern of naked DNA in this region, which is quite periodic (Szent–Györgyi et al., 1987).

Other investigators have observed disruption of arrays of phased nucleosomes on induction of genes in several organisms using nuclease mapping techniques (Wu et al., 1979; Lohr, 1983; Bergman and Kramer, 1983). Morgan and Whitlock (1992) used actinomycin-D to show that the observed disruption was dependent on transcription. The smearing of the nucleosomal maps observed on induction indicates *only* that phasing is disrupted; there might not be any effects on nucleosome structure. In the case of the activated yeast *SUC2* gene (Perez–Ortín et al., 1987), the MNase pattern is not smeared but becomes more like that of naked DNA. In most of these studies the active gene was more sensitive to nuclease digestion than the repressed gene, presumably indicating a more accessible chromatin structure. Cavalli and Thoma (1993) used MNase to map the nucleosomal organization of an inducible hybrid yeast *GAL* gene with a high propensity for nucleosome phasing. They concluded that transcription resulted in the loss of some nucleosomes and a re-phasing of the rest (indicated by a shortened repeat length), showing that nucleosomes must have been redistributed on the gene, which is not consistent with a simple unfolding model.

In conclusion, split nucleosomes might be present at the 3′ end of the yeast *HSP82* gene and on the *Drosophila hsp26* gene, but it has proved difficult to find other examples. Transcription-induced supercoiling is associated with structural changes in the chromatin of active genes, as probed by DNase I, but it is not required for the maintenance of these changes. On the other hand, transcription of other yeast genes results in a reorganization of nucleosomes on the gene which is difficult to explain with a simple unfolding model and is most consistent with a displacement/transfer model.

D. The Structure of Nucleosomes on Active Genes

Do nucleosomes split as a result of transcription? If so, does this occur before or during transcription? Direct evidence for stable split nucleosomes containing actively transcribed sequences ("lexosomes"; see Prior et al., 1983), was obtained by chemical probing for the normally unreactive (and therefore buried) sulphydryl group of H3. Nucleosome core particles prepared from purified active nucleolar chromatin had a highly extended conformation and contained non-histone proteins as well as core histones, unlike core particles from inactive nucleolar chromatin. However, this study proved to be controversial (see van Holde, 1988). Later studies made use of the accessibility of the H3 sulphydryl to purify reactive core particles on a mercury affinity column, and enrichments in transcribed sequences were demonstrated (Chen et al., 1990). Core particles retained on the column indirectly through the sulphydryl groups of associated non-histone proteins were removed by washing first with 0.5 M NaCl (to dissociate core particles from bound non-histone proteins), and then eluting the remaining core particles with dithiothreitol. Core particles containing DNA from the transcribed regions of the oncogenes *c-fos* and *c-myc* were preferentially retained on the column only when prepared from cells expressing the genes; this retention was inhibited when the cells were treated with α–amanitin, a specific inhibitor of Pol II. This experiment suggests that the reactive nucleosomes are transient intermediates generated during transcription, but this is at odds with their apparent stability during isolation, and with the isolation of reactive core particles derived from the uninduced yeast *GAL1* gene (Chen–Cleland et al., 1993).

The nucleosomal organization of transcribing SV40 minichromosomes was analyzed in nuclei using psoralen cross-linking (de Bernardin et al., 1986). This technique relies on the fact that psoralen cross-links the two strands together in linker DNA and in RNA–DNA hybrids but not in nucleosomal DNA. After cross-linking, the DNA was extracted, denatured and viewed in the electron microscope. Nucleosome locations were deduced from the distribution of single-stranded DNA bubbles. Transcriptionally active minichromosomes were identified through their associated transcripts. They possessed the same number of nucleosomes as inactive ones, suggesting that transcription did not result in a significant loss of nucleosomes. Transcripts were equally distributed between the bubbles (nucleosomes) and linker DNA, suggesting that Pol II can invade the nucleosome core. However, the full significance of this observation will not be clear until a complete structural explanation for the inability of psoralen to cross-link nucleosomal DNA is available.

Does the histone octamer dissociate as a result of transcription? Evidence that transcription induces exchange of H2A/H2B between nucleosomes *in vivo* was obtained using formaldehyde cross-linking at high pH (which could result in artifacts) to distinguish between core histones in nucleosomes and those which are not (Jackson, 1990). Louters and Chalkley (1984) used the same assay to show that exogenous labeled H2A/H2B could exchange with H2A/H2B in chromatin fragments *in vitro* but, in a direct mixing experiment, *no* exchange of H2A/H2B occurred between chromatin fragments of different sizes. Schwager and colleagues (1985) also presented evidence for H2A/H2B exchange *in vivo*, but there was no direct evidence for a link with transcription.

Nacheva and colleagues (1989) used chemical cross-linking to determine which histones were bound to the *Drosophila hsp70* gene before and after heat shock. Using a protocol for cross-linking histones to DNA primarily through their lysine residues, there was no change in the degree of cross-linking for any of the histones, suggesting that transcription did not result in significant loss of histones. However, using a protocol for preferential cross-linking through histidine residues (which are located almost exclusively in the histone globular domains), activation of transcription resulted in reduced yields of cross-linked histones in the order: H1 < H2A/H2B < H3/H4. They suggested that transcription displaces the globular domains from DNA, leaving the histones bound only through their tail domains. Strictly speaking, however, these observations show only that transcription induced the disruption of some contacts between histidine residues and DNA within the nucleosome and therefore complete displacement of the octamer core from DNA need not necessarily be invoked.

In conclusion, there is some evidence for unfolded nucleosomes on active genes, which is inconsistent with a simple displacement/transfer model. An alternative model in which the core histone tails remain bound and the globular core of the octamer is displaced during transcription is an interesting possibility. However, there is a conceptual problem because acetylation of the tail domains reduces their DNA binding affinity *in vitro* (Cary et al., 1982). For this reason, it is more likely that these results reflect disruption of some histone–DNA contacts during transcription, which is consistent with both classes of model. The release of H2A/H2B dimers is inconsistent with a simple unfolding model but could be accounted for by the displacement/ transfer model if it involved octamer dissociation into tetramer and dimers.

E. Summary

Active and potentially active genes are packaged in domains of structurally modified chromatin which probably contain acetylated core histones and might be enriched in HMGs 14/17. Transcribed regions are organized into nucleosomes partially depleted of H1, and the passage of RNA polymerase disrupts the nucleosomal organization of the gene, possibly causing dramatic conformational changes or release of H2A/H2B dimers. The conflicting nature of the evidence does not allow a conclusion to be drawn regarding the mechanism of transcription through the nucleosome *in vivo*. However, it is clear that the structure of active chromatin is dynamic, that is, transcription induces changes in the nucleosomal organization of a gene.

It is tempting to interpret the biochemical studies in the context of the changes in ultrastructure. DNase I sensitivity could be attributed to the specific decondensation of highly condensed chromatin containing the gene (the potentially active state) coincident with core histone acetylation, resulting in increased accessibility to the enzyme. The increased sensitivity of transcribing genes to nuclease could be attributed to transcription-dependent decondensation of the 30 nm filament and nucleosome displacement or unfolding. Changes in the nucleosomal organization of the active gene would be expected to depend on the polymerase density (i.e., the activity of the gene) and variations in activity might explain conflicting results.

V. TRANSCRIPTION THROUGH THE NUCLEOSOME CORE *IN VITRO*

In this section, experiments designed to elucidate the mechanism of transcription through the nucleosome core are discussed. Most of the studies employed prokaryotic RNA polymerases (i.e., those of *E. coli* and the T7 and SP6 bacteriophages) because, unlike eukaroytic RNA polymerases, they are available highly purified and they initiate transcription specifically and very efficiently. In order to observe the effects of transcription on a nucleosomal template, it is vital that most of the templates are transcribed. If only a tiny fraction of the templates are transcribed, as in most studies using eukaryotic RNA polymerases, the effects of transcription on the nucleosome cannot be analyzed. However, prokaryotic RNA polymerases could not have been designed to transcribe through nucleosomes *in vivo*, because bacteria do not have nucleosomes, and therefore these studies might be irrelevant to an understanding of transcription through the nucleosome by eukaryotic RNA polymerases. The evidence available (discussed below) suggests that this is not the case, but more experiments with eukaryotic polymerases are necessary.

A. Nucleosome Cores and Positively Supercoiled DNA

Because of the importance of transcription-induced supercoiling (Figure 2, Part A) in interpretation of many of the experiments described below, the stability of nucleosome cores on positively supercoiled DNA must be addressed. Clark and Felsenfeld (1991) detected normal nucleosome cores on a positively supercoiled plasmid using the topo I relaxation and MNase assays. However, both topo I and MNase remove the superhelical stress, leaving open the possibility that the nucleosome cores were initially unfolded but reverted to the normal structure when the stress was removed. However, assays which do not remove superhelical stress (chemical cross-linking, circular dichroism, and chemical modification to probe for the normally buried H3 sulphydryl group) did not reveal any differences between nucleosome cores on positively and negatively supercoiled DNA. This conclusion was supported by a later study of the effect of positive supercoiling on DNA compaction induced by nucleosome cores (Clark et al., 1993). Calculation of expected changes in supercoiling free energy predicted that the formation of a normal core on positively supercoiled DNA would be much less favorable than on negatively supercoiled DNA. This was confirmed by the almost exclusive preference of the octamer for negatively supercoiled DNA under equilibrium binding conditions (0.8 M NaCl). It was concluded that nucleosome cores do not unfold in response to positive superhelical stress, although a *minor* conformational change might not have been detected, and the possible effects of post-translational modifications such as core histone acetylation were not addressed. There is a strong potential for octamer transfer from positively supercoiled to negatively supercoiled DNA but, to realize this potential at physiological salt concentration, a catalyst would be necessary because the octamer is irreversibly bound. These results led to a hypothetical transfer model for transcription through the nucleosome core, in which positive supercoils preceding a transcribing polymerase would destabilize a nucleosome core in its path, resulting in transfer of the octamer to the negatively supercoiled DNA behind the polymerase (Figure 2, Part B).

Pfaffle and Jackson (1990) studied the rates of nucleosome core formation on relaxed and negatively supercoiled plasmid DNA at physiological salt concentration using poly(glutamate)-core histone complexes. The number of nucleosome cores formed was determined by relaxation with topo I. Negative supercoils were stabilized on DNA that was initially negatively supercoiled but not on DNA that was initially relaxed. They suggested that normal nucleosome cores could not be formed on the relaxed DNA because this would require the introduction of positive supercoils in the plasmid. Jackson (1993) showed that core histone-DNA complexes containing mildly positively supercoiled DNA were relaxed more slowly by topo I than those containing negatively supercoiled DNA. It was suggested that unfolded nucleosome cores on positively supercoiled DNA refold only as the plasmid DNA is relaxed by topo I. Prior fixation of reconstitutes with formaldehyde almost completely prevented relaxation, suggesting that an unfolded nucleosome structure had been preserved.

These two sets of results are not easily reconciled. If the histone octamer was unable to form a nucleosome core on positively supercoiled DNA, the analysis of Clark and colleagues (1993) would certainly have detected this. In fact, a difference in relaxation rates would be *expected* if normal nucleosome cores had been formed on the positively supercoiled DNA used by Jackson (1993), because of differences in the number of supercoils requiring relaxation and in the degree of condensation of the reconstitute, which might affect accessibility to topo I. (In the experiment of Jackson [1993]: if 12 normal nucleosome cores were formed on a plasmid initially with +3 supercoils, the plasmid would become condensed with +15 supercoils; conversely, 12 cores formed on a plasmid with −14 supercoils would result in a relatively uncondensed plasmid with only −2 supercoils to be relaxed.) Fixation might have prevented relaxation of the condensed structure by topo I. Alternatively, different sources of core histones might account for the differences between these studies (those of Jackson [1993] probably carried more post-translational modifications).

B. Studies with Prokaryotic RNA Polymerases

1. Transcription through Nucleosome Cores

It was appreciated early on that the packaging of genes into chromatin might present steric and accessibility problems for transcription. Williamson and Felsenfeld (1978) reconstituted purified T7 DNA into nucleosome cores and transcribed it with *E. coli* RNA polymerase. Transcription initiation was inhibited by reconstitution to an extent determined by the nucleosome core density. At salt concentrations below physiological, transcript elongation was severely inhibited by nucleosome cores, resulting in much shorter transcripts than those obtained from naked T7 DNA. However, the average transcript lengths indicated that the polymerase must be capable of transcribing through at least some nucleosome cores. Addition of NaCl to 0.45 M removed the obstacle to transcription and polymerase elongated transcripts at a rate similar to that on naked DNA. Similar results were obtained by Wasylyk and colleagues (1979). This obstacle must reflect salt-sensitive interactions within the nucleosome core, but this is not just core histone dissociation because they remain bound at this salt concentration. Thus, *E. coli* RNA polymerase can transcribe through nucleosome cores, albeit at reduced rates.

What happens to the nucleosome core on transcription? This problem was addressed in two studies (Lorch et al., 1987; Losa and Brown, 1987), both using short DNA fragments containing a promoter for SP6 RNA polymerase, reconstituted into a single nucleosome core. Unfortunately, they came to opposite conclusions concerning the fate of the octamer. Lorch and colleagues (1987) showed that SP6 RNA polymerase is incapable of initiating transcription from a promoter in a nucleosome core. To circumvent this problem, they ligated a DNA fragment containing an SP6 promoter to another fragment already reconstituted into a nucleosome core. Transcription resulted in naked templates and they concluded that the first round of transcription caused displacement of the octamer (Lorch et al., 1988) (the fate of the histones was not determined). Losa and Brown (1987) obtained a nucleosomal template with an open promoter using a 360 bp fragment containing a *Xenopus* 5S RNA gene which precisely positions a nucleosome core (reviewed by Simpson, 1991). The nucleosome core was stable to multiple rounds of transcription and its position was unaffected. The same conclusion was reached by Wolffe and Drew (1989). Lorch and colleagues (1988) attempted to resolve this disagreement by comparing their templates and buffers with those of Losa and Brown and concluded that the different fates of the octamer were due to different DNA sequences. The 5S gene used by Losa and Brown retained most (but not all) of its nucleosome cores whereas all other nucleosome cores tested by Lorch and colleagues dissociated as a result of transcription. This implies that there are two different mechanisms for transcription through the nucleosome core, depending on DNA sequence, which seems inherently unlikely (see below for further discussion).

Morse (1989) devised an assay to determine whether an RNA polymerase can transcribe through an array of nucleosome cores: A plasmid containing an SP6 promoter was reconstituted into nucleosome cores and digested with a restriction enzyme. If its site is within a nucleosome core, the DNA is resistant to digestion but if it is in linker DNA, it is digested. If there is no core at this site, a run-off transcript will be produced, but if there is a core there *and* the polymerase can transcribe through it, longer transcripts should be observed. This was the case and Morse (1989) concluded that SP6 polymerase can transcribe through an array of nucleosome cores.

Pfaffle and colleagues (1990) used poly(glutamate)-histone complexes to reconstitute cores on a relaxed plasmid containing a T7 promoter on which polymerase initiation complexes had already been formed (thus protecting the promoter from blockage by nucleosome core formation). However, a very large excess of histone was used, much more than that required to precipitate all the DNA (some precipitation did occur) and yet the soluble templates had no more than one-half the maximum number of nucleosome cores, and during further incubation, nucleosome core formation apparently continued. Transcription resulted in the disruption of nucleosome cores, as measured by loss of negative supercoils in the plasmid during a transcription time course, although nucleosome cores were detected before and after transcription using MNase. This effect almost disappeared when transcribed in the presence of RNase (removal of the RNA is expected to reduce transcription-induced supercoiling by reducing viscous drag), even though a similar amount of RNA was synthesized (measured as RNase digestion product). They suggested that positive supercoils preceding the polymerase caused unfolding of nucleosome cores in its path, and that negative supercoils in the wake of the polymerase caused them to refold. However, neither unfolding nor transcription-induced supercoiling were *necessary* for the passage of the polymerase through the nucleosome core. Therefore they argued that transcription through the nucleosome core required disruption only of histone-DNA contacts and that the disruption of histone–histone contacts required for unfolding is caused by transcription-induced supercoiling. However, it is not clear by what mechanism the polymerase transcribes through the nucleosome core in the absence of unfolding.

O'Neill and colleagues (1992) transcribed a plasmid containing tandem repeats of a nucleosome-positioning sequence linked to a T7 promoter. Transcript elongation was inhibited by nucleosome cores: the average transcript length was less than that for naked DNA and the efficiency of transcription through a nucleosome core was estimated at 85 percent or less. Kirov and colleagues (1992) concluded that transcription did not result in significant loss of nucleosome cores from a linear template with a T7 promoter.

2. The Fate of the Histone Octamer

The fate of the octamer on transcription was investigated by transcribing a single nucleosome core at a known site on a plasmid (Clark and Felsenfeld, 1992). A nucleosome core was reconstituted on a purified 260 bp DNA fragment and ligated into a plasmid containing a promoter and two terminators for SP6 RNA polymerase. The ligated reconstitute was transcribed and then digested with MNase to yield core particles, and core DNA was extracted and labeled. Similar yields of core particles were obtained with or without transcription indicating that octamers were not lost on transcription. Two lines of evidence demonstrated that transcription induced displacement of the octamer and reformation of the nucleosome core elsewhere on the plasmid: (1) Core DNA from transcribed templates was mostly resistant to digestion by restriction enzymes with sites unique to the 260 bp fragment, but core DNA from untranscribed templates was almost all sensitive to digestion. (2) Untranscribed core DNA hybridized only to plasmid fragments containing the 260 bp region in a southern blot of different restriction digests of the plasmid, but transcribed core DNA hybridized to all of the DNA fragments. Cores were reformed with a modest preference (about 2.5-fold) in the region behind the promoter.

These observations rule out models (Figure 4) in which the octamer is lost or remains in position (unfolded) as a result of transcription. A model in which the polymerase slides the nucleosome core along in front of it is also excluded because identical results were obtained with a linearized template. In the "release/recapture" model, the octamer is released into solution and then recaptured on DNA to reform a nucleosome core. Recapture must be rapid because of the instability of the octamer in the absence of DNA (Feng et al., 1993). The octamer is likely to be recaptured on the plasmid molecule it was displaced from due to the high local concentration of the latter, consistent with the inability of competitor DNA at moderate concentrations to capture the octamer during transcription. However, the nonrandom distribution of cores on the plasmid after transcription is more easily explained by a "direct transfer" model in which the octamer is not released but transferred directly from one DNA segment to another.

The mechanism by which nucleosome cores are displaced and reformed during transcription was investigated further (Studitsky et al., 1994). A nucleosome core was reconstituted on a short linear DNA template (227 bp) containing a nucleosome-positioning sequence linked to an SP6 promoter. Remarkably, transcription resulted in transfer of the octamer from one end of the template to the other, a distance of only 75–80 bp, and blocked the promoter. Two lines of evidence indicated that the octamer was not released during this transfer: (1) The octamer could be transferred to competitor DNA in the presence of 0.5 mM ribonucleoside

FATE OF HISTONE OCTAMERS AFTER TRANSCRIPTION

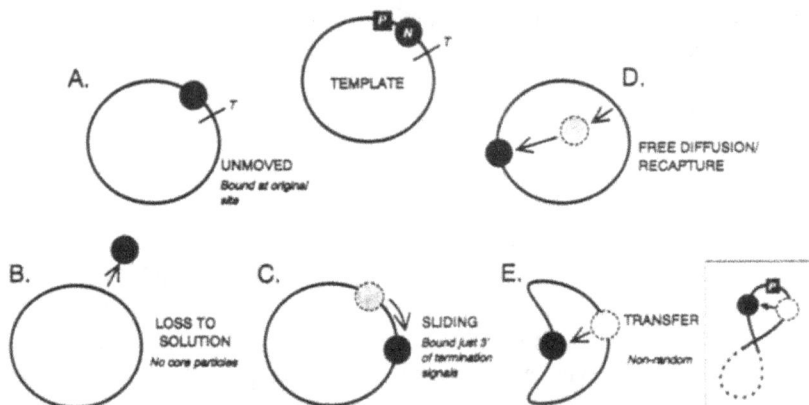

Figure 4. Models for transcription through a nucleosome core *in vitro*. A. The nucleosome core remains in place after transcription (a prediction of the unfolding model). B. The histone octamer is displaced into solution. C. The polymerase slides the nucleosome core along in front of it. D. The octamer is displaced and re-bound on the same plasmid molecule (the release/recapture model). E. The octamer is directly transferred from one DNA segment to another on the same plasmid molecule without release of the octamer (the direct transfer model). The inset shows an alternative, figure-eight configuration which could account for the modest preference for transfer to the region behind the promoter on the circular template (bold and dashed line) which might also be adopted by the linearized template at the moment of transfer (bold line) (from Clark and Felsenfeld, 1992).

triphosphates (rNTPs) but not at 0.1 mM rNTPs, even though transfer still occurred on the template. To account for this result using the release/recapture model, it must be argued that the released octamer binds to the same template molecule because of its high local concentration, and therefore its availability to competitor DNA would depend on the relative concentrations of competitor and template DNAs. This model cannot account for the observed dependence on rNTP concentration. It is not clear how the effect of low rNTP concentration is mediated. (2) The position of a nucleosome core after transcription depended on its position before transcription. This was shown using a slightly longer template on which a single nucleosome core could be formed at either of two positions. In both cases, transcription resulted in octamer transfer to the promoter end of the template, but the final positions were different. This observation is inconsistent with the release/recapture model because, once released, the octamer would lose all information concerning its initial position. In conclusion, the evidence is consistent with direct transfer of the octamer from in front of the transcribing polymerase to behind it and is inconsistent with release/recapture.

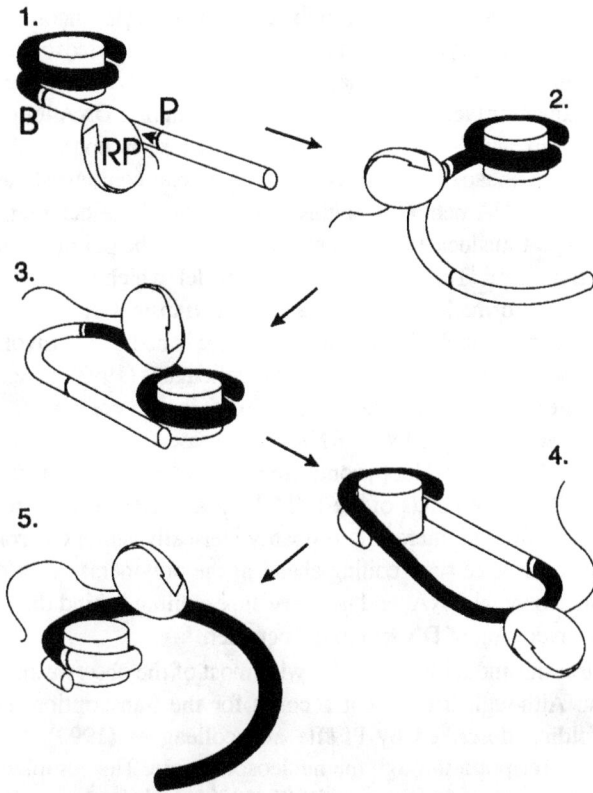

Figure 5. A spooling mechanism for transcription through a nucleosome core. (1) RNA polymerase (RP) initiates transcription at the promoter (P) on the 227 bp template and begins to synthesize a transcript; the border (B) of the nucleosome core is indicated by a bar, and nucleosomal DNA is shaded. The drawing is roughly to scale. (2) As the polymerase approaches the core it induces the dissociation of proximal DNA from the core, exposing part of the octamer surface. (3) The DNA behind the polymerase binds to the exposed octamer surface to form a loop *within the nucleosome core* to which the transcribing polymerase is bound; this loop is a topologically isolated domain subject to superhelical stress. (4) The DNA ahead of the transcribing polymerase continues to uncoil from the octamer and the DNA behind begins to coil around the octamer. (5) The nucleosome core is reformed behind the polymerase (blocking the promoter), and the polymerase completes the transcript (from Studitsky et al., 1994).

At first sight, this would appear to make the problem worse—the transcribing polymerase and the octamer are traveling in opposite directions and neither leaves the template! A "bridging" intermediate must be formed in which the octamer makes contact with the DNA behind the transcribing polymerase before its contacts with DNA ahead of the polymerase are broken, thus transferring itself around the polymerase. These considerations led to a "spooling" model (Figure 5) in which the approaching polymerase causes proximal DNA to begin uncoiling from the octamer surface. As the polymerase continues to invade the nucleosome core, more DNA uncoils from the octamer, and the DNA behind the polymerase is captured on the exposed octamer surface, resulting in the formation of a DNA loop *within the nucleosome core* to which the transcribing polymerase is bound (the bridging complex). The DNA ahead of the transcribing polymerase continues to uncoil from the octamer as the DNA behind it begins to coil around the octamer. Thus, the nucleosome core is translocated to its new position and the polymerase is able to complete the transcript. This is the simplest model which accounts for these observations. A conformational change and even transient splitting of the nucleosome core might occur during transfer, but there is no need to invoke either to explain these observations. In fact, O'Neill and colleagues (1993) have shown that dissociation of the octamer into tetramer and dimers is not necessary for transcription through nucleosome cores by T7 RNA polymerase.

The formation of an internal loop would create ideal conditions for transcription-induced supercoiling: both ends of the DNA loop are tethered (and so unable to release stress) and the polymerase is probably sterically inhibited from rotating around the DNA. Positive supercoiling ahead of the polymerase would drive the uncoiling of nucleosomal DNA, and negative supercoiling behind the polymerase would drive the recoiling of DNA onto the octamer.

This direct transfer model is consistent with most of the observations described in this section. Although it does not account for the transcription-induced nucleosome unfolding described by Pfaffle and colleagues (1990), this was not necessary for transcription through the nucleosome core. The complete displacement observed by Lorch and colleagues (1987, 1988) might be the result of transfer to competitor DNA and the apparent stability of the transcribed nucleosome core observed by Losa and Brown (1987) and Wolffe and Drew (1989) might reflect translocation of the octamer on the same template (in these experiments, competitor DNA was absent). We have confirmed that this is the case (Studitsky et al., unpublished) for a nucleosome core formed on the template used by Losa and Brown (1987), indicating that the mechanism of transcription through the nucleosome core is not sequence-dependent.

C. Studies with Eukaryotic RNA Polymerases

Wasylyk and Chambon (1979) compared the transcription of reconstituted SV40 DNA by Pol I and Pol II with that of naked SV40 DNA. Both polymerases initiated transcription nonspecifically at AT-rich sequences (specific initiation requires other factors and is very inefficient *in vitro*). Transcript initiation and elongation were both inhibited by nucleosome cores. The average length of the transcripts increased with time and some were long enough to indicate that transcription had occurred at least half-way around the template. The analysis was complicated by the presence of some naked DNA and some poorly reconstituted DNA in their preparation, but the observed inhibition of elongation probably was due to transcription through nucleosome cores because it was largely relieved by addition of NH_4^+ ions to 0.4 M, even though core histones were stably bound to DNA at this salt concentration (i.e., they did not exchange to competitor DNA).

The binding of purified Pol II to nucleosome core particles was investigated by Baer and Rhodes (1983). Only a subset of core particles (15%) from mouse myeloma cells could bind Pol II. These cores were deficient in H2A/H2B (one dimer missing) and enriched in transcriptionally active sequences. It is not obvious why only one Pol II molecule was able to bind to these cores since there are apparently two binding sites. It was suggested that these nucleosomes might be deficient in H2A/H2B as a result of transcription *in vivo*, but this cannot be the case because the particles contained DNA of core particle length. Alternatively, it was suggested that the dimer was lost during isolation, or that Pol II displaced the dimer on binding to the core particle. Thus, nucleosome cores derived from transcriptionally active genes apparently tend to lose a dimer. González and Palacián (1989) showed that Pol II can bind to intact core particles and even synthesize short transcripts but could not transcribe the entire length of the core. In contrast, Pol II was able to transcribe through particles with a dimer missing (prepared *in vitro*). Removal of the core histone tail domains (which contain all the acetylation sites) with trypsin facilitated binding of Pol II but had no effect on transcription. Identical results were obtained with *E. coli* RNA polymerase. These results imply that Pol II can transcribe through a nucleosome core deficient in a dimer more easily than an intact nucleosome core. Alternatively, the approximately 40 bp exposed by removing a dimer might be required to form a stable Pol II elongation complex.

The fate of the nucleosome core after transcription by Pol II was addressed by Lorch and colleagues (1987). Transcription could not be initiated by purified Pol II from the adenovirus-2 major late promoter within a nucleosome core, confirming the observations of Knezetic and Luse (1986). This promoter block was overcome by ligating a promoter-containing DNA fragment to a nucleosome core on another DNA fragment. Pol II was able to transcribe through this nucleosome core, but very inefficiently (producing less than 0.05 transcripts/ template). Therefore, to investigate the fate of the nucleosome core, a nucleosomal template with a poly(dC) tail rather than a promoter was used to increase the fraction of transcribed templates. Templates with a poly(dC) tail are transcribed much more efficiently and are useful for studying elongation, but they have a tendency to form stable RNA–DNA hybrids (Dedrick and Chamberlin, 1985). The result was the same as for SP6 polymerase (see preceding): naked templates were produced indicating that octamers were displaced (the histones presumably ended up on the plasmid DNA in the transcription buffer). However, whether the template renatured after transcription or whether a stable RNA–DNA hybrid was formed was not discussed.

Izban and Luse (1991) investigated the mechanism of transcription through the nucleosome core by Pol II using a HeLa extract to assemble paused Pol II elongation complexes on the adenovirus-2 major late promoter in a plasmid. All other proteins were stripped from the template with sarcosyl, the complexes were purified by gel filtration, assembled into nucleosome cores using a *Xenopus* egg extract, rinsed with salt to remove other DNA-binding proteins, and finally purified by gel filtration. The missing rNTP was added to allow Pol II to resume elongation. The average length of transcripts was much less than for naked DNA, indicating that elongation was inhibited by nucleosome core(s) downstream of the promoter. Nucleosome cores accentuated pause sites for the polymerase, and the paused transcripts could be extended on addition of sarcosyl (which dissociates histones and prevents reinitiation, but leaves Pol II elongation complexes intact), indicating that Pol II had paused on nucleosomal DNA but had not dissociated from the template. Addition of 0.4 M KCl, which does not dissociate core histones, also alleviated nucleosome-induced pausing. Elongation factors IIS, TFIIF, or TFIIX improved elongation through the nucleosome core, but the effects were not dramatic (Izban and Luse, 1992). Thus, Pol II has some difficulty transcribing through nucleosome cores *in vitro*.

Specific initiation by Pol III in *Xenopus* oocyte extracts is highly efficient (e.g., Clark and Wolffe, 1991), probably due to the abundance of Pol III transcription factors, but extracts contain many other proteins, the effects of which are difficult to control for. The 5S RNA gene has been used as a model template for Pol III. It is only 120 nucleotides long and contains an internal control region (ICR) to which factors must bind for transcription to occur. Assembly of the ICR into a nucleosome core prevented initiation by Pol III *in vitro* (Morse, 1989; Felts et al., 1990; Clark and Wolffe, 1991). To determine whether Pol III can transcribe through a nucleosome core, artificial "maxi-genes" were used to avoid blocking the ICR (extra DNA was inserted just before the Pol III termination sequence). Morse (1989) found that Pol III could not transcribe an 1185 bp maxi-gene reconstituted into nucleosome cores. However, Felts and colleagues (1990) showed that Pol III could transcribe through four nucleosome cores quite readily, but transcription through an array of 10 cores was relatively inefficient. Hansen and Wolffe (1992) also found that Pol III could transcribe through nucleosome cores. These results suggest that Pol III, like Pol II, has a tendency to pause or terminate transcription at each nucleosome core it encounters.

D. Summary

Most of the evidence from experiments using prokaryotic RNA polymerases is consistent with the displacement/transfer mechanism depicted in Figure 5. Eukaryotic RNA polymerases are similar in size to *E. coli* RNA polymerase (about 500 kD) but they are about five times larger than T7 and SP6 polymerases. However, there is considerable circumstantial evidence suggesting that prokaryotic polymerases are good models for understanding the mechanism of transcription through the nucleosome core by eukaryotic RNA polymerases: (1) If the promoter is within a nucleosome core, transcription cannot be initiated. (2) Transcript elongation is inhibited by nucleosome cores and transcription through an array of nucleosomes is progressively inhibited. (3) Inhibition of elongation is relieved by addition of 0.4 M monovalent cations. It may be concluded that Pol II can transcribe through nucleosome cores *in vitro*, but the contradictory nature of the evidence makes it unclear whether Pol II transcribes through the nucleosome core by the displacement/ transfer mechanism.

Elongation through the nucleosome core is inhibited by a salt-sensitive obstacle, the nature of which is likely to be critically important in understanding the mechanism. It might reflect the interaction of the core histone tail domains with DNA because these dissociate from DNA in 0.3–0.6 M NaCl (Walker, 1984). This is well above physiological ionic strength but core histone acetylation might have the same effect as increased salt concentration because it reduces the affinity of the tail domains for core DNA, at least for H4 (Cary et al., 1982), and elongation factors might also be important. Analysis of the effects of acetylation on transcription through the nucleosome core using the defined systems we now possess might well be the best way forward. Other factors worthy of consideration are HMGs 14 /17 (Crippa et al., 1993), other post-translational modifications of the histones, and possible inhibitory effects of H1.

VI. THE DYNAMIC NUCLEOSOMAL ORGANIZATION OF ACTIVE GENES

A. Overview

It is fair to say that there is no conclusive evidence in favor of either nucleosome unfolding or displacement/transfer *in vivo*. There is evidence for stable unfolded nucleosomes derived from active genes. On the other hand, electron micrographs of active chromatin provide evidence only for *transient* unfolding, and this only in the vicinity of the polymerase (and these unfolded nucleosomes were not clearly distinguishable from naked DNA). Nuclease digestion studies have provided evidence to support all models and are not all accounted for by one model. If we make the reasonable assumption that the mechanism of transcription through nucleosomes is conserved, then it must be concluded that all of these studies should be interpreted with caution. *In vitro*, displacement/transfer accounts for most of the observations with prokaryotic polymerases, but whether this is also true for eukaryotic RNA polymerases is not yet clear. Although it is not possible to reach a conclusion on the mechanism of transcription through nucleosomes *in vivo* at the present time, there is now a consensus that active genes have a dynamic chromatin structure (Kornberg and Lorch, 1992; Morse, 1992; Hansen and Ausio, 1992; van Holde et al., 1992).

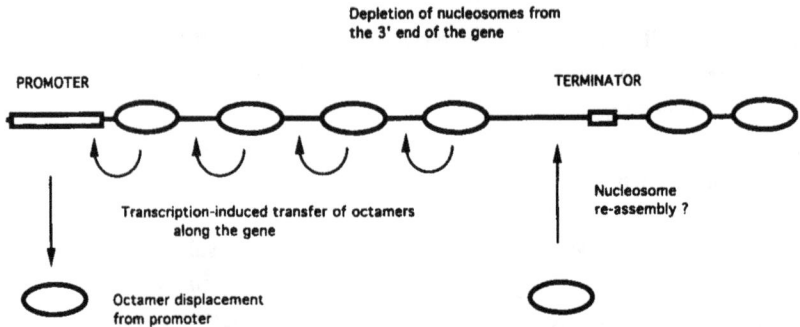

Figure 6. A hypothetical octamer shuttling model for the nucleosomal organization of an active gene. A transcriptionally competent eukaryotic gene is typically organized into a more or less regularly spaced array of nucleosomes. In this model, every transcribing polymerase (not shown) which passes down the gene induces the translocation of every nucleosome in the array a limited distance up the gene towards the promoter, by the spooling mechanism (Figure 5). This would tend to deplete nucleosomes from the 3′ end of the gene, but nucleosomes might be re-assembled here. Each octamer arriving at the transcript initiation site would have to be displaced into solution (or transferred elsewhere). The result is a "shuttling" of octamers which form nucleosomes at the 3′ end, pass stepwise up the gene and are eventually displaced, one by one, at the transcript initiation site.

B. A Model for the Nucleosomal Organization of an Active Gene

The consequences of applying the nucleosome transfer model (Figure 5) to an array of nucleosomes on a gene are suggested in Figure 6. Each RNA polymerase molecule passing down the gene would induce the translocation of every nucleosome in the array stepwise up the gene towards the promoter, resulting in progressive depletion of nucleosomes from the 3′ end of the gene. This suggests an explanation for the inverse dependence of the number of nucleosomes on polymerase density. Nucleosome depletion might be countered by concomitant nucleosome assembly. At the promoter, nucleosomes would be sequentially deposited at the transcript initiation site and a mechanism for their removal would be required. Thus, in this "shuttling" model for the dynamic organization of active chromatin, histone octamers are transferred stepwise up the gene as it is transcribed and are eventually displaced at the promoter, and might then reform nucleosomes at the 3′ end. Of course, this model is just one of many possible models, but it is based on evidence *in vitro*, and it suggests some interesting experiments.

ACKNOWLEDGMENT

I thank Vasily Studitsky for much helpful discussion and comments on the manuscript.

REFERENCES

Andersson, K., Björkroth, B., & Daneholt, B. (1984). Packing of a specific gene into higher order structures following repression of RNA synthesis. J. Cell Biol. 98, 1296–1303.

Arents, G., & Moudrianakis, E.N. (1991). The nucleosomal core histone octamer at 3.1 Å resolution: A tripartite protein assembly and a left-handed superhelix. Proc. Natl. Acad. Sci. USA 88, 10148–10152.

Arents, G., & Moudrianakis, E.N. (1993). Topography of the histone octamer surface: Repeating structural motifs utilized in the docking of nucleosomal DNA. Proc. Natl. Acad. Sci. USA 90, 10489–10493.

Baer, B.W., & Rhodes, D. (1983). Eukaryotic RNA polymerase II binds to nucleosome cores from transcribed genes. Nature 301, 482–488.

Benezra, R., Cantor, C.R., & Axel, R. (1986). Nucleosomes are phased along the mouse β-major globin gene in erythroid and non-erythroid cells. Cell 44, 697–704.

Bergman, L.W., & Kramer, R.A. (1983). Modulation of chromatin structure associated with derepression of the acid phosphatase gene of *Saccharomyces cerevisiae*. J. Biol. Chem. 258, 7223–7227.

Björkroth, B., Ericsson, C., Lamb, M.M., & Daneholt, B. (1988). Structure of the chromatin axis during transcription. Chromosoma 96, 333–340.

Camerini-Otero, R.D., Sollner-Webb, B., & Felsenfeld, G. (1976). The organization of histones and DNA in chromatin: Evidence for an arginine-rich histone kernel. Cell 8, 333–347.

Caplan, A., Kimura, T., Gould, H., & Allan, J. (1987). Pertubation of chromatin structure in the region of the adult beta-globin gene in chicken erythrocyte chromatin. J. Mol. Biol. 193, 57–70.

Cartwright, I.L., & Elgin, S.C.R. (1986). Nucleosomal instability and induction of new upstream protein-DNA associations accompany activation of four small heat shock protein genes in *Drosophila melanogaster*. Mol. Cell. Biol. 6, 779–791.

Cary, P.D., Crane-Robinson, C., Bradbury, E.M., & Dixon, G.H. (1982). Effect of acetylation on the binding of N–terminal peptides of histone H4 to DNA. Eur. J. Biochem. 127, 137–140.

Cavalli, G., & Thoma, F. (1993). Chromatin transitions during activation and repression of galactose-regulated genes in yeast. EMBO J. 12, 4603–4613.

Chen-Cleland, T.A., Smith, M.M., Le, S., Sternglanz, R., & Allfrey, V.G. (1993). Nucleosome structural changes during derepression of silent mating-type loci in yeast. J. Biol. Chem. 268, 1118–1124.

Chen, T.A., Sterner, R., Cozzolino, A., & Allfrey, V.G. (1990). Reversible and irreversible changes in nucleosome structure along the c-fos and c-myc oncogenes following inhibition of transcription. J. Mol. Biol. 212, 481–493.

Clark, D.J., & Felsenfeld, G. (1991). Formation of nucleosomes on positively supercoiled DNA. EMBO J. 10, 387–395.

Clark, D.J., & Felsenfeld, G. (1992). A nucleosome core is transferred out of the path of a transcribing polymerase. Cell 71, 11–22.

Clark, D.J., Ghirlando, R., Felsenfeld, G., & Eisenberg, H. (1993). Effect of positive supercoiling on DNA compaction by nucleosome cores. J. Mol. Biol. 234, 297–301.

Clark, D.J., & Kimura, T. (1990). Electrostatic mechanism of chromatin folding. J. Mol. Biol. 211, 883–896.

Clark, D.J., & Wolffe, A.P. (1991). Superhelical stress and nucleosome-mediated repression of 5S RNA gene transcription *in vitro*. EMBO J. 10, 3419–3428.

Cohen, R.B., & Sheffery, M. (1985). Nucleosome disruption precedes transcription and is largely limited to the transcribed domain of globin genes in murine erythroleukemia cells. J. Mol. Biol. 182, 109–129.

Crippa, M.P., Treischmann, L., Alfonso, P.J., Wolffe, A.P., & Bustin, M. (1993). Deposition of chromosomal protein HMG–17 during replication affects the nucleosomal ladder and transcriptional potential of nascent chromatin. EMBO J. 12, 3855–3864.

Davie, J.R., & Murphy, L.C. (1990). Level of ubiquitinated histone H2B in chromatin is coupled to ongoing transcription. Biochemistry 29, 4752–4757.

de Bernardin, W., Koller, T., & Sogo, J.M. (1986). Structure of *in vivo* transcribing chromatin as studied in Simian virus 40 minichromosomes. J. Mol. Biol. 191, 469–482.

Dedrick, R.L., & Chamberlin, M.J. (1985). Studies on transcription of 3′-extended templates by mammalian RNA polymerase II. Parameters that affect the initiation and elongation reactions. Biochemistry 24, 2245–2253.

Eickbush, T.H., & Moudrianakis, E.N. (1978). The histone core complex: An octamer assembled by two sets of protein-protein interactions. Biochemistry 17, 4955–4964.

Ericsson, C., Grossbach, U., Björkroth, B., & Daneholt, B. (1990). Presence of histone H1 on an active Balbiani ring gene. Cell 60, 73–83.

Ericsson, C., Mehlin, H., Björkroth, B., Lamb, M.M., & Daneholt, B. (1989). The ultrastructure of upstream and downstream regions of an active Balbiani ring gene. Cell 56, 631–639.

Felsenfeld, G. (1992). Chromatin as an essential part of the transcriptional mechanism. Nature 355, 219–224.

Felsenfeld, G., & McGhee, J.D. (1986). Structure of the 30 nm chromatin fiber. Cell 44, 375–377.

Felts, S.J., Weil, P.A., & Chalkley, R. (1990). Transcription factor requirements for *in vitro* formation of transcriptionally competent 5S rRNA gene chromatin. Mol. Cell. Biol. 10, 2390–2401.

Feng, H.P., Scherl, D.S., & Widom, J. (1993). Lifetime of the histone octamer studied by continuous-flow quasielastic light scattering: Test of a model for nucleosome transcription. Biochemistry 32, 7824–7831.

Fisher, E.A., & Felsenfeld, G. (1986). Comparison of the folding of β-globin and ovalbumin gene containing chromatin isolated from chicken oviduct and erythrocytes. Biochemistry 25, 8010–8016.

Fotedar, R., & Roberts, J.M. (1989). Multistep pathway for replication-dependent nucleosomal assembly. Proc. Nat. Acad. Sci. U.S.A. 86, 6459–6463.

González, P.J., & Palacián, E. (1989). Interaction of RNA polymerase II with structurally altered nucleosomal particles. J. Biol. Chem. 264, 18457–18462.

Gross, D.S., & Garrard, W.T. (1987). Poising chromatin for transcription. Trends Biochem. Sci. 12, 293–297.

Hansen, J.C., & Ausio, J. (1992). Chromatin dynamics and the modulation of gene activity. Trends Biochem. Sci. 17, 187–191.

Hansen, J.C., & Wolffe, A.P. (1992). Influence of chromatin folding on transcription initiation and elongation by RNA polymerase III. Biochemistry 31, 7977–7988.

Hebbes, T.R., Clayton, A.L., Thorne, A.W., & Crane-Robinson, C. (1994). Core histone hyperacetylation co-maps with generalized DNase I sensitivity in the chicken β-globulin chromosomal domain. EMBO J. 13, 1823–1830.

Hebbes, T.R., Thorne, A.W., & Crane-Robinson, C. (1988). A direct link between core histone acetylation and transcriptionally active chromatin. EMBO J. 7, 1395–1402.

Hill, R.J., Watt, F., Wilson, C.M., Fifis, T., Underwood, P.A., Tribbick, G., Geysen, H.M., & Thomas, J.O. (1989). Bands, interbands and puffs in native *Drosophila* polytene chromosomes are recognized by a monoclonal antibody to an epitope in the carboxy-terminal tail of histone H1. Chromosoma 98, 411–421.

Izban, M.G., & Luse, D.S. (1991). Transcription on nucleosomal templates by RNA polymerase II *in vitro*: Inhibition of elongation with enhancement of sequence-specific pausing. Genes & Dev. 5, 683–696.

Izban, M.G., & Luse, D.S. (1992). Factor-stimulated RNA polymerase II transcribes at physiological elongation rates on naked DNA but very poorly on chromatin templates. J. Biol. Chem. 267, 13647–13655.

Jackson, V. (1990). *In vivo* studies on the dynamics of histone-DNA interaction: Evidence for nucleosome dissolution during replication and transcription and a low level of dissolution independent of both. Biochemistry 29, 719–731.

Jackson, V. (1993). Influence of positive stress on nucleosome assembly. Biochemistry 32, 5901–5912.

Kamakaka, R.T., & Thomas, J.O. (1990). Chromatin structure of transcriptionally competent and repressed genes. EMBO J. 9, 3997–4006.

Kimura, T., Mills, F.C., Allan, J., & Gould, H. (1983). Selective unfolding of erythroid chromatin in the region of the active β-globin gene. Nature 306, 709–712.

Kirov, N., Tsaneva, I., Einbinder, E., & Tsanev, R. (1992). *In vitro* transcription through nucleosomes by T7 RNA polymerase. EMBO J. 11, 1941–1947.

Knezetic, J.A., & Luse, D.S. (1986). The presence of nucleosomes on a DNA template prevents initiation by RNA polymerase II *in vitro*. Cell 45, 95–104.

Kornberg, R.D., & Lorch, Y. (1991). Irresistible force meets immovable object: transcription through the nucleosome. Cell 67, 833–836.

Kornberg, R.D., & Lorch, Y. (1992). Chromatin structure and transcription. Ann. Rev. Cell Biol. 8, 563–587.

Lee, M.-S., & Garrard, W.T. (1991a). Transcription-induced nucleosome 'splitting': an underlying structure for DNase I sensitive chromatin. EMBO J. 10, 607–615.

Lee, M.-S., & Garrard, W.T. (1991b). Positive DNA supercoiling generates a chromatin conformation characteristic of highly active genes. Proc. Natl. Acad. Sci. USA 88, 9675–9679.

Liu, L.C., & Wang, J.C. (1987). Supercoiling of the DNA template during transcription. Proc. Natl. Acad. Sci. U.S.A. 84, 7024–7027.

Lohr, D. (1983). The chromatin structure of an actively expressed, single copy gene. Nucl. Acids Res. 11, 6755–6773.

Lorch, Y., LaPointe, J.W., & Kornberg, R.D. (1987). Nucleosomes inhibit the initiation of transcription but allow chain elongation with the displacement of histones. Cell 49, 203–210.

Lorch, Y., LaPointe, J.W., & Kornberg, R.D. (1988). On the displacement of histones from DNA by transcription. Cell 55, 743–744.

Losa, R., & Brown, D.D. (1987). A bacteriophage RNA polymerase transcribes *in vitro* through a nucleosome core without displacing it. Cell 50, 801–808.

Louters, L., & Chalkley, R. (1984). *In vitro* exchange of nucleosomal histones H2A and H2B. Biochemistry 23, 547–552.

Malik, N., Smulson, M., & Bustin, M. (1984). Enrichment of acetylated histones in polynucleosomes containing high mobility group protein 17 revealed by immunoaffinity chromatography. J. Biol. Chem. 259, 699–702.

McGhee, J.D., Nickol, J.M., Felsenfeld, G., & Rau, D.C. (1983). Histone hyperacetylation has little effect on the higher order folding of chromatin. Nucl. Acids Res. 11, 4065–4075.

McGhee, J.D., Rau, D.C., & Felsenfeld, G. (1982). The high mobility group proteins, HMG 14 and 17, do not prevent the formation of chromatin higher order structure. Nucl. Acids Res. 10, 2007–2016.

McKnight, S.L., Bustin, M., & Miller, O.L. (1978). Electron microscopic analysis of chromosome metabolism in the *Drosophila melanogaster* embryo. Cold Spring Harb. Symp. 42, 741–754.

Morgan, J.E., & Whitlock, J.P. (1992). Transcription-dependent and transcription-independent nucleosome disruption induced by dioxin. Proc. Nat. Acad Sci. U.S. A. 89, 11622–11626.

Morse, R.H. (1989). Nucleosomes inhibit both transcriptional initiation and elongation by RNA polymerase III *in vitro*. EMBO J. 8, 2343–2351.

Morse, R.H. (1992). Transcribed chromatin. Trends Biochem. Sci. 17, 23–26.

Nacheva, G.A., Guschin, D.Y., Preobrazhenskaya, O.V., Karpov, V.L., Ebralidse, K.K., & Mirzabekov, A.D. (1989). Change in the pattern of histone binding to DNA upon transcriptional activation. Cell 58, 27–36.

O'Neill, T.E., Roberge, M., & Bradbury, E.M. (1992). Nucleosome arrays inhibit both initiation and elongation of transcripts by bacteriophage T7 RNA polymerase. J. Mol. Biol. 223, 67–78.

O'Neill, T.E., Smith, J.G., & Bradbury, E.M. (1993). Histone octamer dissociation is not required for transcript elongation through arrays of nucleosome cores by phage T7 RNA polymerase *in vitro*. Proc. Natl. Acad. Sci. USA 90, 6203–6207.

Pederson, D.S., & Morse, R.H. (1990). Effect of transcription of yeast chromatin on DNA topology *in vivo*. EMBO J. 9, 1873–1881.

Perez-Ortìn, J.E., Estruch, F., Matallana, E., & Franco, L. (1987). Fine analysis of the chromatin structure of the yeast *SUC2* gene and of its changes upon derepression. Comparison between the chromosomal and plasmid-inserted genes. Nucl. Acids Res. 15, 6937–6956.

Pfaffle, P., Gerlach, V., Bunzel, L., & Jackson, V. (1990). *In vitro* evidence that transcription-induced stress causes nucleosome dissolution and regeneration. J. Biol. Chem. 265, 16830–16840.

Pfaffle, P., & Jackson, V. (1990). Studies on rates of nucleosome formation with DNA under stress. J. Biol. Chem. 265, 16821–16829.

Postnikov, Y.V., Shick, V.V., Belyavsky, A.V., Khrapko, K.R., Brodolin, K.L., Nikolskaya, T.A., & Mirzabekov, A.D. (1991). Distribution of high mobility group proteins 1/2, E and 14/17 and linker histones H1 and H5 on transcribed and non-transcribed regions of chicken erythrocyte chromatin. Nucl. Acids Res. 19, 717–725.

Prior, C.P., Cantor, C.R., Johnson, E.M., Littau, V.C., & Allfrey, V.G. (1983). Reversible changes in nucleosome structure and histone H3 accessibility in transcriptionally active and inactive states of rDNA chromatin. Cell 34, 1033–1042.

Richmond, T.J., Finch, J.T., Rushton, B., & Klug, A. (1984). Structure of the nucleosome core particle at 7 Å resolution. Nature 311, 532–537.

Ridsdale, J.A., Hendzel, M.J., Delcuve, G.P., & Davie, J.R. (1990). Histone acetylation alters the capacity of the H1 histones to condense transcriptionally active/competent chromatin. J. Biol. Chem. 265, 5150–5156.

Scheer, U. (1987). Structure of lampbrush chromosome loops during different states of transcriptional activity as visualized in the presence of physiological salt concentrations. Biol. Cell 59, 33–42.

Schwager, S., Retief, J.D., de Groot, P., & von Holt, C. (1985). Rapid exchange of histones H2A and H2B in sea urchin embryo chromatin. FEBS Lett. 189, 305–309.

Simpson, R.T. (1991). Nucleosome positioning: occurrence, mechanisms and functional consequences. Prog. Nuc. Acids Res. Mol. Biol. 40, 143–184.

Simpson, R.T., Thoma, F., & Brubaker, J.M. (1985). Chromatin reconstituted from tandemly repeated cloned DNA fragments and core histones: A model system for study of higher order structure. Cell 42, 799–808.

Smith, R.D., Seale, R.L., & Yu, J. (1983). Transcribed chromatin exhibits an altered nucleosome spacing. Proc. Nat. Acad. Sci. U.S.A. 80, 5505–5509.

Smith, R.D., Yu, J., Annunziato, A., & Seale, R.L. (1984). β-Globin gene family in murine erythroleukemia cells resides within two chromatin domains differing in higher order structure. Biochemistry 23, 2970–2976.

Studitsky, V.M., Clark, D.J., & Felsenfeld, G. (1994). A histone octamer can step around a transcribing RNA polymerase without leaving the template. Cell 76, 371–382.

Sun, Y.L., Xu, Y.Z., Bellard, M., & Chambon, P. (1986). Digestion of the chicken β-globin gene chromatin with micrococcal nuclease reveals the presence of an altered nucleosomal array characterized by an atypical ladder of DNA fragments. EMBO J. 5, 293–300.

Szent-Györgyi, C., Finkelstein, D.B., & Garrard, W.T. (1987). Sharp boundaries demarcate the chromatin structure of a yeast heat-shock gene. J. Mol. Biol. 193, 71–80.

Thoma, F. (1991). Structural changes in nucleosomes during transcription: strip, split or flip? Trends Genet. 7, 175–177.

Thoma, F., Koller, T., & Klug, A. (1979). Involvement of histone H1 in the organization of the nucleosome and of the salt-dependent superstructures of chromatin. J. Cell Biol. 83, 403–427.

Thomas, J.O., & Kornberg, R.D. (1975). An octamer of histones in chromatin and free in solution. Proc. Nat. Acad. Sci. U.S.A. 72, 2626–2630.

Tsao, Y.P., Wu, H.Y., & Liu, L.F. (1989). Transcription-driven supercoiling of DNA: direct biochemical evidence from *in vitro* studies. Cell 56, 111–118.

Turner, B.M., Franchi, L., & Wallace, H. (1990). Islands of acetylated H4 in polytene chromosomes and their relationship to chromatin packaging and transcriptional activity. J. Cell Sci. 96, 335–346.

van Holde, K.E. (1988). Chromatin. Springer-Verlag, New York.

van Holde, K.E., Lohr, D.E., & Robert, C. (1992). What happens to nucleosomes during transcription? J. Biol. Chem. 267, 2837–2840.

Villeponteau, B., Brawley, J., & Martinson, H.G. (1992). Nucleosome spacing is compressed in active chromatin domains of chick erythroid cells. Biochemistry 31, 1554–1563.

Walker, I.O. (1984). Differential dissociation of histone tails from core chromatin. Biochemistry 23, 5622–5628.

Wasylyk, B., & Chambon, P. (1979). Transcription by eukaryotic RNA polymerases A and B of chromatin assembled *in vitro*. Eur. J. Biochem. 98, 317–327.

Wasylyk, B., Thevenin, G., Oudet, P., & Chambon, P. (1979). Transcription of *in vitro* assembled chromatin *Escherichia coli* RNA polymerase. J. Mol. Biol. 128, 411–440.

Westermann, R., & Grossbach, U. (1984). Localization of nuclear proteins related to high mobility group protein 14 (HMG 14) in polytene chromosomes. Chromosoma 90, 355–365.

Widmer, R.M., Lucchini, R., Lezzi, M., Meyer, B., Sogo, J.M., Edstrom, J.E., & Koller, T. (1984). Chromatin structure of a hyperactive secretory protein gene (in Balbiani ring 2) of *Chironomous*. EMBO J. 3, 1635–1641.

Williamson, P., & Felsenfeld, G. (1978). Transcription of histone-covered T7 DNA by *Escherischia coli* RNA polymerase. Biochemistry 17, 5695–5705.

Wolffe, A.P. (1991). Developmental regulation of chromatin structure and function. Trends Cell Biol. 1, 61–66.

Wolffe, A.P. (1992). Chromatin: Structure and function. Academic Press, London.

Wolffe, A.P., & Drew, H.R. (1989). Initiation of transcription on nucleosomal templates. Proc. Natl. Acad. Sci. USA 86, 9817–9821.

Wood, W.I., & Felsenfeld, G. (1982). Chromatin structure of the chicken β-globin region. J. Biol. Chem. 257, 7730–7736.

Workman, J.L. & Buchman, A.R. (1993). Multiple functions of nucleosomes and regulatory factors in transcription. Trends Biochem. Sci. 18, 90–95.

Wu, C., Wong, Y., & Elgin, S.C.R. (1979). The chromatin structure of specific genes: II. Disruption of chromatin structure during gene activity. Cell 16, 807–814.

INDEX